T0406304

4
Topics in Medicinal Chemistry

Transporters as Targets for Drugs

Volume Editors: Susan Napier · Matilda Bingham

With contributions by

M. D. Andrews · G. Antoni · M. Bingham · R. J. Bridges
A. D. Brown · Z. Chen · S. G. Dahl · P. V. Fish · R. Gilfillan
H. Hall · J. L. Katz · J. Kerr · S. Napier · A. Hauck Newman
S. A. Patel · A. W. Ravna · G. Sager · P. Skolnick · J. Sörensen
A. Stobie · I. Sylte · F. Wakenhut · G. Walker · G. A. Whitlock
G. Wishart · J. Yang

 Springer

Drug research requires interdisciplinary team-work at the interface between chemistry, biology and medicine. Therefore, the new topic-related series should cover all relevant aspects of drug research, e.g. pathobiochemistry of diseases, identification and validation of (emerging) drug targets, structural biology, drugability of targets, drug design approaches, chemogenomics, synthetic chemistry including combinatorial methods, bioorganic chemistry, natural compounds, high-throughput screening, pharmacological in vitro and in vivo investigations, drug-receptor interactions on the molecular level, structure-activity relationships, drug absorption, distribution, metabolism, elimination, toxicology and pharmacogenomics.

In references *Topics in Medicinal Chemistry* is abbreviated *Top Med Chem* and is cited as a journal.

Springer WWW home page: springer.com
Visit the TIMC content at springerlink.com

ISSN 1862-2461   e-ISSN 1862-247X
ISBN 978-3-540-87911-4   e-ISBN 978-3-540-87912-1
DOI 10.1007/978-3-540-87912-1
Springer Dordrecht Heidelberg London New York

Library of Congress Control Number: 2009926137

Cover design: WMXDesign GmbH, Heidelberg
Typesetting and Production: le-tex publishing services oHG, Leipzig

Printed on acid-free paper

Springer is part of Springer Science+Business Media (www.springer.com)

Topics in Medicinal Chemistry
Also Available Electronically

For all customers who have a standing order to Topics in Medicinal Chemistry, we offer the electronic version via SpringerLink free of charge. Please contact your librarian who can receive a password or free access to the full articles by registering at:

springerlink.com

If you do not have a subscription, you can still view the tables of contents of the volumes and the abstract of each article by going to the SpringerLink Homepage, clicking on "Browse by Online Libraries", then "Chemical Sciences", and finally choose Topics in Medicinal Chemistry.

You will find information about the

- Editorial Board
- Aims and Scope
- Instructions for Authors
- Sample Contribution

at springer.com using the search function.

Color figures are published in full color within the electronic version on SpringerLink.

Preface to the Series

Medicinal chemistry is both science and art. The science of medicinal chemistry offers mankind one of its best hopes for improving the quality of life. The art of medicinal chemistry continues to challenge its practitioners with the need for both intuition and experience to discover new drugs. Hence sharing the experience of drug discovery is uniquely beneficial to the field of medicinal chemistry.

The series *Topics in Medicinal Chemistry* is designed to help both novice and experienced medicinal chemists share insights from the drug discovery process. For the novice, the introductory chapter to each volume provides background and valuable perspective on a field of medicinal chemistry not available elsewhere. Succeeding chapters then provide examples of successful drug discovery efforts that describe the most up-to-date work from this field.

The editors have chosen topics from both important therapeutic areas and from work that advances the discipline of medicinal chemistry. For example, cancer, metabolic syndrome and Alzheimer's disease are fields in which academia and industry are heavily invested to discover new drugs because of their considerable unmet medical need. The editors have therefore prioritized covering new developments in medicinal chemistry in these fields. In addition, important advances in the discipline, such as fragment-based drug design and other aspects of new lead-seeking approaches, are also planned for early volumes in this series. Each volume thus offers a unique opportunity to capture the most up-to-date perspective in an area of medicinal chemistry.

Dr. Peter R. Bernstein
Prof. Dr. Armin Buschauer
Prof. Dr. Gunda J. Georg
Dr. John Lowe
Dr. Hans Ulrich Stilz

Preface to Volume 4

Transporters are proteins which span the plasma membrane and regulate the traffic of small molecules in and out of the cell. Transporters play a particularly important role in chemical signalling between neurons in the CNS, where they act to control the concentration of neurotransmitters in the synapse. The majority of transporters which are actively being pursued as targets for drug discovery are CNS located and this reflects the history of the field which began with the tricyclic antidepressants (TCAs) over half a century ago. The use of transporter inhibition to regulate the synaptic concentrations of key neurotransmitters is an established approach in the discovery of psychiatric medications. This volume reviews advances in the field of transporters as targets for drug discovery in the last 10 years. The volume will be of interest to scientists engaged in drug research in the pharmaceutical industry, biotech and academia. Following an overview chapter, seven chapters written by leading experts in their area reflect a range of topics pertinent to the transporter field. General topics include recent advances in the structural biology of transporters and its impact on potential structure-based drug design and the design of ligands for Positron Emission Tomography and the importance of molecular imaging in understanding early clinical data. Medicinal chemistry approaches are described outlining the discovery of selective serotonin, noradrenaline and dopamine reuptake inhibitors, current efforts towards the discovery of mixed re-uptake inhibitors with varied "flavours" of monoamine inhibition, advances in the development of inhibitors for the glycine transporter and the discovery of subtype selective EAAT inhibitors. In addition to being an interesting read, the reader will receive a critical overview of progress made in this rapidly developing field.

January 2009

Matilda Bingham
Susan Napier

Contents

Top Med Chem (2009) 4: 1–13
DOI 10.1007/7355_2009_029
© Springer-Verlag Berlin Heidelberg
Published online: 28 March 2009

Overview: Transporters as Targets for Drug Discovery

Matilda Bingham (✉) · Susan Napier

Schering Plough Corporation, Newhouse, Motherwell ML1 5SH, UK
matilda.bingham@spcorp.com

Abbreviations

ABC	ATP binding cassette
ADHD	Attention deficit hyperactivity disorder
ATP	Adenosine triphosphate
DAACS	Dicarboxlate/amino acid:cation (Na$^+$ or H$^+$) symporter
DAT	Dopamine reuptake transporter
EAAT	Excitatory amino acid transporter
GLUT	Glucose transporter
HTS	High-throughput screening
PET	Positron emission tomography
P-gp	P-glycoprotein
MDD	Major depressive disorder
MFS	Major facilitator superfamily
NET	Noradrenaline reuptake transporter
NRI	Noradrenaline reuptake inhibitor
NSS	Neurotransmitter; sodium symporter
SBDD	Structure-based drug design
SERT	Serotonin reuptake transporter
SNRI	Serotonin noradrenaline inhibitor
SSRI	Selective serotonin reuptake inhibitor
TCA	Tricyclic antidepressant

1
Introduction

Transporters are proteins that span the plasma membrane and regulate the traffic of small molecules in and out of the cell. They play a particularly important role in chemical signalling between neurons in the CNS, where they act to control the concentration of neurotransmitters in the synapse. In most systems the termination of chemical transmission is achieved by rapid uptake of the transmitter molecule from the synapse by transporters located on the synaptic terminal or surrounding glial cells. Another key role for transporters is in excluding undesirable xenobiotics from the cell, whilst allowing key molecules required for the cell life cycle to enter. It is increasingly recognised that these efflux or uptake transporters respectively, play an important role in the disposition of many marketed drugs, and whilst the field of drug transport is yet to attain the level of maturity of drug metabolism, it is certain to be of increasing importance in future drug discovery programmes.

2
Transporter Classification

Transporters can be classed into two main families; the ATP binding cassette (ABC) family, and the solute carrier (SLC) family. The SLC family is a very broad categorisation which encompasses, amongst others, three important families of transporters for organic molecules; the major facilitator superfamily (MFS) and two neurotransmitter transporter families, the neurotransmitter; sodium symporter (NSS, or SLC6) and the dicarboxylate/amino acid:cation (Na^+ or H^+) symporter (DAACS, or SLC1) family.

2.1
The ATP Binding Cassette Family

The ATP binding cassette (ABC) family use a primary active transport mechanism to transport the substrate across the membrane [1, 2]. As the name suggests, the free energy stored in the phosphate bonds of ATP is harnessed directly to move substances through the membrane against a concentration gradient from a lower to a higher concentration. Perhaps the best known example of this class is the efflux transporter P-glycoprotein or P-gp, also known as MDR1 and ABCB1. This protein was identified because of its overexpression in cultured tumor cells associated with acquired resistance to multiple anticancer agents [3]. It has since been recognised that this transporter is expressed in many normal tissues including the gastrointestinal tract and blood–brain barrier and plays a key role in limiting the absorption and CNS penetration of several marketed drugs [4]. Interaction with P-gp can also have significant consequences in terms of drug–drug interactions [5]. Testing for P-gp inhibition in cell-based systems is now a routine in vitro screen incorporated into early phase drug discovery programmes. Following the discovery of P-gp a number of other transporters from the ABC family have also been shown to have important roles in the disposition of drugs including MRP2 (ABCC2) [6], SPGP (BSEP, ABCB11) [7], and BCRP (ABCG2) [8, 9]. Two transporter families from the SLC group, the organic anion transporting polypeptide (OATP or SLC21, SLCO and SLC22) family [10], and the organic cation transporter (OCT, or SLC22) family [11], are also important in the uptake of drugs into the brain and systemic circulation.

2.2
The Solute Carrier Family

2.2.1
The Major Facilitator Superfamily

The major facilitator superfamily (MFS) is the largest group of transporters containing over 15, 000 sequenced members to date [12]. This family is ubiquitous in both eukaryotes and prokaryotes and accepts an enormous diversity of substrates including sugars, sugar phosphates, polyols, nucleosides, amino acids, neurotransmitters, and peptides amongst many others. The MFS proteins operate by "facilitating" the diffusion of a solute from a higher to a lower electrochemical potential, via specific binding between the solute and the transporter. The substrate of interest is therefore transported by one of three mechanisms: (a) uniport where the substrate is the "solute" and is energised solely by its own concentration gradient, (b) symport where the substrate is translocated in the same direction as the solute, and (c) antiport wherein the substrate is transported in the opposite direction to the solute. This facilitated diffusion therefore does not require a supply of energy. The potential of these proteins as targets for drug discovery is beginning to be realised; for example several companies are currently pursuing clinical trials on inhibitors of the glucose transporter SGLT-2 for the treatment of non-insulin dependent diabetes [13].

2.2.2
The Neurotransmitter: Sodium Symporter Family

The neurotransmitter: sodium symporter (NSS) family is arguably the best exploited in terms of targets for drug discovery. Many marketed drugs act on NSS transporters including blockbusters such as fluoxetine **1** (Prozac), sertraline **2** (Zoloft), paroxetine **3** (Paxil), buproprion **4** (Wellbutrin) and venlafaxine **5** (Effexor), as well as the newer duloxetine **6** (Cymbalta) and

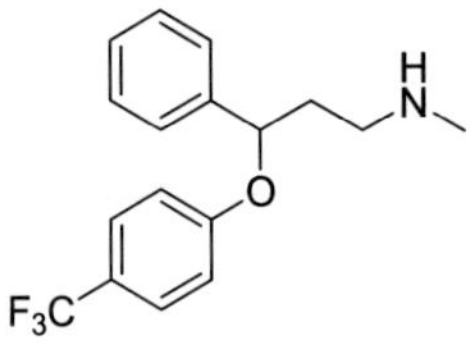

Fluoxetine (Prozac®)
1

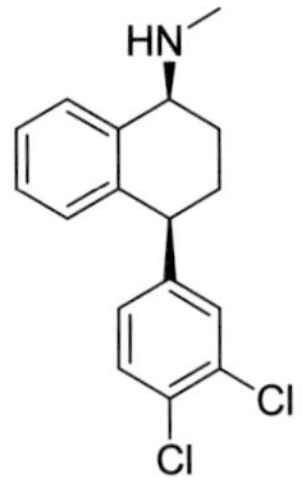

Sertraline (Zoloft®)
2

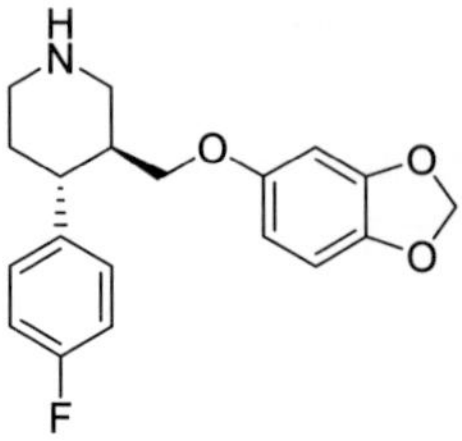

Paroxetine (Paxil®)
3

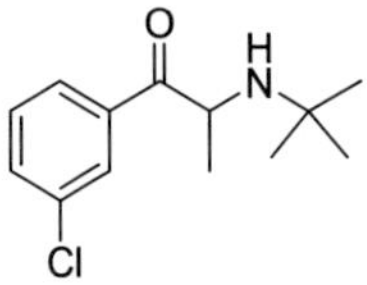

Bupropion (Wellbutrin®)
4

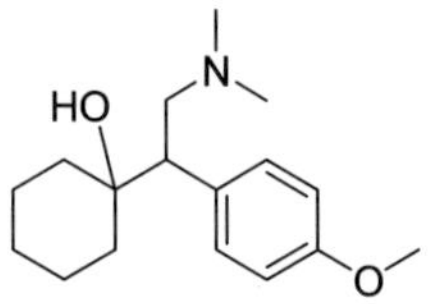

Venlafaxine (Effexor®)
5

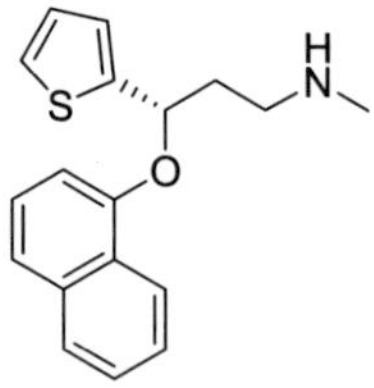

Duloxetine (Cymbalta®)
6

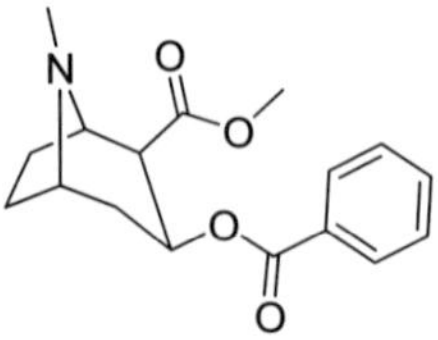

Cocaine
7

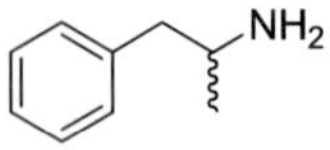

Amphetamine
8

Fig. 1

a number of psychostimulant drugs of abuse such as cocaine 7, and amphetamine 8 (Fig. 1) . The NSS family plays a particularly important role in neurotransmitter signalling in the CNS and hence NSS transporter inhibition has found widespread application in psychiatry. The NSS family of transporters, which includes SERT, NET, DAT, GAT and GlyT therefore forms the focus for this volume.

The NSS transporters use a secondary active transport mechanism to translocate the substrate across the membrane; in an analogous fashion to the ABC family, the energy stored in ATP phosphate bonds is harnessed to drive an ion, eg. H^+ or Na^+, down a concentration gradient. At the same time the substrate is transported against its concentration gradient. The process can be termed symport or antiport as the substrate is transported in the same direction, or the opposite direction to the ion current, respectively.

2.2.3
The Dicarboxylate/Amino Acid:Cation (Na$^+$ or H$^+$) Symporter (DAACS) Family

The most prominent members of this family are the sodium-dependent excitatory amino acid transporters EAATs which have been implicated in a number of neurodegenerative and neuropsychiatric disorders. The EAATs use a secondary active transport mechanism in which the transport of the excitatory amino acid glutamate across the membrane is accompanied by the transport of three sodium ions and a proton.

3
NSS Transporter Assays

The two most common methods for evaluating transporter inhibitors in vitro are equilibrium binding assays and functional reuptake assays, reported as K_is or IC_{50}s. It is this data which is most commonly used in medicinal chemistry programmes to compare compounds of interest and to select compounds for progression to later stage drug development. The classic methods for both approaches use a radiolabelled compound; either a radiolabelled transporter inhibitor for the binding assay, or a radiolabelled neurotransmitter for measuring the reuptake inhibition. Both binding and functional assays can be carried out in either cultured cell systems (where available) or in native tissue such as rat synaptosomes. Care needs to be taken in extrapolating data from these assays, it is not uncommon to observe potency differences between binding and functional assays and between overexpressing systems and native preparations. This is of particular importance when a mixed inhibitor profile is desired, as is increasingly the case in the monoamine transporter field (vide infra Chen et al. and Whitlock et al.). The situation is further complicated by the fact that there is significant "cross talk" between neurotransmitter neurons in the brain such that a given inhibitor profile in vitro may not correlate with the expected profile in vivo. In vivo neurochemistry studies are usually required to confirm the inhibition of the desired neurotransmitter/s in a suitable animal model.

More recently, there has been progress in developing fluorescence-based functional uptake assays, notably for the monoamine transporters [14–16] GlyT [17, 18] and the EAATs [19]. This allows for the possibility of large scale high-throughput screening (HTS) campaigns. The costs associated with the use and disposal of radiolabelled compounds make the classical binding and functional assays less amenable to HTS, and hence the development of viable fluorescent substrates for other transporters is eagerly anticipated.

More detailed pharmacology of transporter inhibitors can be evaluated in functional assays to look at whether the compound is a substrate, or inhibitor, a reversible or non-reversible binder and to investigate inhibition kinetics. However, these assays are not routinely carried out and are typically reserved for compounds that have been selected as potential drug candidates.

4
Transporters and Structure-Based Drug Design (SBDD)

Obtaining suitable crystals for X-ray structure determination is a problem common to all classes of integral membrane proteins, and transporters are no exception. The central portion of the transporter which is embedded in the plasma membrane usually exists as a bundle of α-helices which has evolved to be stabilised by a lipid environment, conversely, the extracellular and intracellular loops which link the α-helices are hydrophilic in character. This environment is very difficult to replicate in suitable crystallisation conditions, and as a result it has only been in the last decade that the first high-resolution X-ray crystal structures have started to emerge. Recently, representative transporters from each of the categories have been successfully crystallised to afford high-resolution 3D structures; three from the MFS family LacY [20], EMrD [21] and GlpT [22], the ABC importers MetNI [23], BtuCD [24, 25], HI1470/71 [26], ModBC [27], and MalFGK [28] and exporters Sav1866 [29] and MsbA [30], and two high-resolution X-ray structures from the neurotransmitter family, the NSS transporter LeuT [31] and the DAACs family archaeal glutamate transporter homologue from *Pyrococcus horikoshii* Glt$_{Ph}$ [32].

The availability of these structures has the potential to revolutionise our understanding of the complex kinetics and mechanisms involved in the selective recognition and transport of molecules across the plasma membrane. The potential impact of the publication of these structures on SBDD is also significant. The implications and additional potential for homology modeling, NMR studies and electron microscopy are discussed in detail by Sylte et al. (vide infra).

Of particular interest, given the focus herein on the NSS transporters, is recent work from Gouaux et al. They identified the tricyclic antidepressant (TCA) clomipramine (Fig. 2) as a weak $\sim 2\,\mu$M non-competitive inhibitor of LeuT, and were able to obtain high-resolution crystal structures of clomipramine in complex with LeuT, the Leu substrate, and a sodium ion [33]. The 3D structure of LeuT, thought to share a common architecture with other NSS transporters, consists of 12 α-helical transmembrane domains, with TMs 1–5 and 6–10 related to each other by a pseudo 2-fold axis in the membrane plane (Fig. 2). The Leucine and sodium ions are bound in the centre of the protein, half way across the membrane bilayer [31]. The clomipramine bound structure shows the clomipramine positioned at the extracellular side of the transporter, just above the substrate binding pocket.

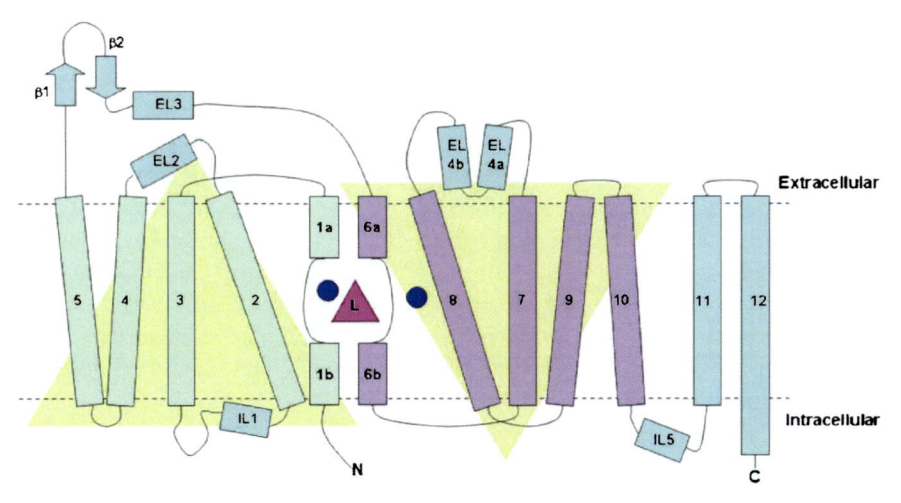

Clomipramine

Fig. 2 a Structure of Clomipramine

Fig. 2 b The LeuT$_{Aa}$ topology reported by Gouaux et al. [23], as determined by X-ray crystallography. TM1–TM5 and TM6–TM10 are related by a pseudo-two-fold axis located in the plane of the membrane. The leucine substrate is represented by a *magenta triangle* and the 2 sodium ions as *blue circles*

Given that clomipramine is known to inhibit SERT and NET in humans, this clomipramine-bound structure has the potential to further our understanding of the structural basis of inhibition at the monoamine transporters. However, it is important to note that the sequences of SERT and NET are significantly different from LeuT, and recent data on SERT [34] and NET [35] suggest that the TCAs may bind in a region closer to the substrate binding pocket.

A contemporaneous study by Wang and Reith et al. reported the crystal structure of LeuT in complex with the leucine substrate and the antidepressant desipramine [36]. They also reported gain-of-function mutagenesis experiments on hSERT and hDAT which suggested that the binding mode observed in the LeuT-desipramine structure is conserved in the monoamine transporters.

5
Transporters as Targets for Drug Discovery

This volume focusses on transporters as targets for drug discovery. A diverse range of topics are covered; from recent advances in the structural biology of transporters and its impact on potential structure-based drug design (Sylte et al.), through a case study of medicinal chemistry drug design (Newman et al.), to the design of PET ligands and their importance in understanding early clinical trial data (Antoni et al.).

The monoamine transporters are the most established drug targets, and as such there are many excellent reviews covering the launched antidepressant selective serotonin uptake inhibitors (SSRIs), selective serotonin and noradrenaline reuptake inhibitors (SNRIs) and TCAs [37]. In this volume we focus rather on more recent developments in the search for second generation monoamine reuptake inhibitors which address the deficits in current marketed drugs. The SSRIs have a good side effect profile however, as antidepressants they suffer from a slow onset of action and significantly, 30–40% of patients do not respond satisfactorily to them. Conversely, although the TCAs are effective antidepressants they have poor selectivity over muscarinic, histaminic and adrenergic receptors, resulting in cardiovascular, anticholinergic and sedative side effects.

Whitlock et al. describe progress in the discovery of SSRIs, noradrenaline reuptake inhibitors (NRIs), and SNRIs from 2000 to the present day. Whilst Chen et al. focus on recent developments in the search for triple SERT, NET and DAT reuptake inhibitors. The interest in these areas stems not only from the potential for improved antidepressant efficacy and side effect profiles, as has been proposed for the triple reuptake inhibitors [38], but the recognition that by tweaking the transporter profile potential therapies for other diseases associated with neurotransmitter imbalance can be developed. For example, although duloxetine 6 (Fig. 1), a dual SNRI, was initially launched in 2004 for the treatment of major depressive disorder (MDD) [39], since 2004 additional approvals have been granted for pain associated with diabetic neuropathy [40] and fibromyalgia [41], for stress urinary incontinence [42] and generalised anxiety disorder [43]. NRIs have been licensed for the treatment of attention deficit hyperactivity disorder (ADHD), as well as being of interest for the treatment of neuropathic pain. The continuing increase in the number of patents being filed for monoamine reuptake inhibitors reflects the ongoing interest in these transporters as targets for drug discovery.

The monoamine reuptake transporters, and DAT in particular, are also responsible for the stimulant properties of several drugs of abuse such as cocaine and amphetamine. The mechanisms underlying the abuse potential of these drugs remain the subject of debate. The dopamine transporter (DAT) hypothesis of cocaine's behavioral effects was first proposed by Ritz et al. [44] following their observation that there was a positive correlation between the binding affinity at DAT and the potency for self-administration of a variety of monoamine uptake inhibitors. Many studies have subsequently supported the DAT hypothesis, however it is becoming more evident that the elegant simplicity of this hypothesis may hide a more complex reality. For example, DAT knockout mice still exhibit place preference and self-administration of cocaine [45, 46]. The interdependency of the monoamine neurotransmission systems and the fact that cocaine also inhibits SERT and NET further complicate interpretation of the in vivo data. Finally, there are non-cocaine based DAT inhibitors which are used clinically which do not show abuse potential and it is one class of these inhibitors, the benztropines (Fig. 3), which forms the subject of a case study in drug discovery by Newman et al. In their chapter they review the current state-of-the-art in our understanding of the mechanisms underlying the abuse potential of this class of drugs.

It is remarkable that despite nearly half a century of use of monoamine reuptake inhibitors our understanding of their mode of action, cocaine being a case in point, is still limited. Despite advances in our knowledge of the neurotransmitter systems at the molecular level, translating this information into predictions for the effect of a given compound in man is still fraught with problems. It is not possible to develop a true animal model for a disease such as MDD since it is questionable whether animals can suffer from a similar illness and we cannot ask the animal for a subjective opinion as to how it feels! This is a problem common to all psychiatric illnesses and hence the challenges associated with developing suitable animal models mean that it is often not until late stage clinical trials that the hypotheses for the therapeutic benefit of reuptake inhibition can be tested. The cost of a compound failing late in clinical development is significant and companies are increasingly looking to incorporate imaging techniques into early clinical trials for CNS drugs to limit late stage attrition rates. In their review Antoni et al. introduce the PET imaging technique and discuss the current state-of-the-art with respect to imaging the transporters. They cover not only the neurotransmitter trans-

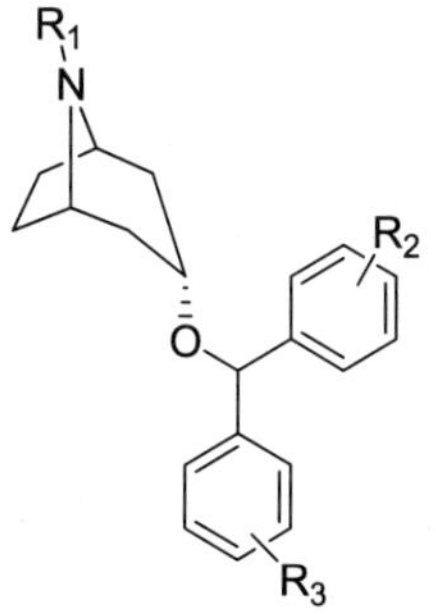

Benztropine R_1=CH$_3$, R_2, R_3=H

Fig. 3

porters but also the ABC transporters, the glucose transporter GLUT and the vesicular monoamine transporter-2. PET imaging cannot increase the likelihood of a compound succeeding in the clinic, but rather allows companies to make faster decisions to stop the development of ineffective compounds. PET imaging can be used to relate drug pharmacokinetics in plasma to receptor occupancy, and subsequently to relate this to clinical efficacy. In the absence of the receptor occupancy information from imaging studies it is difficult to assess whether a lack of clinical efficacy is due to insufficient drug at the desired site of action or due to failure of the mechanistic hypothesis. The translatability of this non-invasive technique provides a link with PET imaging in animal models, thus allowing preclinical evaluation and ranking of new chemical entities to select the most promising to progress into the clinic.

Although there are PET tracers available, there is still much work to be done to identify "ideal" PET ligands for the transporters suitable for use in the clinic. As highlighted by Antoni et al. in their review, "the stringent criteria required for a suitable PET tracer mean that the process of identifying a suitable PET ligand presents as many challenges as the discovery of a new drug".

In addition to regulating monoaminergic chemical transmission, transporters also play a role in controlling synaptic concentrations of amino acid neurotransmitters. Two such transporters for the CNS active amino acid glycine, GlyT1 and GlyT2 from the NSS family were identified in the early 1990s [47–49]. Since that time there has been significant interest in GlyT1 inhibition as a therapy for schizophrenia with the proposed additional benefit of improved cognition. Although there is substantial pharmacological evidence to support this therapeutic hypothesis, clinical proof of concept is yet to be determined. However, a number of interesting compounds have now progressed to clinical trials and hence the validity of GlyT1 as a target for schizophrenia, as well as differentiation between the different structural classes of inhibitors, is likely to be clarified in the next decade. In their review Walker et al. (vide infra) review the current medicinal chemistry landscape for the glycine transporters and report progress towards the identification of subtype selective GlyT inhibitors. Particular emphasis is given to developments in the last 2 years towards the identification of non-amino acid based inhibitors.

A volume on neurotransmitter transporters would not be complete without inclusion of the EAATs. Glutamate is now recognised as the primary excitatory neurotransmitter in the CNS where glutamate synapses mediate the majority of fast excitatory neurotransmission. The glutamatergic synapses are essential for normal development and are involved in synaptic plasticity, learning and memory. Attention has tended to focus on the ionotrophic and metabotrophic glutamate receptors (iGluRs and mGluRs, respectively) as targets for therapy however, the EAATs also play an important physiological role in glutamate neurotransmission by clearing glutamate from the synapse. Bridges et al. describe recent research towards the development of subtype selective EAAT inhibitors and subsequent attempts to better understand the contributions of the different EAAT transporters. The studies to date highlight the neuroprotective role of the EAATs and point towards the use of compounds which enhance glutamate uptake for therapy. The feasibility of this as an approach is yet to be determined, since there is little precedent for developing agents which act as positive modulators of the transporters.

6
Summary

The majority of transporters that are actively being pursued as targets for drug discovery are CNS located, and this probably reflects the history of the field which began with the TCAs over half a century ago. The use of transporter inhibition to regulate the synaptic concentrations of key neurotransmitters is an established approach in the discovery of psychiatric medications and the continued interest in the area is manifest in current efforts towards the discovery of mixed reuptake inhibitors with varied 'flavours' of monoamine inhibition. As highlighted by Whitlock et al. in their review of SNRIs, targeting multiple receptors poses significant challenges for the medicinal chemist and also complicates the biological evaluation of these novel ligands. Advances in PET ligand development (Antoni et al.), may help differentiate these second generation antidepressants in the clinic, especially if a suitable NET ligand is identified in the near future.

Until recently, discovery of new inhibitors has typically been through screening compounds prioritised via a pharmacophore-rational design based approach. However, advances in assay technologies such as the development of fluorescence-based reuptake assays, and the recently published X-ray crystal structures of transporters suggest that the impact of HTS and SBDD may change the way that medicinal chemists approach the discovery of transporter inhibitors in the future.

References

1. Linton KJ (2007) Physiology 22:122
2. Davidson AL, Maloney PC (2007) Trends Microbiol 15:448
3. Juliano RL, Ling V (1976) Biochim Biophys Acta 455:152
4. Szákacs G, Váradi A, Özvegy-Laczka C, Sarkadi B (2008) Drug Discov Today 13:379
5. Yu DK (1999) J Clin Pharmacol 39:1203
6. Keppler D, König J, Büchler M (1997) Adv Enzym Regul 37:321
7. Childs S, Yeh RL, Georges E, Ling V (1995) Cancer Res 55:2029
8. Doyle L, Yang W, Abruzzo L, Krogmann T, Gao Y, Rishi AK, Ross DD (1998) Proc Natl Acad Sci USA 95:15665

9. Miyake K, Mickley L, Litman T, Zhan Z, Robey R, Cristensen B, Brangi M, Greenberger L, Dean M, Fojo T, Bates SE (1999) Cancer Res 59:8
10. Rizwan AN, Burckhardt G (2007) Pharmaceut Res 24:450
11. Choi M-K, Song I-S (2008) Drug Metab Pharmacokin 23:243
12. Law CJ, Maloney PC, Wang D-N (2008) Annu Rev Microbiol 62:289
13. www.Thomson-pharma.com
14. Haunso A, Buchanan D (2007) J Biomol Screen 12:378
15. Mason JN, Farmar H, Tomlinson ID, Schwartz JW, Savchenko V, DeFelice LJ, Rosenthal SJ, Blakely RD (2005) J Neurosci Methods 143:3
16. Jørgensen S, Østergaard N, Peters D, Dyhring T (2008) 169:168
17. Allan L, Leith JL, Papakosta M, Morrow JA, Irving NG, McFerran BW, Clark AG (2006) Comb Chem High Throughput Screen 9:9
18. Benjamin ER, Skelton J, Hanway D, Olanrewaju S, Pruthi F, Ilyin VI, Lavery D, Victory SF, Valenzano KJ (2005) J Biomol Screen 10:365
19. Jensen AA, Brauner-Osborne H (2004) Biochem Pharmacol 67:2115
20. Abramson J, Smirnova I, Kasho V, Verner G, Kaback HR, Iwata S (2003) 301:610
21. Yin Y, He X, Szewczyk P, Nguyen T, Chang G (2006) Science 312:714
22. Huang Y, Lemieux MJ, Song J, Auer M, Wang D-N (2003) Science 301:616
23. Kadaba NS, Kaiser JT, Johnson E, Lee A, Rees DC (2008) Science 321:250
24. Locher KP, Lee AT, Rees DC (2002) Science 296:1091
25. Hvorup RN, Goetz BA, Niederer M, Hollenstein K, Perozo E, Locher KP (2007) Science 317:1387
26. Pinkett HW, Lee AT, Lum P, Locher KP, Rees DC (2007) Science 315:373
27. Hollenstein K, Frei DC, Locher KP (2007) Nature 446:213
28. Oldham ML, Khare D, Quiocho FA, Davidson AL, Chen J (2007) Nature 450:515
29. Dawson RJP, Locher KP (2006) Nature 443:180
30. Ward A, Reyes CL, Yu J, Roth CB, Chang G (2007) Proc Natl Acad Sci USA 104:19005
31. Yamashita A, Singh SK, Kuwate T, Jin Y, Gouaux E (2005) Nature 437:215
32. Yernool D, Boudker O, Jin Y, Gouaux E (2004) Nature 431:811
33. Singh SK, Yamashita A, Gouaux E (2007) Nature 448:952
34. Henry LK, Field JR, Adkins EM, Parnas ML, Vaughan RA, Zou M-F, Newman AH, Blakely RD (2006) J Biol Chem 281:2012
35. Paczkowski FA, Sharpe IA, Dutertre S, Lewis RJ (2007) J Biol Chem 282:17837
36. Zhou Z, Zhen J, Karpowich NK, Goetz RM, Law CJ, Reith MEA, Wang D-N (2007) Science 317:1390
37. Butler SG, Meegan MJ (2008) Curr Med Chem 15:1737
38. Chen Z, Skolnick P (2007) Expert Opin Investig Drugs 16:1365
39. Frampton JE, Plosker GL (2007) CNS Drugs 21:581
40. Smith T, Nicholson RA (2007) Vasc Health Risk Manag 3:833
41. Üçeleyer N, Offenbächer M, Petzke F, Häuser W, Sommer C (2008) Neuropsych Dis Treat 4:525
42. Agur W, Abrams P (2007) Expert Rev Obstet Gynecol 2:133
43. Hartford JA, Kornstein SB, Liebowitz MC, Pigott TD, Russell JE, Detke MEFG, Walker DE, Ball SEF, Dunayevich EE, Dinkel JE, Erickson JE (2007) Int Clin Psychopharmacol 22:167
44. Ritz MC, Lamb RJ, Goldberg SR, Kuhar MJ (1987) Science 237:1219
45. Rocha BA, Fumagalli F, Gainetdinov RR, Jones SR, Ator R, Giros B, Miller GW, Caron MG (1998) Nat Neurosci 1:132
46. Sora I, Wichems C, Takahashi N, Li XF, Zeng Z, Revay R, Lesch KP, Murphy DL, Uhl GR (1998) Proc Natl Acad Sci USA 95:7699

47. Guastella J, Brecha N, Weigmann C, Lester HA, Davidson N (1992) Proc Natl Acad Sci USA 89:7189
48. Liu QR, López-Corcuera B, Mandiyan S, Nelson H, Nelson N (1993) J Biol Chem 268:22802
49. Smith KE, Borden LA, Hartig PR, Branchek T, Weinshank RL (1992) Neuron 8:927

Top Med Chem (2009) 4: 15–51
DOI 10.1007/7355_2008_023
© Springer-Verlag Berlin Heidelberg
Published online: 22 May 2008

Membrane Transporters: Structure, Function and Targets for Drug Design

Aina W. Ravna · Georg Sager · Svein G. Dahl · Ingebrigt Sylte (✉)

Department of Pharmacology, Institute of Medical Biology, Faculty of Medicine,
University of Tromsø, 9037 Tromsø, Norway
sylte@fagmed.uit.no

Abstract Current therapeutic drugs act on four main types of molecular targets: enzymes, receptors, ion channels and transporters, among which a major part (60–70%) are membrane proteins. This review discusses the molecular structures and potential impact of membrane transporter proteins on new drug discovery. The three-dimensional (3D) molecular structure of a protein contains information about the active site and possible ligand binding, and about evolutionary relationships within the protein family. Transporters have a recognition site for a particular substrate, which may be used as a target for drugs inhibiting the transporter or acting as a false substrate. Three groups of transporters have particular interest as drug targets: the major facilitator superfamily, which includes almost 4000 different proteins transporting sugars, polyols, drugs, neurotransmitters, metabolites, amino acids, peptides, organic and inorganic anions and many other substrates; the ATP-binding cassette superfamily, which plays an important role in multidrug resistance in cancer chemotherapy; and the neurotransmitter:sodium symporter family, which includes the molecular targets for some of the most widely used psychotropic drugs. Recent technical advances have increased the number of known 3D structures of membrane transporters, and demonstrated that they form a divergent group of proteins with large conformational flexibility which facilitates transport of the substrate.

Keywords Three-dimensional structure · Drug discovery · Drug targets · Membrane proteins · Transporters

Abbreviations

ABC	ATP-binding cassette
ATP	Adenosine triphosphate
cGMP	Cyclic guanosine monophosphate
CNS	Central nervous system
2D	Two dimensional
3D	Three dimensional
DAACS	Dicarboxylate/amino acid:cation symporter
DAT	Dopamine transporter
DHA1	Drug:H^+ antiporter-1
DMT	Drug/metabolite transporter
DNA	Deoxyribonucleic acid
EAAT	Excitatory amino acid transporter
E-MeP	European Membrane Protein Consortium
EU	European Union
GABA	Gamma-aminobutyric acid
GAT	GABA transporter
GLUT	Glucose transporter
HAE	Hydrophobe/amphiphile efflux
HIV	Human immunodeficiency virus
5-HT	5-Hydroxytryptamine (serotonin)
MFS	Major facilitator superfamily
MS	Mass spectrometry

NARI	Noradrenaline reuptake inhibitor
NBD	Nucleotide binding domain
NET	Noradrenaline transporter
NMR	Nuclear magnetic resonance
NSS	Neurotransmitter:sodium symporter
OAT	Organic anion transporter
PDB	Protein data bank
PEPT	Dipeptide transporter
PEPT1	H^+/dipeptide symporter
PfHT	Parasite-encoded facilitative hexose transporter
RMSD	Root mean square difference
RNA	Ribonucleic acid
RND	Resistance-nodulation-cell division
SERT	Serotonin transporter
SMR	Small multidrug resistance
SSRI	Selective serotonin reuptake inhibitor
TC	Transporter classification
TCDB	Transport classification database
TMD	Trans-membrane domain
TMH	Trans-membrane helix
VMAT	Vesicular monoamine transporter

1
Introduction

A number of consortia bringing together researchers from academic research institutions and companies have been established to determine the three-dimensional (3D) structures of proteins, rapidly and cost-effectively using modern methodologies [1]. At the end of April 2007, the number of entities in the PDB database (http://www.rcsb.org/pdb/) was greater than 42 000. The number of entities in the PDB database increased by more than 5000 during 2006, which is equivalent to the total number of entities in the database 10 years ago. Genome sequencing together with significant advances in process automation and informatics have aided the development of high-throughput X-ray crystallography, and are the main reasons for the large increase in the number of available 3D structures.

Atomic-resolution 3D structures provide important knowledge on biologically active molecules. The molecular structure of a protein contains information about the active site architecture, possible ligand or antigen binding sites, and evolutionary relationships within the protein family, and may serve as a basis for designing protein engineering experiments. The shape and electrostatic properties obtained from the molecular structure are also important for predicting possible interaction partners involved in regulation and complexation. Knowledge of the 3D structures of drug-target complexes defines the topography of the complementary surface between the drug target and the ligands, and provides the possibility of virtual screening experiments searching for possible new molecules binding to the target [2] and for structure-aided drug design [3].

When detailed structural data for the target protein are available, computer programs can be used for ligand docking and virtual screening of compound libraries, and to predict protein–ligand binding affinities in the search for possible lead compounds. The obtained information can help the synthetic chemist to optimize compounds by including chemical groups that can form better interactions with the target, resulting in improved potency and selectivity [4]. At the moment there are several drugs on the market originating from a structure-based design approach. Examples include the HIV drugs Agenerase and Viracept developed using the X-ray crystal structure of HIV proteinase [5, 6], development of the flu drug Zanamivir based on the X-ray structure of neuraminidase [7], and the angiotensin-converting enzyme inhibitors [8, 9]. Direct structural determination by experimental methods like NMR spectroscopy and X-ray crystallography, and indirect structural knowledge obtained by different biophysical and molecular biology studies, together with bioinformatics and computational chemistry are of pivotal importance in the discovery and development of biologically active molecules, and of more effective and safer drugs.

During the last few years progress in genome sequencing has provided, and still provides, important information about the genetic map of different organisms. Modern technologies, such as microarray technology and 2D electrophoresis/mass spectrometry (MS), have provided insight into regulatory mechanisms at the DNA, RNA and protein levels. In the post-genomic era, focus will be on understanding the cellular machinery for regulation and communication, and how proteins and other gene products cooperate on a detailed atomic level. Such information provides insight into biological mechanisms and disease processes, and is important for the discovery and development of new drugs. However, knowledge of the detailed 3D structure of molecules involved in cellular communication will also be important in order to understand the cellular machinery. Structural information about central macromolecules and their regulation and interaction partners will most probably contribute to the discovery of new targets for therapeutic intervention, and may also give new insight into how drug targets can be therapeutically exploited. The drugs of the future may not only be traditional ligands functioning as an agonist, antagonist, substrate or inhibitor, but also act as scaffolding ligands by promoting protein–protein association, by preventing protein–protein association, or by enhancing or preventing degradation, internalization, etc. Future drugs may even be able to interfere with the specific signalling pathway(s) of a receptor without interfering with the other pathway(s) of the same receptor. Structural biology techniques, including theoretical calculations, and 3D structural information may therefore become even more important in the future.

Membrane transporter proteins are crucial co-players in cellular processes, and are known molecular components of many disease processes. The membrane transporter proteins are targeted by several presently used drugs, and have a large potential as targets for new drug development. In this review we discuss the current structural knowledge of membrane transporter proteins and its impact on new drug discovery.

2
Membrane Protein Structures

The protein targets for drug action on mammalian cells can broadly be divided into four main types: receptors, enzymes, ion channels and transporters. Integral membrane proteins are involved in a variety of processes governing cellular functions, and provide a plethora of molecular targets for pharmacological intervention. A large number (60–70%) of the presently known drug targets are proteins embedded in a cellular membrane, and membrane proteins are among the most interesting macromolecules to study by structural biology techniques. High-resolution structural information about proteins embedded in a cellular membrane is of pivotal importance for developing new drugs with therapeutic potential, but is also important for the understanding of the molecular mechanisms of cellular communication and function.

During the last few years, several international structural genomics networks have been established focusing on whole genomes [10], and some networks are focusing uniquely on membrane proteins. One of these networks is the EU-funded E-MeP consortium (http://www.ebi.ac.uk/e-mep/) that was established in 2005 with the goal of developing novel technologies to facilitate the purification and crystallization of membrane proteins. Currently around 20 European laboratories are members of the consortium, while additional laboratories are associate members. E-MeP is exploring several expression systems for 100 different prokaryotic and 200 different eukaryotic membrane proteins.

Crystallization and structure determination of membrane proteins are still not straightforward processes, and current knowledge of the detailed 3D structures of membrane proteins is limited. Out of the more than 42 000 entities deposited in the PDB database, only around 0.3% are unique structures of membrane proteins, although membrane proteins are estimated to represent approximately one third of the proteins coded for in the human and other genomes [11, 12]. Some of the most important questions in the fields of biology, chemistry and medicine remain unsolved as a result of the currently limited understanding of the structure, behaviour and molecular interactions of membrane proteins.

Integral membrane proteins of known 3D structure basically have two different types of architecture: α-helical bundles or β-barrels. Up to now, eukaryotic plasma and reticulum membrane proteins have been shown to be α-helical, while the β-barrel membrane proteins are mainly found in the outer membrane of Gram-negative bacteria and in mitochondria and chloroplast membranes [13]. The helix bundle proteins contain quite long transmembrane hydrophobic α-helices that are packed together into bundles with relatively complicated structure, while the β-barrel proteins are large proteins consisting of anti-parallel β-sheets that fold into a barrel closed by the first and last strands of the sheet [14, 15]. In amino acid sequences of proteins with unknown 3D structure and function, the long hydrophobic transmembrane α-helices are easier to recognize in the sequence than the less hydrophobic trans-membrane β-strands. Bioinformatics studies are therefore generally easier to perform for α-helical bundle trans-membrane proteins than for β-barrel trans-membrane proteins, and have produced much more information about α-helical bundle proteins. Since the 3D structure of integral membrane proteins is not easily determined experimentally, prediction of the secondary structure from the amino acid sequence is important for annotating protein sequences to membrane protein families. This, together with recognition of structural motifs by bioinformatics, provides structural information of value for determining the function and predicting the 3D structure of trans-membrane proteins [16–18].

3
Membrane Transporter Proteins

Ions and small organic molecules are often too polar to penetrate the cellular membrane on their own, and require a transport protein. Trans-membrane solute transporters may be divided into channels that function as selective pores opening in response to a chemical or electrophysiological stimulus, thus allowing movement of a solute down an electrochemical gradient, and active carrier proteins which use an energy-producing process to translocate a substrate against a concentration gradient [19].

Transporter proteins have a recognition site making them specific for a particular solute. The human genome contains many different transporters, including those responsible for the transport of glucose and amino acids into cells, transport of ions and organic molecules by the renal tubules, transport of Ca^{2+} and Na^+ out of cells, uptake of neurotransmitters and neurotransmitter precursors into nerve terminals and vesicles, and transporters involved in multidrug resistance. Drugs may exert their effect by binding to transporters and either inhibiting transport of the solute or functioning as a false substrate for the transport process.

Examples of such drugs include the antidepressant drugs that inhibit the neuronal transporters for noradrenaline and serotonin [20, 21], probenecid which inhibits the weak acid transporter protein in the renal tubule [22], loop diuretics inhibiting the $Na^+/K^+/2Cl^-$ co-transporter of the loop of Henle [23], and the irreversible inhibitor of the H^+/K^+ ATPase (proton pump) of the gastric mucosa, omeprazole [24]. The lack of atomic-resolution 3D structures of membrane transporter proteins limits the design of new ligands interfering with the structure and function of the transporter. Only a few membrane transporter proteins from bacterial species have been crystallized and examined by X-ray diffraction experiments [25]. This makes molecular modelling by biocomputing an interesting methodological alternative, and in many cases the only method available for structural studies of membrane transporter proteins. However, such methods depend on a combination of computational techniques and experimental structural information to guide the molecular modelling process.

3.1
Classification of Membrane Transport Proteins

According to the classification approved by the transporter nomenclature panel of the International Union of Biochemistry and Molecular Biology [19], transporters belong to six categories:

1. Channels and pores
2. Electrochemical potential-driven transporters (secondary and tertiary transporters)
3. Primary active transporters
4. Group translocators
8. Accessory factors involved in transport
9. Incompletely characterized transport proteins

Categories 2, 3 and 4 are carriers. In contrast to most channels, carriers exhibit stereospecific substrate specificities, and their rates of transport are several orders of magnitude lower than those of other channels [19]. Mammalian species have carriers for peptides, nucleosides, sugars, bile acids, amino acids, organic anions, organic cations, vitamins, fatty acids, bicarbonate, phosphates and neurotransmitters. Numerous transporters of interest as drug targets belong to subclasses 2A (porters) and 3A (diphosphate bond hydrolysis-driven transporters).

Porters are either uniporters, symporters or antiporters. Uniporters are facilitated diffusion carriers that transport single molecules, symporters transport two or more molecules in the same direction, while antiporters transport two or more molecules in opposite directions [19].

Carrier mechanisms are distinguished by the source of energy used to activate the transporter, which may be either one of two:

- Facilitated diffusion
- Active transport

3.1.1
Facilitated Diffusion

Facilitated diffusion is accelerated by specific binding between the solute and the transporter. The solute flows from a higher to a lower electrochemical potential, so-called passive transport, via a uniporter, and facilitated diffusion therefore does not require a supply of energy. Examples of uniporters, or facilitated diffusion transporters, are glucose transporters (GLUTs), as indicated in Fig. 1, and the parasite-encoded facilitative hexose transporter (PfHT) of the major facilitator superfamily (MFS). Examples of GLUTs are GLUT1 and GLUT2. GLUT1 is expressed in highest concentrations in erythrocytes and in endothelial cells of barrier tissues, such as the blood–brain barrier. GLUT2 is expressed in liver cells, pancreatic beta-cells, renal tubular cells and intestinal epithelial cells that transport glucose. GLUT1 is responsible for the basal glucose uptake required to maintain respiration in all cells, and GLUT1 levels are decreased by increased glucose levels and increased by decreased glucose levels. PfHT is used by the malaria parasite to absorb glucose, which it needs to grow and multiply in red blood cells.

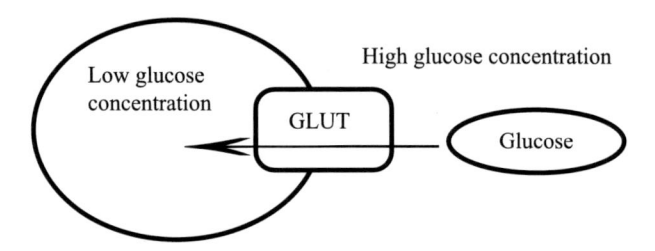

Fig. 1 Facilitated diffusion of glucose through GLUT down the concentration gradient

3.1.2
Active Transport Mechanisms

Active transport uses the free energy stored in the high-energy phosphate bonds of adenosine triphosphate (ATP) as energy source to activate the transporter. There are three types of active transport mechanisms: primary active transport, secondary active transport and tertiary active transport.

Primary active transporters (Fig. 2) use the energy from ATP directly. They exhibit ATPase activity to cleave ATP's terminal phosphate, and move substances from regions of low concentration to regions of high concentration. The ATP-binding cassette (ABC) transporters are primary active transporters comprising a family of structurally related membrane proteins that share a common intracellular structural motif in the domain that binds and hydrolyses ATP. ABC transporters are molecular pumps that regulate the movement of diverse molecules across cellular membranes and represent an

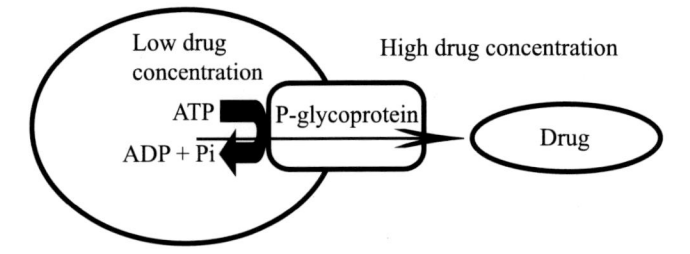

Fig. 2 Primary active transport of drug via P-glycoprotein. The energy from ATP is used to expel the drug out of the cell

important class of targets for discovery of novel small-molecule drugs for treatment of a broad range of human diseases. ABC transporters have both trans-membrane domains (TMDs) and nucleotide binding domains (NBDs). The domain arrangement of these transporters is generally TMD-NBD-TMD-NBD, but domain arrangements such as TMD-TMD-NBD-TMD-NBD, NBD-TMD-NBD-TMD, TMD-NBD and NBD-TMD have also been demonstrated [26, 27]. ABC transporters can be either exporters or importers. A well-characterized ABC exporter is P-glycoprotein, or ABCB1, which is widely distributed in normal cells, such as liver cells, renal proximal tubular cells, cells lining the intestine and the capillary endothelial cells of the blood–brain barrier. P-glycoprotein has broad substrate specificity and may have evolved as a defence mechanism against toxic substances. It actively pumps chemotherapeutic agents out of cancer cells, resulting in multidrug resistance to such drugs (Fig. 2).

Secondary active transporters (Fig. 3) use the energy from a concentration gradient previously established by a primary active transport process. Thus, secondary active transport indirectly uses the energy derived from the hydrolysis of ATP. The driving force of secondary active transport is an ion, for instance H^+ or Na^+, transported down its concentration gradient. Simultaneously, a substrate is transported against its concentration gradient.

There are two types of secondary active transport processes: antiport and symport. In antiport, the driving force ion and the substrate are transported

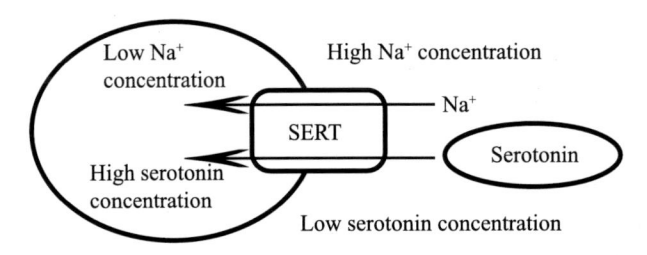

Fig. 3 Secondary active transport. The energy established by the Na^+ gradient is used to transport serotonin against its concentration gradient

in opposite directions, while in symport, they are transported in the same direction. Examples of secondary transporters are the H^+/dipeptide symporter (PEPT1) mainly involved in absorption of di- and tripeptides across plasma membranes in the small intestine and kidney proximal tubules, and central nervous system (CNS) transporters such as the serotonin (5-HT) transporter (SERT), noradrenaline transporter (NET), dopamine transporter (DAT), GABA transporter (GAT) and excitatory amino acid (glutamate) transporter (EAAT). By pumping neurotransmitters back into presynaptic nerve terminals, these CNS transporters play central roles in maintaining the homeostasis of neutrotransmitter levels in neuronal synapses.

Tertiary active transporters like the organic anion transporters (OATs). Tertiary active transporters utilize a gradient generated by secondary active transport. OATs use the outwardly directed dicarboxylate gradient to move (exchange) the organic substrate into the cell. The dicarboxylate gradient is generated by the sodium dicarboxylate co-transporter (secondary active transporter) which is using the inwardly directed sodium gradient initially generated by the Na^+/K^+-ATPase (primary active transporter) [28].

4
Structure Determination of Membrane Proteins

Although structural determination of membrane proteins is not a trivial task, improvements in membrane protein molecular biology and biochemistry, technical advances in structural data collection, notably using synchrotron X-ray beamlines, and the availability of several sequenced genomes have contributed to progress in the number of trans-membrane proteins determined by X-ray crystallography [29–31]. The difficulties in experimental structure determination of trans-membrane proteins arise from their amphiphilic nature. The hydrophilic surfaces are exposed to the aqueous medium, while the hydrophobic surfaces interact with non-polar alkyl chains of phospholipids. The amphiphilic nature makes it difficult to obtain stable and homogeneous protein preparations, and during crystallization, crystal contacts are formed between hydrophilic and hydrophobic surfaces.

Key issues that need to be considered before the structure of a trans-membrane drug target can be determined are [10]:

- How to produce a sufficient amount of the membrane protein.
- How to solubilize and purify the membrane protein without destroying the active 3D conformation of the protein. For membrane transporter proteins this is not trivial, due to the hydrophobic nature of the membrane-spanning region of the protein.
- How to crystallize the membrane transport protein, and what can be done in order to study the 3D membrane protein structure in solution.

4.1
Expression and Purification of Membrane Proteins

In order to determine a protein structure at high resolution, at least milligram quantities of the protein are required. In spite of recombinant protein production techniques and a variety of available expression systems, it has been difficult to provide membrane proteins in a quantity and quality for X-ray crystallographic structure determination. Membrane proteins are often expressed in low abundance in native tissues, and it is therefore necessary to produce the proteins in heterologous expression systems. However, heterologous membrane protein expression may produce toxic effects on host cells, contributing to poor stability and low yields [1]. This problem can be reduced by introducing deletions and mutations into the proteins and by generating fusion constructs. It is also important to use an expression system that does not significantly affect the activity of the mammalian membrane protein, compared with the activity in the native tissue [10]. Prokaryotes may lack many post-translational modification systems of importance for the native activity of the membrane protein. Many different types of recombinant expression systems have been tested for membrane proteins.

The most widely used system for recombinant protein expression of trans-membrane proteins has been *Escherichia coli*, due to the simple and inexpensive scale-up [32], which has so far also been the most successful approach. The expression has been directed to the bacterial membrane or inclusion bodies. Suitable expression vectors are available, and proteins can be labelled metabolically with heavy-atom-labelled amino acids for X-ray crystallography or with stable isotopes for NMR spectroscopy [33]. In addition to *E. coli*, other bacteria have also been tested for membrane protein expression, but have usually given lower yields [34].

Different yeast strains have been used for recombinant expression of a number of trans-membrane proteins [10]. Insect cells have a close resemblance to mammalian cells and have been used for membrane protein expression [10, 35]. Expression in mammalian cells has also been performed, resulting in both transient and stable expression. A general drawback with the use of mammalian cell lines has been that it has given quite low yields compared with bacterial expression systems, and it also involves a more time-consuming procedure [36]. Expression in COS cells and HEK293 cells has successfully been done for membrane transporter proteins including the glutamate transporter [37, 38].

After expression, the protein is solubilized and separated from the lipid components by the use of detergents. This process very often requires an intensive screening process, since different detergents have to be used for different trans-membrane proteins [10]. After solubilization, the recombinant protein is often purified by affinity chromography methods, after insertion of histidine tags into the N- or C-terminal of the protein.

4.2
Structure Determination of Membrane Proteins

The methods used to determine high-resolution atomic structures of proteins are nuclear magnetic resonance (NMR) spectroscopy and X-ray crystallography. Structural determination by X-ray crystallography is so far the method with largest success for trans-membrane proteins. X-ray crystallography and NMR have complementary features in elucidating the structure–functional relationships of proteins and protein–ligand complexes. If a protein forms suitable crystals, X-ray crystallography may represent a convenient and rapid approach, while NMR spectroscopy may have advantages when the structure is partly distorted, exists in several stable conformations in solution or does not crystallize. Solution and solid-phase NMR are also alternatives for structure determination, especially for smaller proteins, but also for protein domains where the electron density is not observed by X-ray crystallography. This is exemplified by the solution NMR structure of the periplasmic signalling domains of the TonB-dependent outer membrane transporter FecA from *E. coli* [39]. Electron cryomicroscopy also contributes valuable structural information about membrane proteins, although at much lower resolution than that obtained by X-ray crystallography [40].

Since structure determination of membrane proteins by experimental methods has so far proven very challenging, structure prediction by homology modelling [25] using modern bioinformatics techniques may represent an alternative, and very often the only alternative, to obtain insight into the atomic structure of membrane transporters and other membrane proteins.

4.2.1
X-Ray Crystallography

The use of advanced protein expression and purification procedures, crystallization robots and powerful synchrotron radiation sources has enabled high-throughput structure determination using X-ray crystallographic techniques. Crystallization techniques and structure determination have become "high-throughput" for several protein families, but for membrane proteins including transporter proteins, the available crystallization and structure solution methods are not regarded as high throughput.

A high-resolution X-ray crystallographic structure provides structural information at an atomic level and is a powerful method for studying the structure of drug targets and their ligands. X-ray crystal structures represent time and space averages of all atoms present within the protein molecule, and may also provide information about the structural movements of the protein. The process of X-ray structure determination of trans-membrane proteins has different steps including crystallization of the purified membrane protein, measurements of crystal diffractions, calculation of electron density and model building [1, 10].

A major challenge of X-ray crystallography of trans-membrane proteins is to obtain suitable 3D crystals. Homogeneity and stability at high protein concentrations are important to obtain good results. Different strategies have been used for producing suitable crystals. These strategies include the use of detergents that replace the native membrane lipids and form mixed detergent–membrane protein micelles, crystallization using vapour diffusion, and crystallization using lipid cubic phases and bicelles [29]. The rationale behind the methods using cubic phases [41–43] or bicelles [44, 45] is that the solubilized membrane protein is inserted into a native-like environment that is believed to improve the chances of crystallization.

4.2.2
NMR Spectroscopy

NMR spectroscopy investigates transition between spin states of magnetically active nuclei in a magnetic field. Determination of the solution structure of trans-membrane proteins by NMR requires that well-resolved 2D ^{1}H/^{15}N chemical shift correlation spectra can be obtained. For helical trans-membrane proteins, spectral resolution is complicated by the limited amide ^{1}H chemical shift dispersion in α-helixes and the slow correlation time for many micelle-bound proteins [46].

In general, NMR methods have advanced to the point where small to medium sized protein domain structures can be determined in a quite routine manner, and solution NMR spectroscopy has emerged as an eminent tool in studies of protein structure [47] and intermolecular interactions [48]. Of about 42 000 entities (April 2007) deposited in the PDB database (http://www.rcsb.org/pdb/), about 15% were determined by NMR techniques. NMR spectroscopy of proteins contributes with important information about the kinetics, thermodynamics, conformational equilibria, molecular motions and ligand binding equilibria of the protein, since the signals observed in solution by NMR show the chemical properties of atomic nuclei, including the their relative motions [49].

If over-expressed proteins are inserted efficiently into membranes, they might also be studied by solid-state NMR spectroscopy without prior dissociation. When this method can handle larger proteins, the method holds promise for 3D structure determination of membrane proteins [50, 51].

4.2.3
Electron Microscopy

The basic idea of electron microscopic 3D structure determination is to produce 2D projection images (2D crystals) from a 3D object. These 2D projection images can then be used to reconstruct the 3D structure of the original object by applying back-projection algorithms [1]. The method can be used to study large macromolecular machines like the ribosome or spliceosome which undergo massive structural rearrangements [40]. A number of membrane proteins have been reconstituted to form 2D crystals. The quality of the diffraction in the best direction of optimum crystals typically ranges from about 6–7 Å resolution up to 3 Å. At around 6 Å resolution trans-membrane α-helices can be revealed [52], while at 3 Å the protein backbone and larger side chains can be modelled.

4.2.4
Three-Dimensional Structure Prediction

Comparative modelling or homology modelling can be used to generate 3D structural models of proteins with unknown structure [53]. In homology modelling or comparative modelling, molecular modelling techniques are used to construct 3D models of the protein of interest (the target protein) using structural information from a protein with known 3D structure (the template protein), based on a postulated structural conservation between the template and target proteins. The homology modelling approach is based on the observation that the 3D structure of homologous proteins is more conserved than the amino acid sequence. Combined with structural information from molecular biology studies (e.g. site-directed mutagenesis experiments) and ligand binding studies, homology modelling provides indirect structural knowledge about the target protein and its interactions with drugs and other interaction partners.

When the structural similarities between the target and the template protein are high, the homology modelling approach may give structural models of sufficient accuracy for virtual screening of compound libraries and target-based ligand design. The accuracy of a model constructed by homology modelling depends on the conservation of secondary structure between the template and the target [54]. Sequence similarities larger than 50% between the template and the target are assumed to produce quite accurate structural models. Sequence similarities of 50% are expected to give a root mean square difference (RMSD) of about 1 Å between the backbone atoms of the template structure and the model. However, even at an overall sequence identity of <20% between the template and the target, the active sites and the secondary structure elements necessary for building the protein scaffold may have very similar geometries [55].

Several automatic homology modelling methods are available on the internet [56–58]. Such automatic modelling methods may provide models of high accuracy when the structural conservation between the template and the target is high. The most important single determinant for the quality of the homology-based model is the accuracy of the amino acid sequence alignment between the template and the target [54]. For membrane transporters, the sequence identity between a bacterial template and a modelled mammalian membrane transporter is often low, and the alignment often has to be manually adjusted based on experimental observations, particularly from site-directed mutagenesis experiments [59]. The interpretation of site-directed mutagenesis results is therefore very important in the process of modelling membrane proteins. Models with low sequence similarity to the template structure are valuable working tools for generating hypotheses about the structure and function of the target protein, for designing new experimental studies, and along with structural information would contribute value to ligand design.

5
Transporters of Known 3D Structure

Three-dimensional crystal structures of several bacterial transporters for organic molecules have been determined by X-ray crystallography at atomic resolution, as shown in Table 1.

5.1
The Major Facilitator Superfamily

The major facilitator superfamily (MFS) includes almost 4000 different transporter proteins. The MFS family members transport diverse substrates including sugars, polyols, drugs, neurotransmitters, Krebs cycle metabolites, phosphorylated glycolytic intermediates, amino acids, peptides, organic and inorganic anions and many more (http://www.tcdb.org/) [60–63]. These transporters function by uniport, symport or antiport mechanisms, and may possess either 12 [64], 14 [65] or 24 [66] trans-membrane helices (TMHs), with a common evolutionary ancestor [67]. Examples of human MFS transporters are glucose uniporters (GLUTs), the vesicular monoamine transporter 1 and 2 (VMAT1 and VMAT2), the thyroid hormone transporter (MCT8) and the organic anion transporter (OAT) family.

The 3D structures of three *E. coli* transporter proteins of the MFS family have been determined by X-ray crystallography at atomic resolution: EmrD [68] (Fig. 4a), GlpT [69] and LacY [70] (Fig. 4b), at 3.5, 3.3 and 3.5 Å, respectively. These structures indicate that MFS proteins with 12 TMHs share a common architecture of the membrane spanning region, organized in symmetrical N- and C-terminal domains each of six TMHs, with overall structural topologies resembling each other. TMHs 3, 6, 9 and 12 are facing away from the interior of the transporters [68–70] (Fig. 4). The orientations of the TMHs of EmrD are different from those of GlpT and LacY, presumably

A.W. Ravna et al.

Table 1 Transporters for organic substrates with 3D structure determined at atomic resolution by X-ray crystallography. TC: transporter classification number (http://www.tcdb.org/)

Name	Family/superfamily	Function	Organism	TC number	PDB ident.	Refs.
EmrD (hydrophobic uncoupler, e.g. CCCP, benzalkonium, SDS):H^+ antiporter	Major facilitator superfamily (MFS)	H^+ antiport (multiple drugs)	*Escherichia coli*	2.A.1.2.9	2gfp	[68]
GlpT (glycerol-P:Pi antiporter)	Major facilitator superfamily (MFS)	Pi/glycerol-3-phosphate antiport	*Escherichia coli*	2.A.1.4.3	1PW4	[69]
LacY (lactose:H^+ symporter)	Major facilitator superfamily (MFS)	Symport (lactose/H^+)	*Escherichia coli*	2.A.1.5.1	1PV6 1PV7	[70]
AcrB (multidrug/dye/detergent resistance pump)	Resistance-nodulation-cell division (RND) superfamily	H^+ antiport	*Escherichia coli*	2.A.6.2.2	1OY6 etc	[73]
EmrE (small multidrug efflux pump)	Drug/metabolite transporter (DMT) superfamily	H^+ antiport	*Escherichia coli*	2.A.7.1.3	1S7B 2F2M	[82]
LeuT$_{Aa}$ (amino acid (leucine):2 Na^+ symporter)	Neurotransmitter:sodium symporter (NSS) family	Na^+ symport	*Aquifex aeolicus*	2.A.22.4.2	2A65	[64]
Glt$_{ph}$ (archaeal glutamate transporter homologue)	Dicarboxylate/amino acid:cation (Na^+ or H^+) symporter (DAACS) family	H^+ symport	*Pyrococcus horikoshii*	2.A.23.1.5	1IXFH	[86]
Sav1866 (multidrug exporter)	ABC superfamily		*Staphylococcus aureus*	3.A.1.106.2	2HYD	[88]
BtuCD (bacterial ABC transporter involved in B_{12} uptake)	Vitamin B_{12} uptake transporter (B_{12}T) family (ABC superfamily)	Vitamin B_{12} uptake (ATP)	*Escherichia coli*	3.A.1.13.1	1L7V	[93]
HI1470/1 (bacterial ABC transporter involved in haem and B_{12} uptake)	Vitamin B_{12} uptake transporter (B_{12}T) family (ABC superfamily)	Haem and B_{12} uptake	*Haemophilus influenzae*	–	2NQ2	[94]

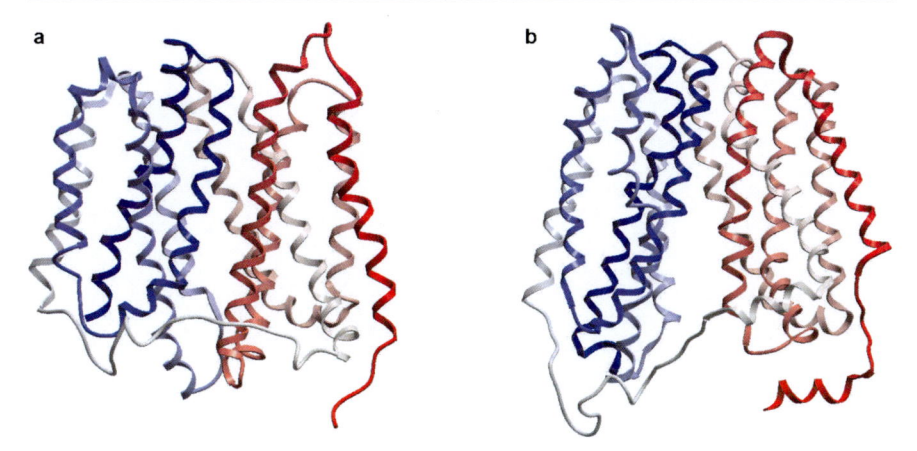

Fig. 4 Backbone Cα trace of the X-ray crystallographic structure of EmrD (**a**) [68] (PDB code 2GFP) and LacY (**b**) [70] (PDB code 1PV6), viewed in the membrane plane (cytoplasm downwards). *Colour coding* of the structures: *blue* via *white* to *red* from N-terminal to C-terminal

because GlpT and LacY are facing the cytoplasm in a V-shaped conformation [69, 70], while EmrD probably represents an intermediate state [68]. It has been proposed that the substrates for GlpT, LacY and EmrD are translocated across the membrane by an alternating-access mechanism [68–70].

5.2
The Resistance-Nodulation-Cell Division Superfamily

All known members of the resistance-nodulation-cell division (RND) superfamily catalyse substrate efflux via and H^+ antiporter mechanisms. These transporters are found in bacteria, archaea and eukaryotes and are organized in eight phylogenic families (http://www.tcdb.org/) [71, 72]. Up to now, 18 different X-ray crystallographic structures of the proton:drug antiporter Acriflavine resistance protein B (AcrB) have been reported [73–77] [78] (Fig. 5). This transporter is a major drug-resistance pump of the RND superfamily that belongs to the largely Gram-negative bacterial hydrophobe/amphiphile efflux-1 (HAE1) family (http://www.tcdb.org/). In *E. coli* the protein cooperates with a membrane fusion protein AcrA, and the outer membrane channel Tol C [73–75]. The substrate specificity of this large protein complex is broad, transporting cationic neutral and anionic substrates [79]. The AcrB protomer is organized as a homotrimer with a jellyfish-like structure [73], and the crystal structures of AcrB with and without substrates indicate that drugs are exported by a functionally rotating mechanism [74] or an alternating access peristaltic mechanism [75].

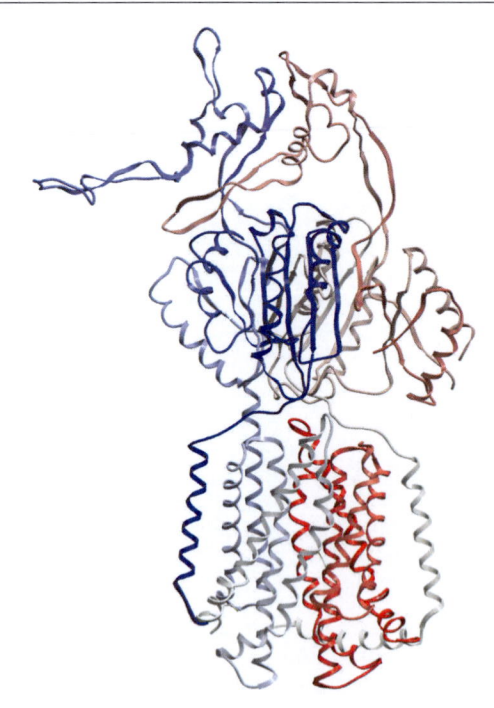

Fig. 5 Backbone Cα trace of the X-ray crystallographic structure of a single AcrB protomer [73] (PDB code 1IWG) viewed in the membrane plane. *Colour coding* as in Fig. 4

5.3
The Drug/Metabolite Transporter Superfamily

The drug/metabolite transporter (DMT) superfamily consists of 18 recognized families, each with a characteristic function, size and topology [80]. The multidrug transporter EmrE belongs to this superfamily, and to the four TMH small multidrug resistance (SMR) family (http://www.tcdb.org/). The SMR family members are prokaryotic transport systems consisting of homodimeric or heterodimeric structures [81]. Two *E. coli* EmrE X-ray crystal structures have been reported at 3.7 [82] and 3.8 Å [83]. The EmrE transporter is a proton drug:antiporter [82], and two EmrE subunits form a homodimer that binds substrate at the interface [83] (Fig. 6).

5.4
The Neurotransmitter:Sodium Symporter Family

Members of the neurotransmitter:sodium symporter (NSS) family catalyse uptake of a variety of neurotransmitters, amino acids, osmolytes and related nitrogenous substances by a solute:Na^+ symport mechanism [84]. In 2005 the crystal structure of the *Aquifex aeolicus* $LeuT_{Aa}$ determined by X-ray diffrac-

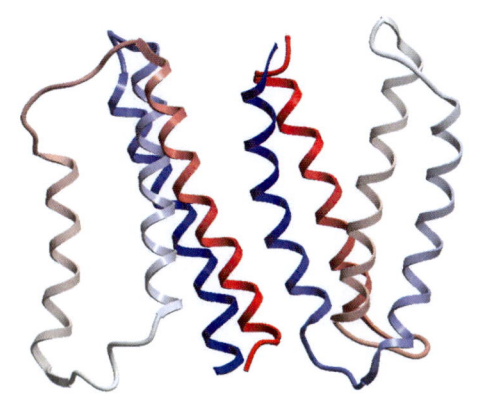

Fig. 6 Cα trace of the X-ray crystal structure of the EmrE dimer X [83] (PDB code 2F2M) viewed in the membrane plane. *Colour coding* as in Fig. 4

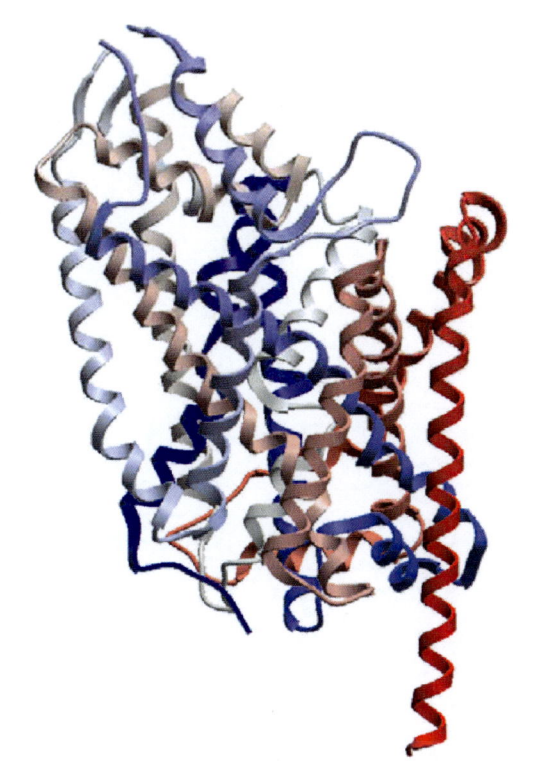

Fig. 7 Cα trace of the LeuT$_{Aa}$ X-ray crystallographic structure [64] (PDB code 2A65) viewed in the membrane plane. *Colour coding* as in Fig. 4

tion at 1.65 Å resolution was reported [64] (Fig. 7). LeuT$_{Aa}$ belongs to the NSS family (http://www.tcdb.org/), and is a bacterial homologue of SERT, DAT, NET and GAT-1. It is a 12 TMH sodium/leucine symporter, where TMHs 1–

5 are related to TMHs 6–10 by a pseudo-twofold axis in the membrane plane. The structure resembles a shallow "shot glass" [64], with leucine and sodium ions bound within the protein core. The substrate and sodium ion binding sites are comprised of TMHs 1, 3, 6 and 8. An alternating access model for transport, where all symported substrates must bind simultaneously before translocation, has been confirmed by experimental studies on SERT [85].

5.5
The Dicarboxylate/Amino Acid:Cation (Na$^+$ or H$^+$) Symporter Family

The members of the dicarboxylate/amino acid:cation symporter (DAACS) family catalyse Na$^+$ and/or H$^+$ symport together with either a Krebs cycle dicarboxylate (malate, succinate or fumarate), a dicarboxylic amino acid (glutamate or aspartate), a small, semipolar, neutral amino acid (Ala, Ser, Cys, Thr), neutral and acidic amino acids or most zwitterionic and dibasic amino acids (http://www.tcdb.org/).

The *Pyrococcus horikoshii* Glt$_{ph}$ (archaeal glutamate transporter homologue) structure has been determined by X-ray crystallography at 3.5 Å resolution [86] (Fig. 8) and 2.96 Å resolution [87]. This is a proton symporter belonging to the DAACS family (http://www.tcdb.org/). The transporter is organized as a trimer, with each protomer having eight TMHs, two re-entrant helical hairpins, and independent substrate translocation pathways [87]. It has been proposed that glutamate transport is achieved by movements of the

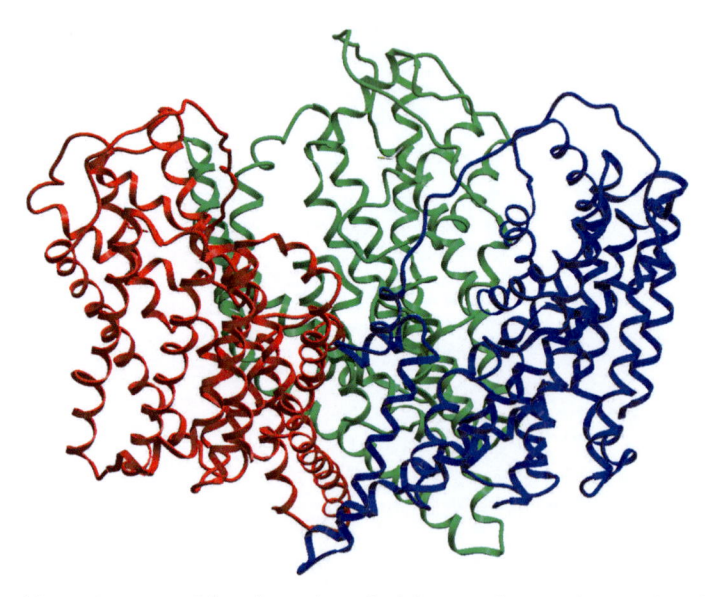

Fig. 8 Backbone Cα trace of the Glt$_{ph}$ trimer [86] (PDB code 1XFH) viewed in the membrane plane. The three protein subunits are shown in different colours (*red, green, blue*)

hairpins allowing alternating access to either side of the membrane [86]. Glt_{ph} is a bacterial homologue of the human EAAT1–5.

5.6
The ATP-Binding Cassette Superfamily

The ABC superfamily contains both uptake and efflux transport systems. Phylogenically, the members of these two porter groups generally cluster loosely together with just a few exceptions. There are dozens of families within the ABC superfamily, and family classification generally correlates with substrate specificity (http://www.tcdb.org/).

In 2006, the multidrug transporter *Staphylococcus aureus* Sav1866 was determined by X-ray crystallography at 3.0 Å resolution in an outward-facing conformation, reflecting the ATP-bound state [88]. Sav1866 belongs to the ABC superfamily and shows sequence similarity to human P-glycoprotein (P-gp). The transporter consists of two subunits, each with a trans-membrane domain–nuclear binding domain (TMD-NBD) topology, with six TMHs in each TMD. The two subunits are twisted and embracing each other, and both the TMDs and NBDs are tightly interacting. Towards the extracellular side, bundles of TMHs diverge into two "wings", with each wing consisting of TMH1 and TMH2 from one subunit and TMH3–TMH6 from the other subunit. The crystal structure of Sav1866 indicates that ABC transporters may use an "alternating access and release" mechanism where ATP binding and hydrolysis control the conversion of one state into the other, and that domain swapping and subunit twisting takes place in the transport cycle [88].

The MsbA structures from three different bacteria have been determined by X-ray crystallography at 4.5 Å resolution (*E. coli*) [89], 3.8 Å resolution (*Vibrio cholerae*) [90], and 4.2 Å resolution (*Salmonella typhimurium*) [91]. The MsbA transporters belong to the ABC superfamily, and the prokaryotic ABC-type efflux permeases or the lipid exporter (LipidE) family (http://www.tcdb.org/). The three different structures were thought to represent different conformational stages of the transport cycle, an open conformation [89], a closed conformation [90], and a post-hydrolysis conformation [91]. If true, the MsbA structures would represent interesting targets for computer modelling of human ABC transporters. However, these structures were retracted [92] after the publication of the Sav1866 structure [88]. The Sav1866 structure indicated that the MsbA structures were incorrect and that the biological interpretations based on the MsbA structures were invalid.

The *E. coli* BtuCD protein is involved in B_{12} uptake, and the structure was determined by X-ray crystallography at 3.2 Å resolution [93]. BtuCD also belongs to the ABC superfamily, prokaryotic ABC-type uptake permeases, and the vitamin B_{12} uptake transporter ($B_{12}T$) family. It consists of four subunits, two NBDs (BtuD) and two TMDs (BtuC). Each of the two BtuC subunits contains ten TMHs [93].

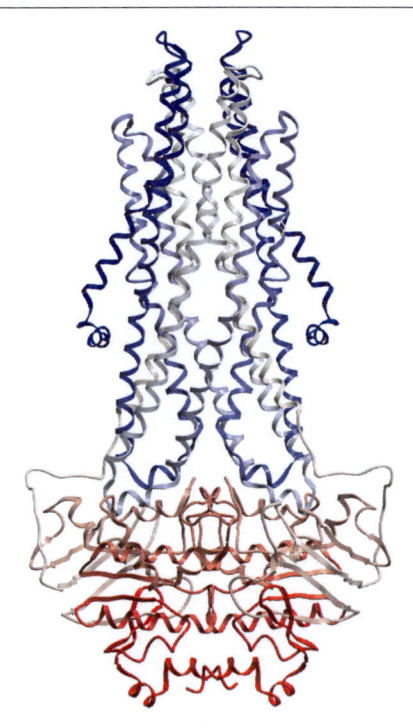

Fig. 9 Backbone Cα trace of the Sav1866 dimer [88] (PDB code 2HYD) viewed in the membrane plane. The NBDs are in *red*, while the TMDs are coloured in *blue* to *white* from the N- to the C-terminal

The 3D structure of the HI1470/1 transporter from *Haemophilus influenzae*, which is also a bacterial ABC transporter mediating the uptake of metal-chelate species including haem and vitamin B_{12}, has been determined at 2.4 Å resolution [94]. It exhibits an inward-facing conformation, which is in contrast to the outward-facing state observed for the homologous vitamin B_{12} importer BtuCD [94]. The 3D structures indicate that the substrate translocation requires large conformational changes, and the differences between the BtuCD and the HI1470/1 transporter from *H. influenzae* may reflect conformations relevant to the alternating access mechanism of substrate translocation.

6
Potential for New Drug Development

6.1
Multidrug Resistance Protein Targets

Cells exposed to toxic compounds can develop resistance by a number of mechanisms, including increased excretion. This may result in multidrug resistance, which is a particular limitation to cancer chemotherapy and antibiotic treatment. Development of inhibitors of drug efflux transporters has been sought for use as a supplement to current therapy in order to overcome multidrug resistance problems [95].

6.1.1
ABC Transporters and Cancer Therapy

ABC transporters play an important role in multidrug resistance in cancer chemotherapy. Human ABC transporters are divided into five different subfamilies ABCA, ABCB, ABCC, ABCD and ABCG, based on phylogenetic analysis. According to the transport classification database (TCDB), these ABC transporter subfamilies (ABC-type efflux permeases) belong to subclasses 3.A.1.201–212 [19] (http://www.tcdb.org/). Transporters in subfamilies ABCA, ABCB, ABCC and ABCG are involved in multidrug resistance [96–99].

ABCB1 (P-Glycoprotein)

The ABCB1 transporter is important in the removal of anticancer agents, such as adriamycin, vincristine and daunorubicin, from cells. ABCB1 is expressed in normal tissues, such as the gastrointestinal epithelium, epithelia of the bronchi, mammary gland, prostate gland, salivary gland, sweat glands of the skin, pancreatic ducts, renal tubules, and in bile canaliculi and ductules, in adrenal and in endothelial cells at blood–brain barrier sites and other blood–tissue barrier sites [100]. ABCB1 expression is highest in tumours from colon, adrenal, pancreatic, mammary and renal tissue, even in the absence of prior chemotherapy [101]. Even though the relationship between ABCB1 expression and response to chemotherapy remains unclear, negative prognostic implications of ABCB1 expression have been established in breast cancer, neuroblastoma, various types of leukaemia, and several sarcomas [101].

Development of ABCB1 inhibitors may help to prevent ABCB1 efflux of anticancer agents. ABCB1 inhibitors are not cytotoxic agents themselves, but when used in combination with cancer drugs which are normally pumped out by the cell by ABCB1, intracellular drug concentrations are maintained, restoring sensitivity to these therapeutics.

Three generations of ABCB1 inhibitors have been developed. The first-generation ABCB1 inhibitors were established therapeutic drugs for diverse targets that were discovered, largely by chance, to also function as ABCB1 inhibitors [101]. In general, these were less potent than later generations of ABCB1 inhibitors; they were not selective, and produced undesirable side effects [95]. The second-generation ABCB1 inhibitors, which were based on the structures of the first-generation compounds and optimized using QSAR [101], were less toxic. However, inhibition of ABCB1 by first- or second-generation compounds has failed to demonstrate the desired clinical benefit.

Dangerously high doses of these agents were needed, and they exhibited toxicity due to an increased availability of the co-administered chemotherapy [95].

The third-generation ABCB1 inhibitors, which were discovered by combinatorial chemistry screening [101], are more potent and more selective than earlier compounds, and are currently in clinical trials [101, 102]. The therapeutic benefit of ABCB1 inhibition is yet to be firmly established, but the continued development of these agents may establish the true therapeutic potential of ABCB1-mediated multidrug resistance reversal. A suggested approach would achieve a balance between the positive effects of ABCB1 inhibition at the tumour site and the negative potential toxic side effects outcome of reducing elimination of the chemotherapy.

ABCC5 (MRP5)

ABCC5 belongs to the ABC superfamily, transports cGMP and is also involved in multidrug resistance [103]. ABCC5 is expressed in most tissues, such as in skeletal muscle, kidney, testis, heart and brain [104–106], in smooth muscle cells of the corpus cavernosum, ureter and bladder, and mucosa in ureter and urethra [107, 108], in vascular smooth muscle cells, cardiomyocytes, and vascular endothelial cells in the heart [109], in placenta [110], and in human erythrocytes [103]. Clinical studies have shown that extracellular cGMP levels are elevated in various types of cancer. Significant elevation of urinary cGMP excretion has been observed in patients with untreatable adenocarcinomas from ovary, stomach or large bowel, whereas a normal range was found in patients where tumours had been removed [111]. Elevated urine cGMP concentrations have also been demonstrated among patients with cancer of the uterine cervix [112], and measurements of urinary cGMP levels after treatment of ovarian cancer has been reported to be a very sensitive tool in therapeutic monitoring [113–115]. Increased cGMP efflux by ABCC5 may be one mechanism whereby cancer cells can develop resistance against endogenous growth control, and also against antineoplastic drugs which are substrates for ABCC5.

ABCB1/ABCC5 Structural Considerations/Molecular Modelling Approach

Knowledge of the ABCB1 and ABCC5 structures may be used to develop membrane transport modulating agents which, in turn, may be helpful in overcoming resistance to chemotherapeutic agents. These transporters feature both TMDs and NBDs, with a TMD-NBD-TMD-NBD domain arrangement. The NBD contains the Walker A and B motifs [116] and a signature C motif, and the substrate specificity of the transporters is provided by the TMDs.

The 3D structures of ABCB1 and ABCC5 have not been experimentally determined, but molecular modelling by homology may be used to gain structural insight into their potential as drug targets. In particular, homology modelling may be used to study substrate difference between ABCB1 and ABCC5, since ABCB1 transports cationic amphiphilic and lipophilic substrates [117–120], while ABCC5 transports organic anions [103, 121].

In order to understand the molecular concepts underlying the substrate difference between ABCB1 and ABCC5, we have used the *Staphylococcus aureus* Sav1866 X-ray crystal structure [88] to construct models of ABCB1 and ABCC5 [122]. Modelling indicated that the electrostatic potential surface of the substrate translocation area of ABCB1 is neutral with negative and weakly positive areas, while the electrostatic potential surface of the ABCC5 substrate translocation chamber generally is positive. These results indicate that ABCB1, transporting cationic amphiphilic and lipophilic substrates, has a more neutral substrate translocation chamber than ABCC5, which has a positive chamber transporting organic anions. Structural information about the ABCB1 and ABCC5 substrate binding sites might be useful in the design of inhibitor multidrug efflux by these transporters.

6.2
Multidrug Resistance and Antibiotic Treatment

Treatment of infections may be limited by the emergence of bacteria that are resistant to multiple antibiotics. Bacterial antibiotic resistance may be caused by intrinsic mechanisms, such as efflux systems, or by acquired mechanisms, such as mutations in genes targeted by the antibiotic [123].

The major mechanism of resistance to tetracycline in Gram-negative bacteria is drug-specific efflux. Drug efflux pumps are involved in fluoroquinolone resistance of *Staphylococcus aureus* and *Streptococcus pneumoniae*, and the antiseptic resistance of *Staphylococcus aureus*. When multidrug pumps are overexpressed, resistance levels are elevated. Efflux pumps are thus potential antibacterial targets, since inhibitors of bacterial efflux pumps may restore the activity of an antibiotic which otherwise is effluxed [124]. Structural knowledge at the atomic level from X-ray crystallographic studies of bacterial multidrug transporters is rapidly growing. Examples are the *E. coli* EmrD [68], *E. coli* AcrB [73–78], *E. coli* EmrE [82, 83], and *Staphylococcus aureus* Sav1866 [88] crystal structures.

6.3
CNS Drug Targets

6.3.1
Neurotransmitter:Sodium Symporter Family

Some of the most successful CNS drugs selectively target secondary transporters. Transporters at the plasma membrane contribute to the clearance and recycling of neurotransmitters in neural synaptic clefts. When a neuro-transmitter transporter is inhibited, the concentration of neurotransmitter increases in the synapse. The *A. aeolicus* LeuT$_{Aa}$ crystal structure [64] of the NSS family has delivered new insight into the structure of NSS trans-porters. Its homologies to the human transporters SERT, NET and DAT are 20–25% [64]. The high sequence conservation in functionally important re-gions between the *A. aeolicus* LeuT$_{Aa}$ transporter and the other NSS family members suggests that these proteins share a common folding and a com-mon transport mechanism, and that the *A. aeolicus* LeuT$_{Aa}$ crystal structure can be used to model the functionally important regions of other NSS family members with quite high accuracy [59].

Serotonin and Noradrenaline Transporters

Noradrenaline and 5-HT modulate the activity of neural circuits influencing mood and sleep. Antidepressants selectively inhibit 5-HT or noradrenaline reuptake into presynaptic neurons. Selective serotonin reuptake inhibitors (SSRIs) have replaced tricyclic antidepressants as the drugs of choice in the treatment of depressive disorders, mainly because of their improved toler-ability and safety if taken in overdose. Still, 10–30% of patients taking an-tidepressants are partially or totally resistant to the treatment. SSRIs block the reuptake of serotonin into the presynaptic nerve terminals, thereby en-hancing serotonergic neurotransmission, which presumably results in their antidepressant effects. SSRIs are prescribed for conditions such as depression, obsessive-compulsive disorder, social phobia, post-traumatic stress disorder, premenstrual dysphoric disorder and generalized anxiety disorder [125]. Side effects of SSRIs include agitation, insomnia, neuromuscular restlessness, nau-sea, dry mouth, fatigue, decreased libido, diarrhoea, vomiting and headache.

Reboxetine is a specific noradrenaline reuptake inhibitor (NARI). The side effects of NARIs include dry mouth, constipation, insomnia, increased sweating, tachycardia, vertigo, urinary retention and impotence. The gen-eral limitations of SSRIs and NARIs are due to side effects directly related to their effect on the serotonergic and noradrenergic systems. Discontinuation symptoms from SSRIs and NARIs are depression, dizziness, nausea, lethargy, headache, flu-like feelings, panic attacks, numbness, agitation and insomnia. A better understanding of the molecular mechanisms of SERT and NET is important for developing new agents with fewer side effects.

The molecular aspects of SSRI binding to SERT and NET have been the subject of several molecular modelling studies [126–129]. The *A. aeolicus* LeuT$_{Aa}$ crystal structure [64] represented a major advance towards under-standing the structure–function relationships of SERT and NET, since this transporter is quite close both in function and amino acid sequence to human SERT and NET, and thus provides a template for updated models [59, 85, 130, 131]. In order to examine the molecular aspects of the selectivities of SSRIs,

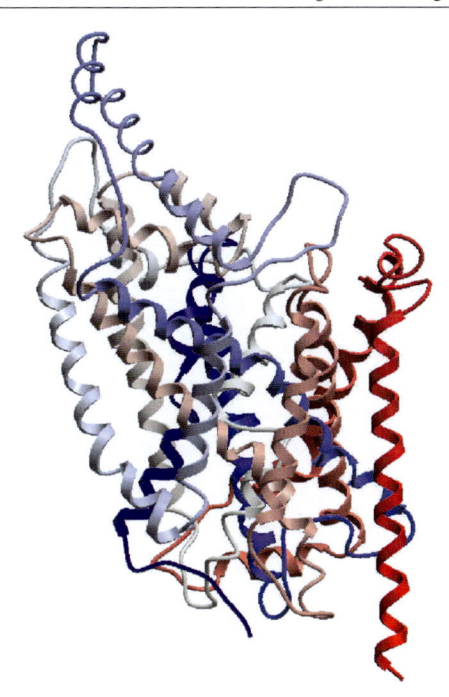

Fig. 10 Cα trace of the homology model of SERT [131] viewed in the membrane plane. *Colour coding* as in Fig. 4

we have constructed molecular models of SERT [131] (Fig. 10) and NET based on the *A. aeolicus* LeuT$_{Aa}$ crystal structure [64]. The ICM pocket finder of the ICM software version 3.4–4 [132] reported amino acids in TMHs 1, 3, 6 and 8 of SERT and NET as being contributors to the putative substrate binding area.

Dopamine Transporter and Drugs of Abuse

Dopamine is involved in the reward system, which is linked to drug abuse. When a person receives positive reinforcement for certain behaviours, which can be both natural rewards and artificial rewards such as addictive drugs, the reward system is activated [133]. When cocaine binds to the dopamine transporter (DAT), the dopamine concentration at the synapse is elevated, resulting in activation of a "reward" mechanism. The binding of cocaine to SERT and NET also contributes to cocaine reward and cocaine aversion [134, 135].

We have previously constructed 3D models of DAT [127–129, 136] based on various low-resolution structural data and transporters with low homology with DAT. The *A. aeolicus* LeuT$_{Aa}$ X-ray crystal structure [64] provides the possibility of updating the previous DAT models. Figure 11 shows a putative binding site of cocaine in DAT (unpublished). Site-directed mutagenesis studies and docking studies of cocaine binding to DAT indicated that cocaine

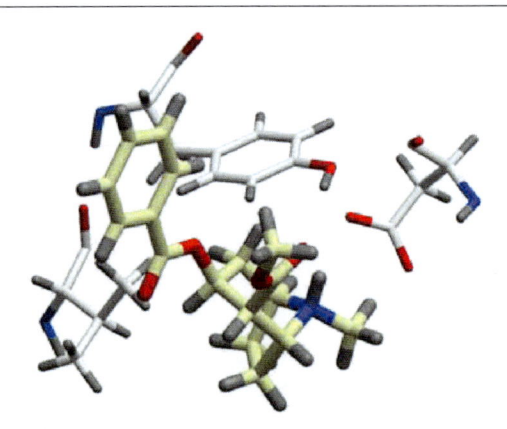

Fig. 11 Cocaine docked into the putative binding site of the DAT model. Amino acids interacting with cocaine displayed in the figure are Asp-79, Val-152 and Tyr-156

interacts with Asp-79 [137], Val-152 [138] and Tyr-156. Tyr-156 corresponds to Tyr-176 in SERT, which has been found by site-directed mutagenesis studies to be important for cocaine binding [139].

Interestingly, cocaine and SSRIs have similar molecular mechanisms of action. However, while SSRIs are therapeutic drugs prescribed for the treatment of depression, cocaine is a local anaesthetic drug and a substance of abuse. Knowledge of cocaine's molecular interactions with DAT may be used to develop agents that block binding of cocaine without inhibiting the reuptake of dopamine. Such agents might be effective in treating cocaine addiction.

GABA Transporter

The neurotransmitter GABA transmits inhibitory signals that reduce excitation and anxiety. A search for selective inhibitors of GABA transporters has led to potent and selective inhibitors of GAT-1 (SwissProt accession number P30531), which is a 12 TMH transporter with homology to LeuT$_{Aa}$ [59]. The only clinically approved GAT-1 inhibitor at present is tiagabine [140, 141]. Tiagabine is a potent and broad spectrum anticonvulsant drug which does not induce tolerance to the anticonvulsant effect [142]. Tiagabine has also shown promise in clinical trials to treat chronic daily headaches with symptoms of migraine [143], and it has been suggested from preclinical studies and human studies that tiagabine also possesses anxiolytic properties [141]. Tiagabine has also been reported to be effective in prophylactic treatment of bipolar disorder [144, 145], but its therapeutic potential in this condition has not been established.

A comprehensive amino acid alignment, including the *A. aeolicus* LeuT$_{Aa}$ sequence, sequences of GABA transporters and the sequences of other NSS family members, would provide the possibility of modelling the ligand binding to GAT-1 and GAT-2, which could be helpful in designing tiagabine derivatives with an improved binding selectivity profile.

Glycine Transporter

Non-competitive N-methyl-D-aspartate (NMDA) blockers induce schizophrenic-like symptoms in humans, presumably by impairing glutamatergic transmission [146]. It has therefore been postulated that compounds potentiating this neurotransmission would increase extracellular levels of glycine, which is a co-agonist of glutamate, and thus possess antipsychotic activity [146, 147].

GlyT1c is a glycine/2Na$^+$/1Cl$^-$ symporter belonging to the NSS family. Blocking of the GlyT1c transporter using a specific inhibitor, SSR504734, resulted in increased extracellular glycine levels in rat prefrontal cortex, and enhanced glutamatergic neurotransmission. The GlyT1c inhibitor has shown activity in several animal models of schizophrenia [146]. A homology model based on the LeuT$_{Aa}$ template will provide the possibility to study the molecular interactions of specific inhibition by SSR504734.

6.3.2
The Drug:H$^+$ Antiporter-1 (DHA1) (12 Spanner) Family

The vesicular monoamine transporter-2 (VMAT2) contributes to regulation of the monoaminergic neuronal function by sequestrating catecholamines and serotonin into synaptic vesicles. VMAT2 pumps dopamine, serotonin, noradrenaline, epinephrine and histamine into storage vesicles against a gradient, powered by the vesicular H-ATPase and the exchange of two protons for one substrate molecule [148]. These transporters also sequester neurotoxins within vesicles, thus playing a role in neuroprotection [149, 150].

Experimental studies have indicated that amphetamine-type agents interact with the VMAT2 and deplete vesicular neurotransmitters via a carrier-mediated exchange mechanism [151]. The cytoplasmic levels of transmitter are increased due to disruption of the vesicular transmitter storage of transmitter, such that more neurotransmitter is available for release by transporter-mediated exchange [151]. It has been proposed, based on its functions as a critical regulator of neurotransmitter disposition within the brain, that the VMAT2 transporter might be a possible target for drugs used in addictive disorders, Parkinson's disease and schizophrenia [152, 153].

VMAT2 is a member of the MFS family of proteins. Three different *E. coli* transporter proteins of the MFS family have been determined by X-ray crystallography at atomic resolution: EmrD [68] (Fig. 4a), GlpT [69], and LacY [70] (Fig. 4b). These structures indicate that the MFS proteins with 12 TMHs share a common architecture of the membrane spanning region, organized in symmetrical N- and C-terminal domains each of six TMHs, with overall structural topologies resembling each other. These observations indicate that the MFS proteins of known 3D structure can be used to predict the structure of VMAT2, and thereby guide the synthesis and evaluation of novel VMAT2 ligands as possible therapeutic agents.

6.3.3
The Dicarboxylate/Amino Acid:Cation (Na$^+$ or H$^+$) Symporter Family

Glutamate Transporters

Glutamate causes excitation, nudging the brain into high alert, affecting cognition and most other fundamental aspects of brain function. It transmits critical instructions between nerve cells involved in sensory perception, learning and memory. Changes in glutamate neurotransmission may contribute to different brain diseases. Glutamate has a dual action: it is an excitatory neurotransmitter under normal conditions, but it is toxic to neuronal cells when present in excess. Selective modulation of the levels of glutamate may produce a therapeutic benefit in disorders where glutamate levels are abnormal, such as in stroke, head trauma, retinal ischaemia, schizophrenia, Alzheimer's disease, Parkinson's disease and other neurodegenerative and psychiatric disorders.

The excitatory amino acid transporter-3 (EAAT-3) is a major neuronal transporter for glutamate in the brain. An EAAT-3 inhibitor may have therapeutic applications in schizophrenia, cognitive impairment (such as that associated with Alzheimer's disease) and other nervous system diseases where glutamate neurotransmission is deficient (http://www.neurocrine.com/html/res_eaats.htm). Increasing the amount of glutamate released from certain nerve cells could improve learning, memory skills and overall cognitive function. A possible limitation to this approach is that glutamate can cause glutamate-induced retinal toxicity, and that long-term inhibition of glutamate transporter activity may cause neuronal damage. Increased knowledge of the structure and function of glutamate transporters is therefore of pivotal importance. So far, the only X-ray crystal structure with similarities in function and sequence with the glutamate transporters is that of the glutamate transporter homologue from *Pyrococcus horikoshii* Glt$_{ph}$ [86].

Sequence alignment of family members followed by homology modelling should take into account that indirect structural knowledge from experimental studies may emphasize the nature of the ligand binding area of the EAAT-3 transporter.

6.4
Transporters Involved in Drug Absorption, Distribution and Elimination

More than 28 different mammalian OATs are expressed in the liver, small intestine, blood–brain barrier endothelial cells, placenta, kidneys and other organs [154, 155]. These carrier proteins, which may transport a wide range of substrates, are major factors in drug absorption, distribution and excretion, and work in concert with the drug metabolism system in order to eliminate drug metabolites from systemic circulation. Their activity is saturable and inducible, and may show polymorphic variation among individuals [156].

P-glycoproteins, Mrps and Oatps/OATPs are transporter proteins playing an important role in drug absorption, distribution and excretion [154, 155, 157]. The organ-specific cellular localization and expression of such transporter proteins in liver, kidneys and intestines has given new insight into the molecular mechanisms of cellular uptake and excretion of drugs and drug metabolites in these organs. Uptake of many different compounds into hepatocytes is mediated by OATs expressed at the sinusoidal membrane (rat Oatp1, Oatp2 and Oatp4, human OATP-B, OATP-C and OATP8) [155]. These Oatps/OATPs mediate sodium-independent uptake of a wide variety of mainly bulky organic anions, neutral compounds and organic cations including drugs [154, 157]. Besides liver specific Oatps/OATPs (Oatp4, OATP-C and OATP8), others are expressed in various tissues including intestine (Oatp3), kidney (Oatp1), brain (Oatp2, Oatp3, OATP-A, OATP-D, OATP-E and OATP-F) and testis (OATP-D and OATP-F) [154]. It has been proposed that Oatps/OATPs may be used to target drugs to certain organs, based on their selective tissue distribution and substrate specificities.

In the future, detailed knowledge of cellular transport mechanisms of drugs in various organs may be taken into account in drug design, in order to develop drug molecules which have both desired pharmacological activities and pharmacokinetic properties which would make them useful therapeutic agents. However, detailed structural knowledge of transporter proteins involved in absorption, distribution and elimination is still too limited to be taken directly into account in target-based drug design projects.

6.5
Prodrug Targets

6.5.1
Dipeptide Transporters

Dipeptide transporters (PEPTs) are involved in transport of di- and tripeptides across plasma membranes in the small intestine and kidney proximal tubules. These transporters are important for efficient absorption of protein ingestion products. Dipeptide transporters are H^+-coupled and localized in brush border membranes. They mediate absorption of certain drugs like cephalosporins, β-lactam antibiotics and ACE inhibitors [158]. For example, whilst methyldopa is poorly absorbed from the intestines, when it is converted to a dipeptyl derivative it is a substrate for dipeptide transporters, and is more efficiently absorbed.

Dipeptide transporters may be exploited for improving intestinal absorption of pharmacologically active amino acids. Therefore, prodrugs targeting the PEPT1 transporter may improve the oral bioavailability of drugs with low intestinal membrane permeability. Structural knowledge of the peptide transporters is important for designing peptidomimetics that may facilitate drug transport across the intestinal epithelium [159–161]. The PEPT transporters belong to the MFS family of transporters, indicating that structural modelling based on the X-ray crystal structure of Lac Permease [70] may provide important structural information about the PEPT transporters.

7
Conclusions

Improved methods in molecular biology, biochemistry, crystallization and X-ray crystallographic data collection and processing have presented the possibility of automation of the different steps of protein expression and structure determination. Together with developments in genomic sequencing, these technical improvements have increased the number of known 3D protein structures. These technical advances have also provided insight into an unprecedented number of potential drug targets and created an environment for the emergence of new strategies for drug discovery. The improvements in technology have also increased the number of available atomic-resolution structures of membrane transporter proteins. However, experimental methods for 3D structure determination of membrane proteins are still difficult and remain a significant challenge for structural biology and new drug discovery. Current knowledge of the 3D structures of membrane proteins, including membrane transporters, is therefore still limited. Almost all known membrane transporter structures have been determined with proteins from bacteria. Several of the bacterial transporters of known 3D structure have mammalian counterparts of therapeutic interest, indicating that molecular modelling approaches may be used to generate 3D models of important drug targets based on structural conservation throughout evolution. Membrane transporters constitute a divergent group of proteins with substrates ranging from ions to relatively large organic molecules, and with large conformational flexibility in order to facilitate substrate transport. This suggests that inhibitors may bind to different conformations of a transporter, and that several conformations of a transporter should be considered in a target-based ligand design approach.

References

1. Walian P, Cross TA, Jap BK (2004) Genome Biol 5:215
2. Klebe G (2006) Drug Discov Today 11:580
3. Blundell TL, Jhoti H, Abell C (2002) Nat Rev Drug Discov 1:45
4. Hajduk PJ, Greer J (2007) Nat Rev Drug Discov 6:211
5. Stoffler D, Sanner MF, Morris GM, Olson AJ, Goodsell DS (2002) Proteins 48:63
6. Tomasselli AG, Heinrikson RL (2000) Biochim Biophys Acta 1477:189
7. Elliott M (2001) Philos Trans R Soc Lond B Biol Sci 356:1885
8. Cushman DW, Ondetti MA, Gordon EM, Natarajan S, Karanewsky DS, Krapcho J, Petrillo EW Jr (1987) J Cardiovasc Pharmacol 10(7):S17
9. Fournie-Zaluski MC, Coric P, Thery V, Gonzalez W, Meudal H, Turcaud S, Michel JB, Roques BP (1996) J Med Chem 39:2594
10. Lundstrom K (2006) Cell Mol Life Sci 63:2597
11. Frishman D, Mewes HW (1997) Nat Struct Biol 4:626
12. Wallin E, von Heijne G (1998) Protein Sci 7:1029
13. Basyn F, Spies B, Bouffioux O, Thomas A, Brasseur R (2003) J Mol Graph Model 22:11
14. Elofsson A, von Heijne G (2007) Annu Rev Biochem
15. Oberai A, Ihm Y, Kim S, Bowie JU (2006) Protein Sci 15:1723
16. Amico M, Finelli M, Rossi I, Zauli A, Elofsson A, Viklund H, von Heijne G, Jones D, Krogh A, Fariselli P, Martelli PL, Casadio R (2006) Nucleic Acids Res 34:W169
17. Granseth E, Viklund H, Elofsson A (2006) Bioinformatics 22:e191
18. Viklund H, Granseth E, Elofsson A (2006) J Mol Biol 361:591
19. Saier MH Jr (2000) Microbiol Mol Biol Rev 64:354
20. Rothman RB, Silverthorn ML, Glowa JR, Matecka D, Rice KC, Carroll FI, Partilla JS, Uhl GR, Vandenbergh DJ, Dersch CM (1998) Synapse 28:322
21. Yamashita M, Fukushima S, Shen HW, Hall FS, Uhl GR, Numachi Y, Kobayashi H, Sora I (2006) Neuropsychopharmacology 31:2132
22. Race JE, Grassl SM, Williams WJ, Holtzman EJ (1999) Biochem Biophys Res Commun 255:508
23. Masereel B, Lohrmann E, Schynts M, Pirotte B, Greger R, Delarge J (1992) J Pharm Pharmacol 44:589
24. Diogo Filho A, Santos PS, Duque AS, Cezario RC, Gontijo Filho PP (2006) Acta Cir Bras 21:279
25. Dahl SG, Sylte I, Ravna AW (2004) J Pharmacol Exp Ther 309:853
26. Dassa E, Bouige P (2001) Res Microbiol 152:211
27. Dean M, Rzhetsky A, Allikmets R (2001) Genome Res 11:1156
28. Zhou F, You G (2007) Pharm Res 24:28
29. Caffrey M (2003) J Struct Biol 142:108
30. Cherezov V, Clogston J, Papiz MZ, Caffrey M (2006) J Mol Biol 357:1605
31. Cherezov V, Peddi A, Muthusubramaniam L, Zheng YF, Caffrey M (2004) Acta Crystallogr D Biol Crystallogr 60:1795
32. Eifler N, Duckely M, Sumanovski LT, Egan TM, Oksche A, Konopka JB, Luthi A, Engel A, Werten PJ (2007) J Struct Biol 159:179
33. Edwards AM, Arrowsmith CH, Christendat D, Dharamsi A, Friesen JD, Greenblatt JF, Vedadi M (2000) Nat Struct Biol 7:970
34. Turner GJ, Reusch R, Winter-Vann AM, Martinez L, Betlach MC (1999) Protein Expr Purif 17:312
35. Summers MD (2006) Adv Virus Res 68:3

36. Eifler N, Duckely M, Sumanovsky LT, Egan TM, Oksche A, Konopka JB, Luthi A, Engel A, Werten PJ (2007) J Struct Biol 159:179
37. Schurmann A, Keller K, Monden I, Brown FM, Wandel S, Shanahan MF, Joost HG (1993) Biochem J 290(Pt 2):497
38. Schurmann A, Monden I, Joost HG, Keller K (1992) Biochim Biophys Acta 1131:245
39. Garcia-Herrero A, Vogel HJ (2005) Mol Microbiol 58:1226
40. Kuhlbrandt W, Wang DN, Fujiyoshi Y (1994) Nature 367:614
41. Chiu ML, Nollert P, Loewen MC, Belrhali H, Pebay-Peyroula E, Rosenbusch JP, Landau EM (2000) Acta Crystallogr D Biol Crystallogr 56:781
42. Landau EM, Rosenbusch JP (1996) Proc Natl Acad Sci USA 93:14532
43. Nollert P, Qiu H, Caffrey M, Rosenbusch JP, Landau EM (2001) FEBS Lett 504:179
44. Faham S, Boulting GL, Massey EA, Yohannan S, Yang D, Bowie JU (2005) Protein Sci 14:836
45. Faham S, Bowie JU (2002) J Mol Biol 316:1
46. Page RC, Moore JD, Nguyen HB, Sharma M, Chase R, Gao FP, Mobley CK, Sanders CR, Ma L, Sonnichsen FD, Lee S, Howell SC, Opella SJ, Cross TA (2006) J Struct Funct Genomics 7:51
47. Kay LE (2005) J Magn Reson 173:193
48. Lian LY, Barsukov IL, Sutcliffe MJ, Sze KH, Roberts GC (1994) Methods Enzymol 239:657
49. Palmer AG III, Kroenke CD, Loria JP (2001) Methods Enzymol 339:204
50. Ratnala VR, Kiihne SR, Buda F, Leurs R, de Groot HJ, DeGrip WJ (2007) J Am Chem Soc 129:867
51. Rosenbusch JP (2001) J Struct Biol 136:144
52. Williams KA (2000) Nature 403:112
53. Wieman H, Tondel K, Anderssen E, Drablos F (2004) Mini Rev Med Chem 4:793
54. Kopp J, Schwede T (2004) Pharmacogenomics 5:405
55. Lundstrom K (2005) Trends Biotechnol 23:103
56. Dalton JA, Jackson RM (2007) Bioinformatics
57. Kopp J, Schwede T (2004) Nucleic Acids Res 32:D230
58. Kopp J, Schwede T (2006) Nucleic Acids Res 34:D315
59. Beuming T, Shi L, Javitch JA, Weinstein H (2006) Mol Pharmacol 70:1630
60. Burckhardt G, Wolff NA (2000) Am J Physiol Renal Physiol 278:F853
61. Lewinson O, Adler J, Sigal N, Bibi E (2006) Mol Microbiol 61:277
62. Smith FW, Rae AL, Hawkesford MJ (2000) Biochim Biophys Acta 1465:236
63. Tamura N, Konishi S, Yamaguchi A (2003) Curr Opin Chem Biol 7:570
64. Yamashita A, Singh SK, Kawate T, Jin Y, Gouaux E (2005) Nature 437:215
65. Jin J, Guffanti AA, Beck C, Krulwich TA (2001) J Bacteriol 183:2667
66. Wood NJ, Alizadeh T, Richardson DJ, Ferguson SJ, Moir JW (2002) Mol Microbiol 44:157
67. Marger MD, Saier MH Jr (1993) Trends Biochem Sci 18:13
68. Yin Y, He X, Szewczyk P, Nguyen T, Chang G (2006) Science 312:741
69. Huang Y, Lemieux MJ, Song J, Auer M, Wang DN (2003) Science 301:616
70. Abramson J, Smirnova I, Kasho V, Verner G, Kaback HR, Iwata S (2003) Science 301:610
71. Nishino K, Yamaguchi A (2001) J Bacteriol 183:5803
72. Saier MH Jr, Tam R, Reizer A, Reizer J (1994) Mol Microbiol 11:841
73. Murakami S, Nakashima R, Yamashita E, Yamaguchi A (2002) Nature 419:587
74. Murakami S, Nakashima R, Yamashita E, Matsumoto T, Yamaguchi A (2006) Nature 443:173

75. Seeger MA, Schiefner A, Eicher T, Verrey F, Diederichs K, Pos KM (2006) Science 313:1295
76. Yu EW, McDermott G, Zgurskaya HI, Nikaido H, Koshland DE Jr (2003) Science 300:976
77. Yu EW, Aires JR, McDermott G, Nikaido H (2005) J Bacteriol 187:6804
78. Sennhauser G, Amstutz P, Briand C, Storchenegger O, Grutter MG (2006) PLoS Biol 5:e7
79. Nikaido H, Zgurskaya HI (2001) J Mol Microbiol Biotechnol 3:215
80. Jack DL, Yang NM, Saier MH Jr (2001) Eur J Biochem 268:3620
81. Chung YJ, Saier MH Jr (2001) Curr Opin Drug Discov Devel 4:237
82. Ma C, Chang G (2004) Proc Natl Acad Sci USA 101:2852
83. Pornillos O, Chen YJ, Chen AP, Chang G (2005) Science 310:1950
84. Clark JA, Amara SG (1993) Bioessays 15:323
85. Zhang YW, Rudnick G (2006) J Biol Chem 281:36213
86. Yernool D, Boudker O, Jin Y, Gouaux E (2004) Nature 431:811
87. Boudker O, Ryan RM, Yernool D, Shimamoto K, Gouaux E (2007) Nature 445:387
88. Dawson RJ, Locher KP (2006) Nature 443:180
89. Chang G, Roth CB (2001) Science 293:1793
90. Chang G (2003) J Mol Biol 330:419
91. Reyes CL, Chang G (2005) Science 308:1028
92. Chang G, Roth CB, Reyes CL, Pornillos O, Chen YJ, Chen AP (2006) Science 314:1875
93. Locher KP, Lee AT, Rees DC (2002) Science 296:1091
94. Pinkett HW, Lee AT, Lum P, Locher KP, Rees DC (2007) Science 315:373
95. Dantzig AH, de Alwis DP, Burgess M (2003) Adv Drug Deliv Rev 55:133
96. Ejendal KF, Hrycyna CA (2002) Curr Protein Pept Sci 3:503
97. Leslie EM, Deeley RG, Cole SP (2005) Toxicol Appl Pharmacol 204:216
98. Ross DD, Doyle LA (2004) Cancer Cell 6:105
99. Sampath J, Adachi M, Hatse S, Naesens L, Balzarini J, Flatley RM, Matherly LH, Schuetz JD (2002) AAPS PharmSci 4:E14
100. Cordon-Cardo C, O'Brien JP, Casals D, Rittman-Grauer L, Biedler JL, Melamed MR, Bertino JR (1989) Proc Natl Acad Sci USA 86:695
101. McDevitt CA, Callaghan R (2007) Pharmacol Ther 113:429
102. Thomas H, Coley HM (2003) Cancer Control 10:159
103. Jedlitschky G, Burchell B, Keppler D (2000) J Biol Chem 275:30069
104. Belinsky MG, Bain LJ, Balsara BB, Testa JR, Kruh GD (1998) J Natl Cancer Inst 90:1735
105. Kool M, de Haas M, Scheffer GL, Scheper RJ, van Eijk MJ, Juijn JA, Baas F, Borst P (1997) Cancer Res 57:3537
106. McAleer MA, Breen MA, White NL, Matthews N (1999) J Biol Chem 274:23541
107. Nies AT, Spring H, Thon WF, Keppler D, Jedlitschky G (2002) J Urol 167:2271
108. Rius M, Thon WF, Keppler D, Nies AT (2005) J Urol 174:2409
109. Dazert P, Meissner K, Vogelgesang S, Heydrich B, Eckel L, Bohm M, Warzok R, Kerb R, Brinkmann U, Schaeffeler E, Schwab M, Cascorbi I, Jedlitschky G, Kroemer HK (2003) Am J Pathol 163:1567
110. Meyer Zu Schwabedissen HE, Grube M, Heydrich B, Linnemann K, Fusch C, Kroemer HK, Jedlitschky G (2005) Am J Pathol 166:39
111. Guthrie D, Isah H, Latner AL, Turner GA (1979) IRCS Med Sci 7:209
112. Turner GA, Ellis RD, Guthrie D, Latner AL, Monaghan JM, Ross WM, Skillen AW, Wilson RG (1982) J Clin Pathol 35:800
113. Luesley DM, Chan KK, Newton JR, Blackledge GR (1987) Br J Obstet Gynaecol 94:461

114. Turner GA, Ellis RD, Guthrie D, Latner AL, Ross WM, Skillen AW (1982) Br J Obstet Gynaecol 89:760
115. Turner GA, Greggi S, Guthrie D, Benedetti Panici P, Ellis RD, Scambia G, Mancuso S (1990) Eur J Gynaecol Oncol 11:421
116. Walker JE, Saraste M, Runswick MJ, Gay NJ (1982) Embo J 1:945
117. Muller M, Mayer R, Hero U, Keppler D (1994) FEBS Lett 343:168
118. Orlowski S, Garrigos M (1999) Anticancer Res 19:3109
119. Smit JW, Duin E, Steen H, Oosting R, Roggeveld J, Meijer DK (1998) Br J Pharmacol 123:361
120. Wang EJ, Lew K, Casciano CN, Clement RP, Johnson WW (2002) Antimicrob Agents Chemother 46:160
121. Sager G (2004) Neurochem Int 45:865
122. Ravna AW, Sylte I, Sager G (2007) Theor Biol Med Model 4:33
123. Alekshun MN, Levy SB (2007) Cell 128:1037
124. Li XZ, Zhang L, Nikaido H (2004) Antimicrob Agents Chemother 48:2415
125. Ables AZ, Baughman OL 3rd (2003) Am Fam Physician 67:547
126. Ravna AW, Edvardsen O (2001) J Mol Graph Model 20:133
127. Ravna AW, Sylte I, Dahl SG (2003) J Pharmacol Exp Ther 307:34
128. Ravna AW (2006) World J Biol Psychiatry 7:99
129. Ravna AW, Sylte I, Kristiansen K, Dahl SG (2006) Bioorg Med Chem 14:666
130. White KJ, Kiser PD, Nichols DE, Barker EL (2006) Protein Sci 15:2411
131. Ravna AW, Jaronczyk M, Sylte I (2006) Bioorg Med Chem Lett 16:5594
132. Abagyan R, Totrov M, Kuznetsov DN (1994) J Comput Chem 15:488
133. Bardo MT (1998) Crit Rev Neurobiol 12:37
134. Hall FS, Li XF, Sora I, Xu F, Caron M, Lesch KP, Murphy DL, Uhl GR (2002) Neuroscience 115:153
135. Uhl GR, Hall FS, Sora I (2002) Mol Psychiatry 7:21
136. Ravna AW, Sylte I, Dahl SG (2003) J Comput Aided Mol Des 17:367
137. Kitayama S, Shimada S, Xu H, Markham L, Donovan DM, Uhl GR (1992) Proc Natl Acad Sci USA 89:7782
138. Lee SH, Chang MY, Lee KH, Park BS, Lee YS, Chin HR (2000) Mol Pharmacol 57:883
139. Chen JG, Sachpatzidis A, Rudnick G (1997) J Biol Chem 272:28321
140. Hog S, Greenwood JR, Madsen KB, Larsson OM, Frolund B, Schousboe A, Krogsgaard-Larsen P, Clausen RP (2006) Curr Top Med Chem 6:1861
141. Schwartz TL, Nihalani N (2006) Expert Opin Pharmacother 7:1977
142. Sills GJ (2003) Epileptic Disord 5:51
143. Capuano A, Vollono C, Mei D, Pierguidi L, Ferraro D, Di Trapani G (2004) Clin Ter 155:79
144. Young AH, Geddes JR, Macritchie K, Rao SN, Vasudev A (2006) Cochrane Database Syst Rev 3:CD005173
145. Young AH, Geddes JR, Macritchie K, Rao SN, Watson S, Vasudev A (2006) Cochrane Database Syst Rev 3:CD004694
146. Depoortere R, Dargazanli G, Estenne-Bouhtou G, Coste A, Lanneau C, Desvignes C, Poncelet M, Heaulme M, Santucci V, Decobert M, Cudennec A, Voltz C, Boulay D, Terranova JP, Stemmelin J, Roger P, Marabout B, Sevrin M, Vige X, Biton B, Steinberg R, Francon D, Alonso R, Avenet P, Oury-Donat F, Perrault G, Griebel G, George P, Soubrie P, Scatton B (2005) Neuropsychopharmacology 30:1963
147. Lechner SM (2006) Curr Opin Pharmacol 6:75
148. Schuldiner S, Steiner-Mordoch S, Yelin R (1998) Adv Pharmacol 42:223

149. Fleckenstein AE, Volz TJ, Riddle EL, Gibb JW, Hanson GR (2007) Annu Rev Pharmacol Toxicol 47:681
150. Volz TJ, Fleckenstein AE, Hanson GR (2007) Addiction 102(1):44
151. Partilla JS, Dempsey AG, Nagpal AS, Blough BE, Baumann MH, Rothman RB (2006) J Pharmacol Exp Ther 319:237
152. Fleckenstein AE, Hanson GR (2003) Eur J Pharmacol 479:283
153. Riddle EL, Fleckenstein AE, Hanson GR (2005) AAPS J 7:E847
154. Hagenbuch B, Meier PJ (2004) Pflugers Arch 447:653
155. Meijer DK, Lennernas H (2005) Eur J Pharm Sci 26:130
156. Tirona RG, Leake BF, Merino G, Kim RB (2001) J Biol Chem 276:35669
157. Huber RD, Gao B, Sidler Pfandler MA, Zhang-Fu W, Leuthold S, Hagenbuch B, Folkers G, Meier PJ, Stieger B (2007) Am J Physiol Cell Physiol 292:C795
158. Brodin B, Nielsen CU, Steffansen B, Frokjaer S (2002) Pharmacol Toxicol 90:285
159. Vabeno J, Lejon T, Nielsen CU, Steffansen B, Chen W, Ouyang H, Borchardt RT, Luthman K (2004) J Med Chem 47:1060
160. Vabeno J, Nielsen CU, Ingebrigtsen T, Lejon T, Steffansen B, Luthman K (2004) J Med Chem 47:4755
161. Vabeno J, Nielsen CU, Steffansen B, Lejon T, Sylte I, Jorgensen FS, Luthman K (2005) Bioorg Med Chem 13:1977

Top Med Chem (2009) 4: 53–94
DOI 10.1007/7355_2008_028
© Springer-Verlag Berlin Heidelberg
Published online: 9 December 2008

Design of Monoamine Reuptake Inhibitors: SSRIs, SNRIs and NRIs

Gavin A. Whitlock (✉) · Mark D. Andrews · Alan D. Brown · Paul V. Fish · Alan Stobie · Florian Wakenhut

Pfizer Global R&D, Ramsgate Road, Sandwich CT13 9NJ, UK
gavin.whitlock@pfizer.com

Abstract This review will detail the medicinal chemistry involved in the design, synthesis and discovery of selective serotonin, noradrenaline reuptake inhibitors and dual serotonin/noradrenaline reuptake inhibitors. In particular, this review will focus exclusively on series and compounds which have been disclosed within the medicinal chemistry literature between January 2000 and June 2008. Background information on previously disclosed clinical agents, such as atomoxetine, milnacipran and reboxetine, is included for comparison purposes with more recently disclosed agents.

Keywords Dual reuptake inhibitor · Design · Monoamine · NRI · SAR · SNRI · SSRI

Abbreviations
5-HT	Serotonin
ADHD	Attention deficit hyperactivity disorder
ATX	Atomoxetine
BBB	Blood–brain barrier
CNS	Central nervous system
DA	Dopamine
DAT	Dopamine transporter
DLM	Dog liver microsome
DLX	Duloxetine
DRI	Dopamine reuptake inhibitor
HLM	Human liver microsome
NA	Noradrenaline
NE	Norepinephrine (= noradrenaline)
NET	Noradrenaline transporter
NRI	Noradrenaline reuptake inhibitor
P-gp	P-glycoprotein
PE	Premature ejaculation
PET	Positron emission tomography
RBX	Reboxetine
RLM	Rat liver microsome
SAR	Structure–activity relationship
SERT	Serotonin transporter
SNRI	Dual serotonin and noradrenaline reuptake inhibitor
SPECT	Single photon emission computed tomography
SSRI	Selective serotonin reuptake inhibitor
TCA	Tricyclic antidepressant

1
Introduction

The understanding that cocaine 1 and the tricyclic antidepressants (TCAs), such as amitriptyline 2, function by blocking the reuptake of monoamines in the brain owes much to Axelrod [1] (Fig. 1). A decade after the clinical introduction of the TCAs, he introduced the concept of reuptake to explain how noradrenaline 3 was taken up by sympathetic nerve terminals. Soon afterwards, similar but distinct uptake mechanisms were shown to exist for dopamine 4 and 5-hydroxytryptamine or serotonin 5 [2]. Subsequently it was

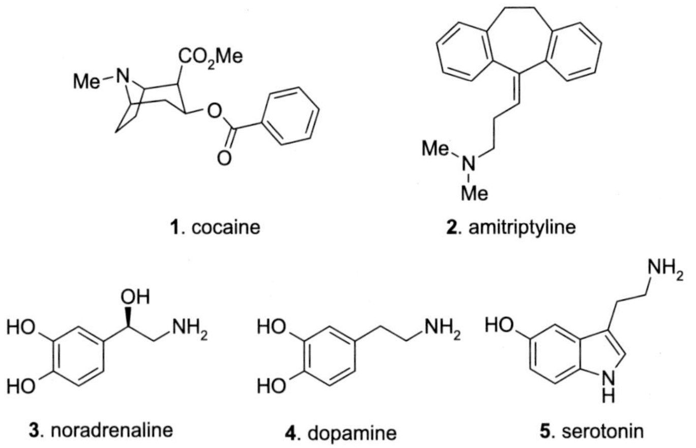

Fig. 1 Structures of cocaine, amitriptyline, noradrenaline, dopamine and serotonin

shown that cocaine blocks reuptake of all three neurotransmitters [3] while the TCAs, such as amitriptyline, block only serotonin and noradrenaline reuptake [4].

Although the TCAs are effective antidepressants, they have poor selectivity over muscarinic, histaminic and adrenergic receptors, resulting in cardiovascular, anticholinergic and sedative side effects [5]. The mechanistic understanding of their side effects, however, led to the development of inhibitors that selectively block the reuptake of serotonin, noradrenaline, or the two neurotransmitters simultaneously, resulting in improved treatment of central nervous system (CNS) disorders including depression, anxiety, obsessive compulsive disorder, attention deficit hyperactivity disorder (ADHD), pain and urinary incontinence.

A number of selective serotonin reuptake inhibitors (SSRIs), such as fluoxetine 6, paroxetine 7, citalopram 8 and sertraline 9, were introduced in the 1980s, and became the most widely used group of antidepressants (Fig. 2). Three mixed serotonin/noradrenaline reuptake inhibitors (SNRIs), venlafaxine 10, duloxetine 11 and milnacipran 12, have been approved for treating depression, while the relatively selective noradrenaline reuptake inhibitor (NRI) atomoxetine 13 has been approved for treating ADHD. Finally, the more selective NRI reboxetine 14 has been approved in Europe for treating depression.

Since Axelrod's initial experiments, thousands of studies have dealt with the pharmacological and functional properties of monoamine reuptake sites in the brain. Details of these are beyond the scope of this review but they include [6]: identification of the genes responsible for encoding the transporters; the localization of monoamine transporters in the brain; the identification of the structural and functional domains of the transporters; and in

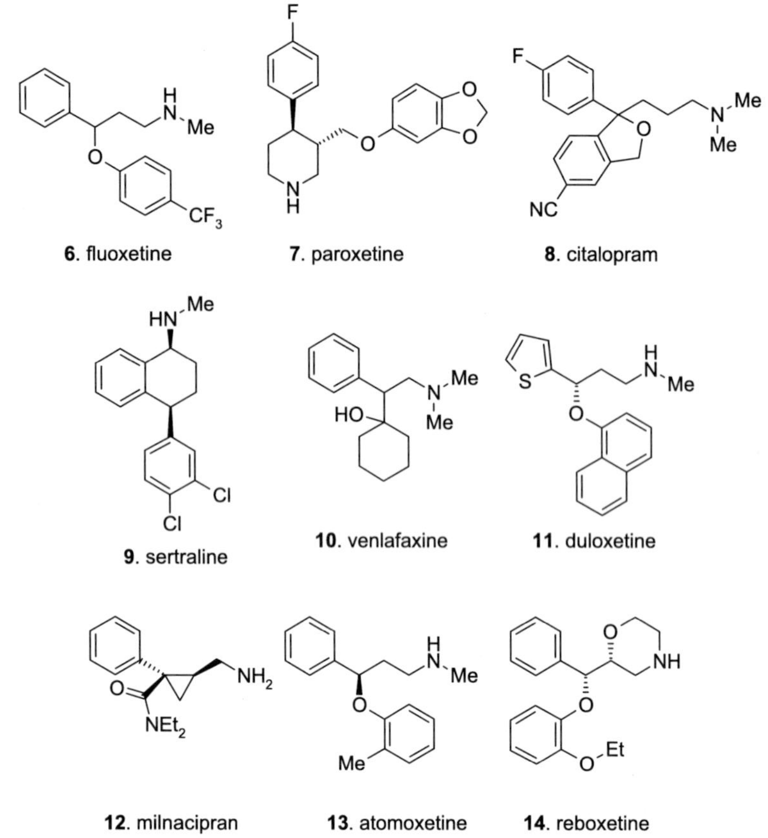

Fig. 2 Structures of the SSRIs fluoxetine, paroxetine, citalopram and sertraline, the SNRIs venlafaxine, duloxetine and milnacipran, and the NRIs atomoxetine and reboxetine

vivo imaging to examine changes in monoamine transporters associated with psychiatric and movement disorders.

As antidepressants, the SSRIs suffer the shortcoming of slow onset of action and 30–40% of patients do not respond satisfactorily to them [7]. As a result, attempts to augment these with other pharmacologies, e.g. 5-HT$_{1A}$, 5-HT$_{2C}$ etc. to "enhance their activity", has been an important aspect of the so-called SSRI+ approach. Although beyond the scope of this review, this subject was covered as recently as 2008 [8]. Mixed noradrenaline/dopamine reuptake, serotonin/noradrenaline/dopamine reuptake, and dopamine reuptake inhibitors are also beyond the scope of this review, but again these have been recently covered [9]. Instead, this review focuses exclusively on the medicinal chemistry involved in the design, synthesis and discovery of SSRIs, NRIs and SNRIs. It covers from January 2000 to June 2008, a period which has seen a steady rise, as measured by patent activity [10], in the SNRI field with 21 patents filed in the first half of 2008, compared to one in the whole of 2000. By the same measure, interest in NRIs continues to grow from 8 patents in 2000 to 85 in 2007, and interest in the more mature field of SSRIs has reduced only marginally from a peak of 68 patents in 2004 to 41 in 2007. Some of this is the result of increased interest in SSRIs for the treatment of sexual dysfunction. Throughout the review, background information on previously disclosed clinical agents, such as atomoxetine, milnacipran and reboxetine, is included for comparison purposes.

2
Selective Serotonin Reuptake Inhibitors

2.1
Tropane Based Selective Serotonin Reuptake Inhibitors: Design and Structure–Activity Relationship

Cocaine (1) is a non-selective transmitter reuptake inhibitor with a slight preference for dopamine transporter (DAT) over noradrenaline transporter (NET) and serotonin transporter (SERT) [1]. Preparation of cocaine analogues has, in the past, largely focused on the development of compounds with high affinity for the DAT [11]. However, more recently cocaine has been investigated by a number of groups as a potential starting point in the design of SSRIs. Ravna et al., working at the University of Tromsø, have constructed molecular models of SERT, NET and DAT based on the structure of the lactose permease symporter [12]. Docking of cocaine and the SSRI S-citalopram suggests that an unconserved amino acid Asp-499 in transmembrane α-helix 10 of NET may contribute to the low affinity of S-citalopram for NET. This analysis has potential utility in the structure based design of cocaine based SRIs although, to date, no such activity has been reported.

Workers at Cairo University have described the preparation and SRI/NRI selectivity profile of a series of tropane based inhibitors 15 (Fig. 3) [13]. Selectivity for SRI over NRI was assessed in vivo in the mouse using 5-HT induced

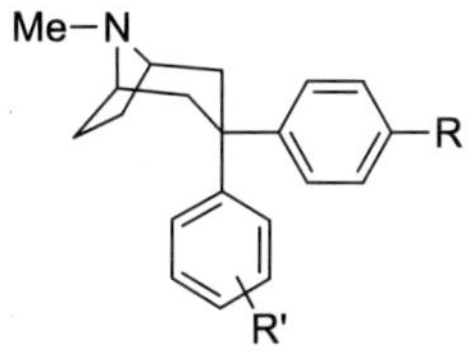

Compounds **15**
15a: R = H, R' = m-CF$_3$

Fig. 3 Compounds prepared by Hanna et al.

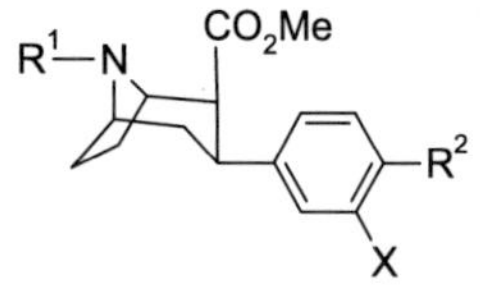

Compounds **16**
16a: R1 = H, R2 = CH(CH$_3$)$_2$; X = I
SERT K$_i$ 25nM; DAT K$_i$ > 500nM

Fig. 4 2β-Carbomethoxy-3β-phenyl tropanes

neurotoxicity and yohimbine based mortality, respectively. All compounds reported had similar SRI and NRI activity, except for **15a** which was reported to have an ED$_{50}$ based selectivity of tenfold for SRI over NRI, a similar selectivity profile to that observed for citalopram in the same models. The structure–activity relationship (SAR) suggested that this selectivity was associated with a *meta*-R′ substituent.

Emond and co-workers at the Laboratoire de Biophysique Medicale et Pharmaceutique have reported a series of 2β-carboxymethoxy-3β-phenyl tropanes **16** [14]. Introduction of a large *para*-alkyl substituent gave reduced DAT affinity but allowed moderate SERT affinity to be maintained (as measured by in vitro binding in rat brain tissue). Further optimization then led to the potent SSRI, **16a** (Fig. 4). The authors noted that this compound, which contains an iodine atom, represents a potential positron emission tomography (PET) or single photon emission computed tomography (SPECT) ligand (through incorporation of ^{123}I). See Sect. 2.4 for further discussions on potential SSRI imaging tools.

Co-workers at Yale University and Harvard Medical School have reported the synthesis and transporter affinity of a range of cocaine analogues bearing a *p*-thiophenyl substituent [15]. One of these, **17**, is an exquisitely potent binder at SERT (rat brain preparation) and shows excellent selectivity over DAT and NET (710-fold and 11 100-fold, respectively) (Fig. 5).

Kozikowski et al. at Georgetown University Medical Center have reported the synthesis of a novel class of tricyclic tropane analogues, with the tropane ring locked in its boat conformation [16]. These compounds were tested for their ability to inhibit monoamine reuptake (rat brain preparation). The

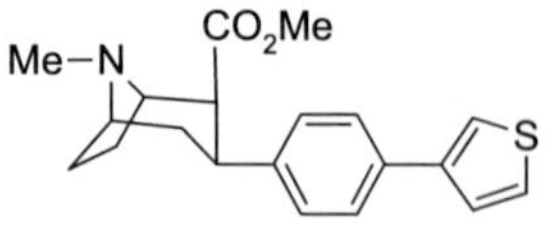

17 SERT K$_i$ 17pM; DAT K$_i$ 12nM; NET K$_i$ 189nM

Fig. 5 Potent and selective thiophenylphenyl tropane reported by Tamagnan and coworkers

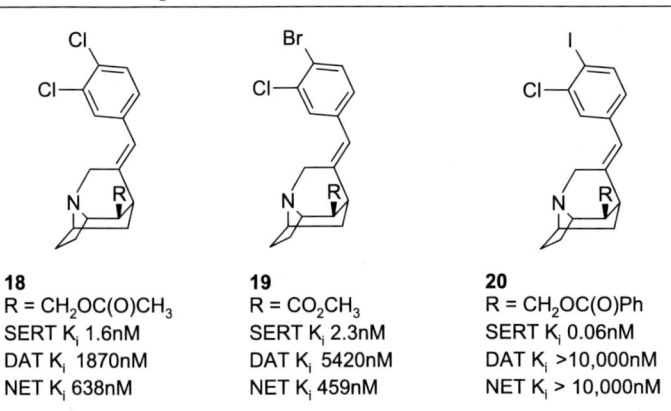

18
R = CH$_2$OC(O)CH$_3$
SERT K$_i$ 1.6nM
DAT K$_i$ 1870nM
NET K$_i$ 638nM

19
R = CO$_2$CH$_3$
SERT K$_i$ 2.3nM
DAT K$_i$ 5420nM
NET K$_i$ 459nM

20
R = CH$_2$OC(O)Ph
SERT K$_i$ 0.06nM
DAT K$_i$ >10,000nM
NET K$_i$ > 10,000nM

Fig. 6 Tricyclic tropanes locked in the boat conformation

authors demonstrated that it was possible to achieve NET or SERT selectivity in systems of this type. Compound **18** demonstrated nanomolar SRI activity and excellent selectivity over NRI and dopamine reuptake inhibitor (DRI) activities. In a related paper [17] the same group describe **19**, a potent and selective SRI. Subsequent optimization of the ester substituent yielded **20**, a 60pM SRI with essentially no DRI or NRI activity (Fig. 6).

2.2
Other Small-Molecule Selective Serotonin Reuptake Inhibitors

Workers at BMS have reported a series of SSRIs based on their initial finding that substituted homotryptamines **21** were relatively potent and selective SRIs (Fig. 7) [18]. They found that a 5-CN substituent was optimal on the indole ring, giving **22** which had a SERT IC$_{50}$ of 2 nM. Constraining the nitrogen in a quinuclidine ring was tolerated giving the equipotent SRI **23**. The group at BMS then went on to investigate constraining the amine side chain in several different ways. Incorporating a cyclopropyl ring gave a series of compounds exemplified by BMS-505130 (**24**) [19]. The *trans*-cyclopropyl ring was found to be significantly more potent than the corresponding *cis* isomer and the (*S,S*) configuration was found to be nearly 50-fold more potent than its enantiomer; interestingly, N alkylation of the indole ring caused a significant drop off in potency. Again 5-CN substitution of the indole ring was found to be optimal, giving compound **24** which was an extremely potent SRI with a SERT K_i of 0.18 nM. Compound **24** was profiled further in wide ligand profiling and was found to have weak activity at 5-HT$_{1A}$ (K_i = 410 nM) and 5-HT$_6$ (K_i = 270 nM). Although no absorption, distribution, metabolism and excretion (ADME) or pharmacokinetic (PK) data were reported, the results of a brain microdialysis study in rats were given. These showed that oral doses of 0.3 or 1 mg/kg gave robust increases in extracellular serotonin levels in the frontal cortex with a t_{max} of approximately 2 h.

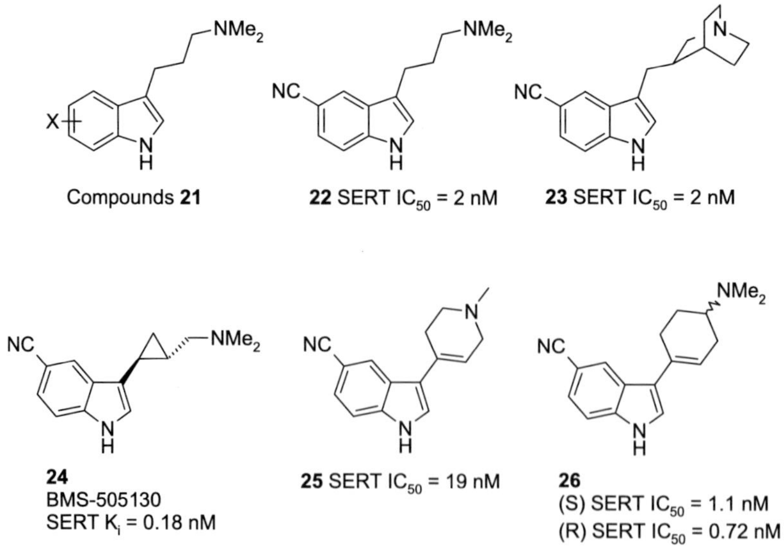

Compounds 21

22 SERT IC$_{50}$ = 2 nM

23 SERT IC$_{50}$ = 2 nM

24
BMS-505130
SERT K$_i$ = 0.18 nM

25 SERT IC$_{50}$ = 19 nM

26
(S) SERT IC$_{50}$ = 1.1 nM
(R) SERT IC$_{50}$ = 0.72 nM

Fig. 7 Homotryptamine based SSRIs

A subsequent paper [20] described the effect on SRI potency of changing the amine side chain for a tetrahydropyridine **25** or aminocyclohexene **26**. While the tetrahydropyridines were weaker (compound **25**, SERT IC$_{50}$ = 19 nM), both the enantiomers of the aminocyclohexene **26** retained excellent potency ((S) SERT IC$_{50}$ = 1.1 nM; (R) SERT IC$_{50}$ = 0.72 nM). Modelling studies suggested that both enantiomers could adopt a similar conformation to the highly potent BMS-505130 (**24**). Brain microdialysis studies in rat with racemic **26** showed a maximal increase in extracellular serotonin levels when it was dosed ip at 1 mg/kg. An oral dose of 10 mg/kg gave a similar effect on extracellular 5-HT levels compared to a 10 mg/kg oral dose of fluoxetine.

The group at BMS also looked at the effect of changing the indole ring for various other ring systems [21], whilst retaining the cyclopropyl side chain of compound **24** (Fig. 8). The indazole analogue of compound **24** was significantly weaker at 71 nM, but the unsubstituted benzothiophene **27**

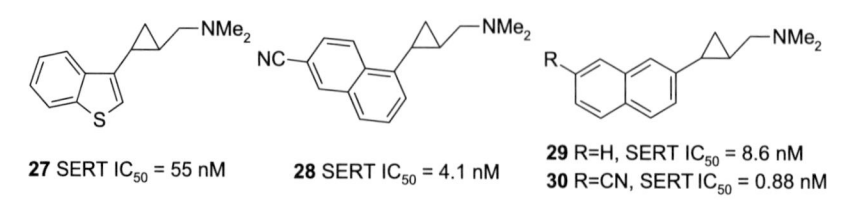

27 SERT IC$_{50}$ = 55 nM

28 SERT IC$_{50}$ = 4.1 nM

29 R=H, SERT IC$_{50}$ = 8.6 nM
30 R=CN, SERT IC$_{50}$ = 0.88 nM

Fig. 8 Analogues of compound **24**

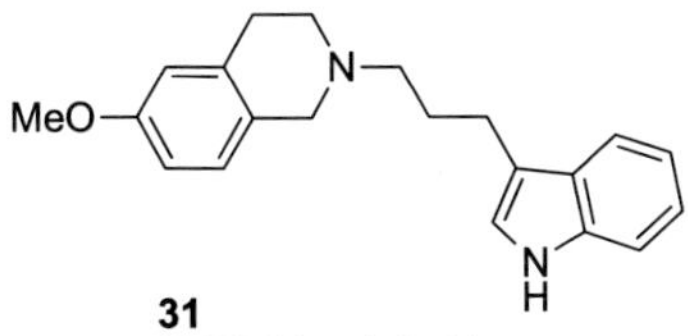

31
SERT K_i = 1.4 nM
$5HT_{1A}$ K_i = 288 nM
$alpha_1$ K_i = 58 nM

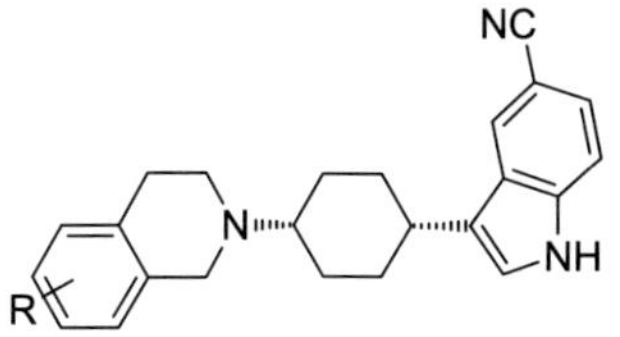

32 R = 6-OMe
SERT K_i = 0.1 nM
$5HT_{1A}$ K_i = 300 nM

Fig. 9 Further indole based SSRIs

was comparable in potency to the unsubstituted indole analogue of compound **24** (55 vs 99 nM). Surprisingly, the 5-cyanobenzothiophene compound was not reported. A number of naphthalene analogues were investigated and attaching the aminoalkyl side chain to both the 1 and 2 positions of the naphthalene led to potent SSRIs. For the 1-aminoalkylnaphthalenes the most potent compound **28** (SERT IC_{50} = 4.1 nM) again incorporated a cyano substituent. For the 2-aminoalkylnaphthalenes even the unsubstituted system **29** (R = H) was relatively potent (SERT IC_{50} = 8.6 nM), and in this case cyano substitution led to an increase in potency (**30**, R = CN, SERT IC_{50} = 0.88 nM) to give activity close to that of the original indole compound **24**. No ADME or in vivo data were reported for any of these compounds.

Workers at Wyeth–Ayerst developed another indole based series in their attempt to develop compounds combining SRI activity with 5-HT_{1A} antagonism (Fig. 9) [22]. Incorporating the basic nitrogen of the indole derivative **21** into a tetrahydroisoquinoline ring led to compounds such as **31** which, in addition to potent SRI activity (SERT K_i = 1.4 nM), had moderate activity at 5-HT_{1A} and $alpha_1$. The most potent SSRI they disclose is compound **32** (SERT K_i = 0.1 nM) where the indole and tetrahydroisoquinoline rings are linked by a *cis*-cyclohexane ring. The corresponding *trans*-cyclohexane isomer was the best dual SRI/5-HT_{1A} antagonist disclosed, with a SERT K_i of 8 nM and a 5-HT_{1A} K_i of 300 nM.

2.3 Rapid Onset Selective Serotonin Reuptake Inhibitors

A development in the last 10 or so years of SSRI research has been the attempt by companies such as Alza and Pfizer to develop rapid onset, short half-life SSRIs, such as dapoxetine **33** and UK-390957 (structure undisclosed), primarily for the treatment of premature ejaculation (PE) [23]. Dapoxetine is a potent SSRI which was in development specifically as a treatment for PE [24]. Despite showing some efficacy in clinical trials, it was given a "not approvable" letter by the FDA in 2005 [25]. The rationale for a rapid onset agent is that it would be suitable for on demand dosing, which may be a preferred dosing regime for some patients. A short half-life minimizes exposure to the drug and so could result in an improved side-effect profile. Both of these features have been questioned in the literature, however. On demand treatment with classical SSRIs such as paroxetine shows reduced efficacy in treating PE relative to chronic (daily) treatment [23], and some studies indicate that patients actually prefer a daily treatment [26].

Workers at Pfizer have reported on some of their efforts to develop rapid onset SSRIs (Fig. 10) [27]. Their strategy to reduce t_{max} (time to maximal compound concentration) was to reduce the volume of distribution which, it was reasoned, would result in a shorter hepatic transit time. Starting from sertraline 9 they introduced polarity in order to reduce lipophilicity and thereby membrane affinity. Electron-withdrawing groups were particularly targeted as a means of reducing pK_a and potentially reducing membrane affinity still further. Remarkably, for a class of drugs characterized by lipophilic amines, it was found that polarity was very well tolerated on the tetrahydronaphthalene ring and incorporation of a large range of polar groups gave very potent SSRIs. Polar heterocycles such as triazole were tolerated, but gave unacceptable levels of CYP2D6 inhibition. C-7 substitution was found to be optimal and the primary sulfonamide UK-373911 (34) was selected for clinical development. It was found that in the rat the volume of distribution was indeed reduced relative to sertraline (19 vs 52 l/kg, unbound volume 660 vs 1900 l/kg) and that the t_{max} had been re-

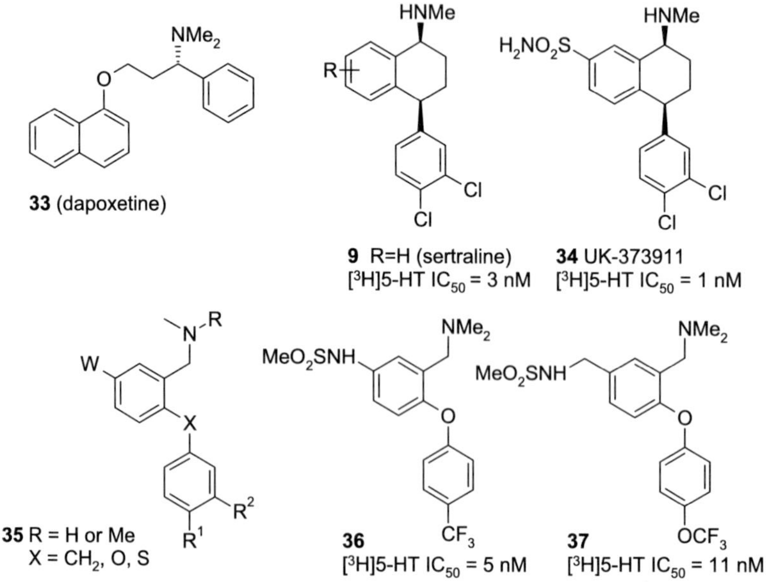

Fig. 10 Rapid onset SSRIs

duced from 4 to 1 h, in keeping with the lower $\log D$ (2.3 vs 3.1) and pK_a (8.4 vs 9.3). Despite the increased polarity, compound **34** retained good flux in a Caco-2 cell line and good brain penetration (CSF to free blood ratio of 2.6 : 1).

The same group at Pfizer then went on to explore the structurally much simpler diphenylmethane, diphenyl sulfide and diphenyl ether systems, **35** [28]. As with the sertraline template, it was found that polar substitution (W) was very well tolerated on one of the aromatic rings. The other aromatic ring was found to give good SRI activity even when substituted with a lipophilic substituent at only the 4-position (e.g. R^1 = Cl, SMe, CF_3, OCF_3). While the diphenylmethane and diphenyl sulfide systems gave potent SRIs, although selectivity was poorer for the diphenylmethanes, the corresponding diphenyl ethers were preferred due to their synthetic accessibility. The polar analogues retained good flux in a Caco-2 cell line and in vivo profiling of analogues **36** and **37** in the dog showed them to have significantly lower volumes of distribution than compound **34** (6 l/kg for both; unbound volumes 60 and 88 l/kg for **36** and **37**, respectively). These compounds also had the rapid t_{max} (0.5 h) that the authors were seeking. The authors also note that the low volume of distribution is structurally driven as the compounds have the same physicochemical properties ($\log D$, pK_a) as sertraline analogue **34**, which has a significantly higher volume of distribution (19 l/kg, unbound volume 1050 l/kg reported in this paper).

2.4
Serotonin Reuptake Inhibitor Ligands for Positron Emission Tomography

Several groups have reported on the identification and synthesis of suitably labelled SRIs for (potential) use in PET and SPECT studies on the in vivo distribution, density and occupancy of the SERT in humans. Work has focused on the optimization of properties in two series, based on ZIENT [29] and on the 3-diphenyl sulfide SRI previously reported by Wilson and Houle [30] (Fig. 11). Key compounds are detailed, along with their in vitro potency and selectivity profiles, in Table 1.

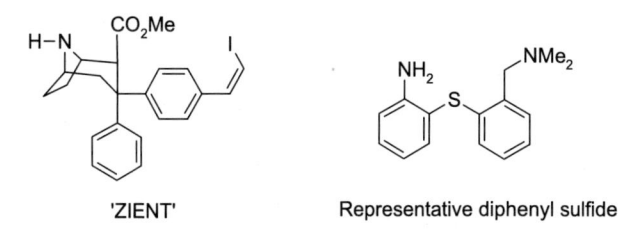

'ZIENT'   Representative diphenyl sulfide

Fig. 11 Compounds used as the basis of new PET ligand design

Table 1 Potential SRI PET and SPECT ligands

Compound[a]	SERT K_i (nM)	NET K_i (nM)	DAT K_i (nM)	Radio-label	In vivo study	Refs.
	0.2	102.2	29.9	^{11}C	Non-human primate	[31]
	0.2	31.7	32.6	^{11}C	Non-human primate	[20]
	0.05	24	3.47	^{123}I	Rat	[18]
	0.08	28	13	^{18}F	Non-human primate	[32]
	0.08	28	13	^{18}F	Non-human primate	[32]
	0.11	450	22	^{11}C	Rat, monkey	[33]
	0.25	7.5	340	^{18}F	Rat	[34]
	1.4	12	299	^{18}F	Rat	[23]

Table 1 (continued)

Compound	SERT K_i (nM)	NET K_i (nM)	DAT K_i (nM)	Radio-label	In vivo study	Refs.
	0.05	650	3020	[18]F	Rat	[35]
	1.1	1350	1423	[11]C	Rat	[36]
	0.25	61	532	[11]C	None	[37, 38]
	1.04	663.8	> 10 000	[19]F	Rat	[39]

[a] Position of radiolabel denoted by *

3
Dual Serotonin and Noradrenaline Reuptake Inhibitors

3.1
Benzylic Amines

Whitlock et al. at Pfizer reported the discovery of both phenyl and pyridyl methanamines and described the SAR for dual SNRI activity [40, 41]. Variation of the B-ring demonstrated that a 2,4-disubstitution pattern afforded potent SNRI activity with good DAT selectivity, exemplified with the 2-OMe 4-Cl analogues **38** and **39** (Fig. 12).

Introduction of polar substitution on the A-ring was then undertaken in order to reduce lipophilicity. The positioning of the polar group was rationalized from the overlap of compound **39** with tetrahydronaphthalene SSRIs such as **34** (Fig. 13) [27].

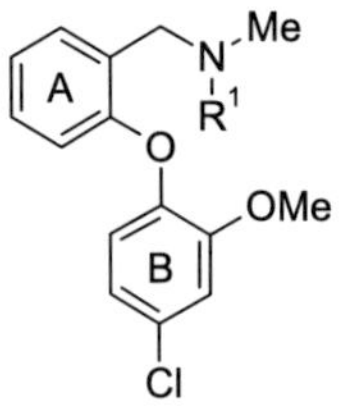

Fig. 12 Structures and uptake activity of benzylic amines **38** and **39**

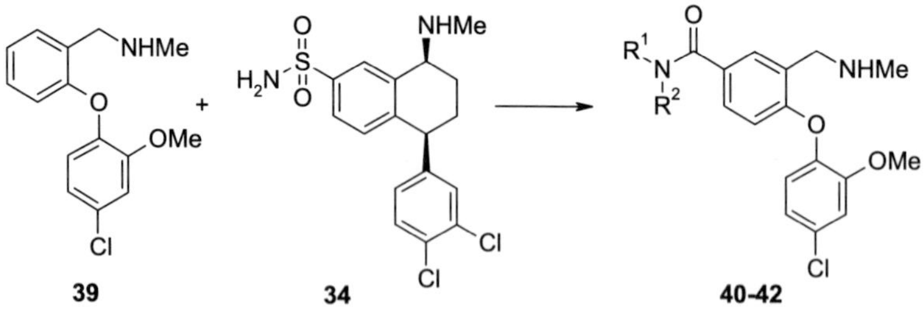

38 R^1 = Me
[^{3}H]5-HT IC$_{50}$ 6nM
[^{3}H]NE IC$_{50}$ 14nM
[^{3}H]DA IC$_{50}$ 4780nM
HLM Cl$_{int}$ 45 μl/min/mg

39 R^1 = H
[^{3}H]5-HT IC$_{50}$ 13nM
[^{3}H]NE IC$_{50}$ 78nM
[^{3}H]DA IC$_{50}$ 19600nM
HLM Cl$_{int}$ 9 μl/min/mg

Fig. 13 Structures of tetrahydronaphthalene **34** and amide targets **40–42**

Polar groups were well tolerated, and the amides **40–42** possessed the desired pharmacological properties (Table 2). Further studies indicated that analogues **40–42** also had good in vitro human liver microsome (HLM) stability, weak hERG activity and good passive membrane permeability. No information on blood–brain barrier (BBB) penetration, in vivo pharmacokinetics or pharmacodynamics has yet been reported.

The authors then discovered that amides such as **42** suffered from non-P450 mediated amide cleavage in human hepatocytes and this could not be improved by modification of the amide substituents. To circumvent this problem, a pyridyl ring was incorporated as a replacement for the benzamide group (Fig. 14) [40]. The 2,4-disubstitution pattern was retained in ring B, and introduction of a methyl group next to the pyridine nitrogen was also investigated to mitigate any potential P450 inhibition issues with unflanked

Table 2 Uptake inhibition and ADME profile of compounds **40–42**

Compound	[^{3}H]5-HT IC$_{50}$ (nM)	[^{3}H]NE IC$_{50}$ (nM)	[^{3}H]DA IC$_{50}$ (nM)	$c\log P$	HLM Cl$_{int}$ (μl/ml/mg)	hERG K_i (μM)	PAMPA Papp (10^{-6} cm/s)
40	12	62	21 000	2.5	8	> 7500	11
41	13	78	15 700	2.8	8	> 7500	14
42	8	32	> 40 000	2.5	15	> 7500	15

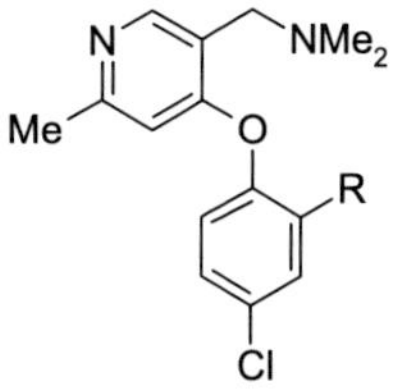

Fig. 14 Structures of pyridyl compounds **43** and **44**

Table 3 Uptake inhibition and ADME profile of compounds 42–44

	42	43	44
[^{3}H]5-HT IC$_{50}$ (nM)	8	13	6
[^{3}H]NE IC$_{50}$ (nM)	32	26	26
[^{3}H]DA IC$_{50}$ (nM)	> 40 000	5040	1680
$c \log P$	2.5	3.5	4.3
HLM Cl$_{int}$ (μl/ml/mg)	15	< 7	< 7
Hheps Cl$_{int}$ (μl min/million cells)	21	< 5	< 5
CYP2D6 IC$_{50}$	NT	> 10 000	> 10 000
CYP3A4 IC$_{50}$	NT	> 10 000	> 10 000
Caco-2 Papp (10^{-6} cm/s) A-B/B-A	NT	34/34	30/34

pyridines. Examples **43** and **44** demonstrated good SNRI activity with excellent selectivity over DAT activity (Table 3). Metabolic stability studies in HLMs and human hepatocytes also indicated that **43** and **44** had improved stability when compared to amide **42**. Additionally, examples **43** and **44** exhibited weak P450 inhibition and good Caco-2 flux with no evidence of P-glycoprotein (P-gp) mediated efflux. These flux data may be suggestive of the potential for good oral absorption and BBB penetration, although no in vivo data were reported in this publication.

The authors then analysed the relationship between SNRI potency and lipophilicity for the pyridyl series. It was found that SRI activity was not influenced by $c \log P$; however NRI, and therefore dual SNRI, activity was heavily influenced by $c \log P$, with good dual potency only being achieved at high $c \log P$ (> 3.5). Due to the potential for high lipophilicity to adversely impact polypharmacology, analogues **43** and **44** were screened against a broad panel of receptors, ion channels and enzymes. It was found that these compounds possessed significant off-target activity and this series targeting dual SNRI pharmacology was halted. The authors highlighted the issues with working in a chemical series where the desired pharmacological activity is highly dominated by lipophilicity.

3.2
Milnacipran Analogues

Milnacipran 12, a dual SNRI, is approved for the treatment of depression [42] and is currently in phase 3 trials for the treatment of fibromyalgia [43] (Fig. 15). Interestingly, milnacipran is a polar compound with a $\log D$ of ~ 0, and as such is different from many other centrally acting monoamine reuptake inhibitors which are far more lipophilic. The polar nature of milnacipran delivers an attractive PK profile in humans, with high bioavailability, long half-life and renal elimination of unchanged drug as the primary route of clearance [44]. Despite its polar nature, milnacipran is still believed to exert its pharmacology via central inhibition of serotonin and noradrenaline reuptake. The lack of interaction of milnacipran with P450 enzymes as either substrate or inhibitor reduces the potential for drug–drug interactions, which are highly prevalent for many other marketed monoamine reuptake inhibitors. This attractive physicochemical and PK profile, coupled with clinical efficacy, has prompted several groups to further investigate the SAR in this template, particularly as 12 has relatively weak SNRI activity.

Roggen et al. from the University of Oslo [45] were the first group to investigate enantiomerically pure milnacipran analogues, and their work focused on variations of the aromatic moiety and the impact of stereochemistry on in vitro monoamine reuptake activity (Fig. 16). They found that the (1R, 2S) enantiomer of milnacipran 12a exhibited weak dual SNRI pharmacology, whilst the (1S, 2R) enantiomer 12b was significantly more potent against the

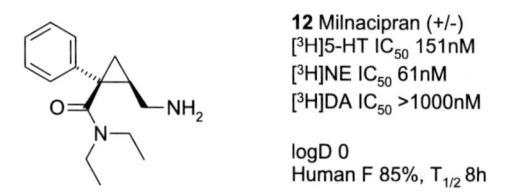

12 Milnacipran (+/-)
[³H]5-HT IC$_{50}$ 151nM
[³H]NE IC$_{50}$ 61nM
[³H]DA IC$_{50}$ >1000nM

logD 0
Human F 85%, T$_{1/2}$ 8h

Fig. 15 Structure and properties of milnacipran 12

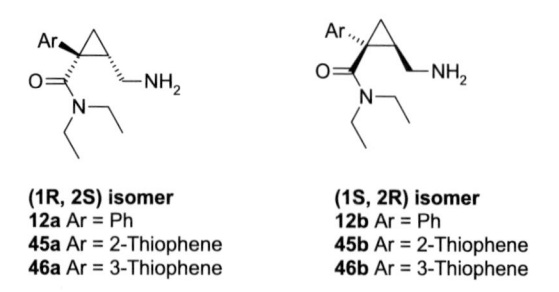

(1R, 2S) isomer
12a Ar = Ph
45a Ar = 2-Thiophene
46a Ar = 3-Thiophene

(1S, 2R) isomer
12b Ar = Ph
45b Ar = 2-Thiophene
46b Ar = 3-Thiophene

Fig. 16 Structures of milnacipran single enantiomers and analogues 12a/b, 45a/b and 46a/b

Table 4 Uptake inhibition of compounds **12a/b**, **45a/b** and **46a/b**

Compound	[^{3}H]5-HT IC$_{50}$ (nM)	[^{3}H]NE IC$_{50}$ (nM)	[^{3}H]DA IC$_{50}$ (nM)
12a	420	200	10000
12b	120	7	9000
45a	520	1500	970
45b	130	29	1400
46a	250	410	10000
46b	190	19	10000

NET. Variations of the aryl group, for example **45** and **46**, generally did not lead to any significant improvements in dual SNRI activity (Table 4).

Chen et al. at Neurocrine Biosciences have investigated the SAR of the milnacipran template still further [46–48]. N alkylation of the primary amine was investigated in the racemic series, but in all cases led to a significant drop in potency. The SAR of secondary and tertiary amides was then probed [46] (Fig. 17, Table 5). In general, secondary amides led to a significant drop in activity, with the exception of anilide **47** which had weak NA activity. The authors rationalized this result by a potential favourable π interaction between the inhibitor and transporter protein. Ethyl-phenyl tertiary amide **48** was tenfold more potent and then cyclization to indoline **49** gave a fur-

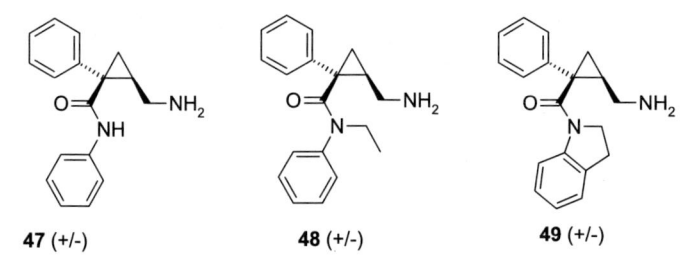

47 (+/-) **48** (+/-) **49** (+/-)

Fig. 17 Structures of milnacipran analogues **47–49**

Table 5 Uptake inhibition of compounds **12** and **47–49**

Compound	[^{3}H]5-HT IC$_{50}$ (nM)	[^{3}H]NE IC$_{50}$ (nM)	[^{3}H]DA IC$_{50}$ (nM)
12	420	77	6100
47	> 10000	570	> 10000
48	4400	63	> 10000
49	13	4.4	3900

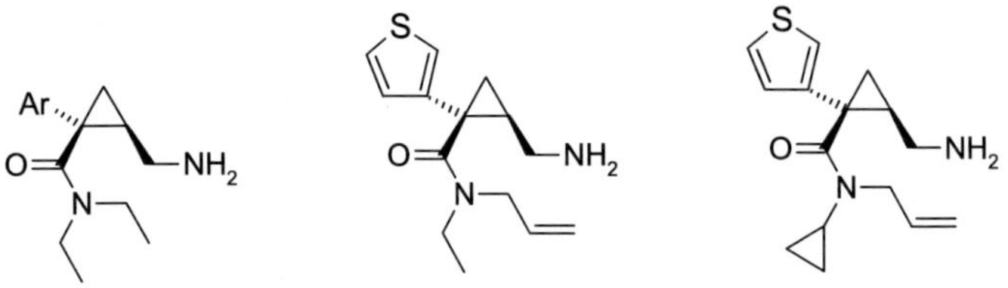

(1S, 2R) isomer
12b Ar = Ph
46b Ar = 3-Thiophene

(1S, 2R) isomer
50

(1S, 2R) isomer
51

Fig. 18 Structures of milnacipran analogues 50 and 51

ther increase in potency at both the SERT and NET, with good selectivity over the DAT. No ADME or in vivo data have yet been published on these compounds.

Chen et al. then extended their work on tertiary amides using a 3-thienyl group as a replacement for the phenyl moiety (Fig. 18), and in this work the compounds were isolated as single enantiomers [47]. This change had originally been reported by Roggen et al. (Table 4) [45]. Although the parent compound 46b had poor 5-HT reuptake activity, introduction of an allylic group (compound 50) increased activity with no significant increase in $c \log P$ (Table 6). Incorporation of a cyclopropyl group in compound 51 increased 5-HT activity further to give a potent and selective SNRI combined with low lipophilicity. In vitro metabolic stability studies were then carried out, and showed that examples 50 and 51 had similar stability in both HLMs and rat liver microsomes (RLMs) to milnacipran. Additionally, these new analogues had similar flux and P-gp mediated efflux in Caco-2 cells when compared to milnacipran. To assess whether this efflux would affect brain penetration, compound 50 was dosed orally to rats alongside milnacipran. The brain/plasma ratios were similar for both compounds, indicating that 50 had the potential to cross the BBB.

Table 6 Uptake inhibition and ADME profile of compounds 12b, 46b, 50 and 51

Compound	$[^3H]$5-HT IC_{50} (nM)	$[^3H]$NE IC_{50} (nM)	$[^3H]$DA IC_{50} (nM)	$c \log P$	Caco-2 A-B/B-A	Rat brain/ plasma ratio (1 h, 4 h)
12b	320	40	3200	1.2	4.5	0.42, 0.86
46b	830	49	> 10 000	0.9	4.8	NT
50	140	5.6	> 10 000	1.2	8.9	0.30, 0.40
51	26	8.7	> 10 000	1.1	7.3	NT

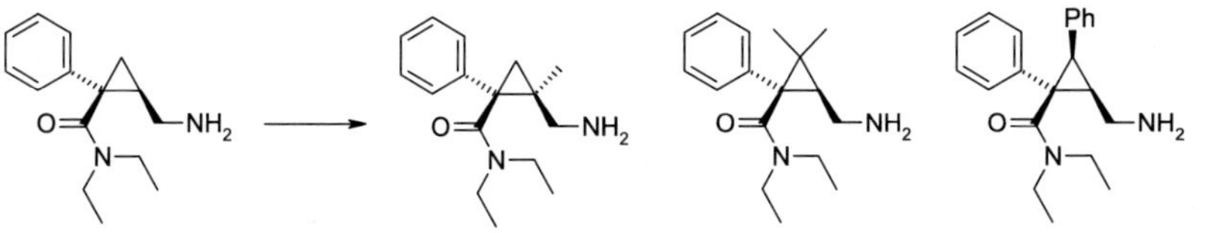

12 Milnacipran (+/-)
[^{3}H]5-HT IC$_{50}$ 420nM
[^{3}H]NE IC$_{50}$ 77nM
[^{3}H]DA IC$_{50}$ 6100nM

52 (+/-)
[^{3}H]5-HT IC$_{50}$ 6900nM
[^{3}H]NE IC$_{50}$ >10000nM
[^{3}H]DA IC$_{50}$ >10000nM

53 (+/-)
[^{3}H]5-HT IC$_{50}$ >10000nM
[^{3}H]NE IC$_{50}$ >10000nM
[^{3}H]DA IC$_{50}$ 4500nM

54 (+/-)
[^{3}H]5-HT IC$_{50}$ 6400nM
[^{3}H]NE IC$_{50}$ >10000nM
[^{3}H]DA IC$_{50}$ >10000nM

Fig. 19 Structures and uptake inhibition of milnacipran analogues 52–54

In their most recent work, Chen et al. have continued to explore SARs in the milnacipran template, with a key goal being to improve on the potency of milnacipran without significantly changing lipophilicity [48]. Incorporation of substitution onto the central cyclopropyl ring was investigated (Fig. 19) without success, with examples 52–54 showing a significant reduction in activity.

Further replacements for the phenyl group were then studied, with a particular focus on oxygen-containing substituted phenyl rings (Fig. 20). This strategy successfully increased SNRI activity, e.g. compounds 55 and 56 (Table 7). Further amide modification by incorporation of N-propargyl groups in 57 increased potency further, whilst keeping $c \log P$ similar to milnacipran. In vitro metabolic stability studies showed that 55–57 had similar stability in HLMs compared to milnacipran. In vitro Caco-2 flux data showed

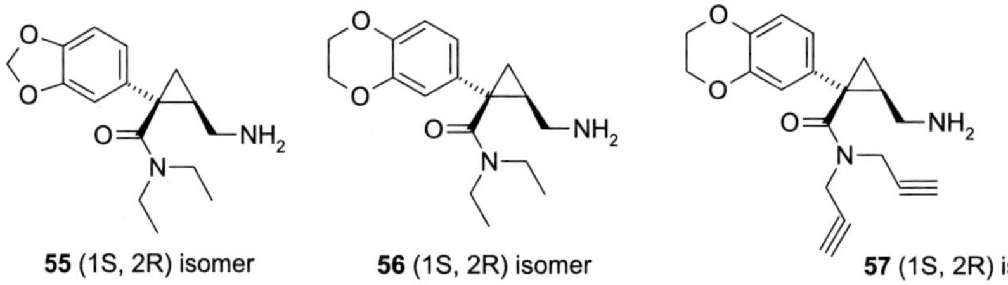

55 (1S, 2R) isomer   **56** (1S, 2R) isomer   **57** (1S, 2R) isomer

Fig. 20 Structures of milnacipran analogues 55–57

Table 7 Uptake inhibition and ADME profile of compounds 12b and 55–57

Compound	[^{3}H]5-HT IC$_{50}$ (nM)	[^{3}H]NE IC$_{50}$ (nM)	[^{3}H]DA IC$_{50}$ (nM)	$c \log P$	Caco-2 A-B/B-A	HLM Cl$_{sys}$ (ml/min/kg)
12b	320	40	3200	1.2	7.4	5.0
55	480	12	>10 000	1.1	6	7.9
56	140	6.3	>10 000	0.9	NT	4.4
57	25	1.7	>10 000	0.7	15	7.9

that **55** had a similar efflux ratio compared to **12**, whereas the efflux ratio of **57** was twofold higher. This P-gp mediated efflux has the potential to adversely affect BBB penetration; however, no in vivo data were reported in this publication.

3.3
Duloxetine and Propanamine Analogues

Duloxetine **11** (DLX), a dual SNRI, is approved for the treatment of depression [49], pain associated with diabetic neuropathy [50], fibromyalgia [51] and stress urinary incontinence [52]. Gallagher et al. at Eli Lilly have recently described the medicinal chemistry programme which led to the discovery of duloxetine [53]. Phenyl-naphthalene lead **58** showed encouraging levels of dual SNRI activity (Fig. 21, Table 8); however, substitution on the phenyl ring generally led to an erosion of NA reuptake activity. Bio-isosteric replacements of the phenyl group were then sought, and the 2-thienyl **59** was identified as a lead compound. The separate enantiomers **60** and **11** showed subtly different pharmacology, and (S) enantiomer **11** (duloxetine) was progressed on the basis of increased 5-HT reuptake potency. Rat microdialysis experiments were then carried out with **11** to determine the increase in synaptic concentrations of 5-HT and NA. Following po dosing at 10 mg/kg, increases over basal levels of 5-HT and NA of $208 \pm 31\%$ and $353 \pm 62\%$, respectively, were observed. The pharmacokinetics, pharmacology and efficacy of duloxetine

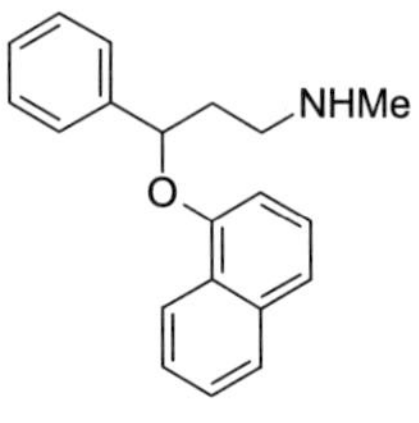

58 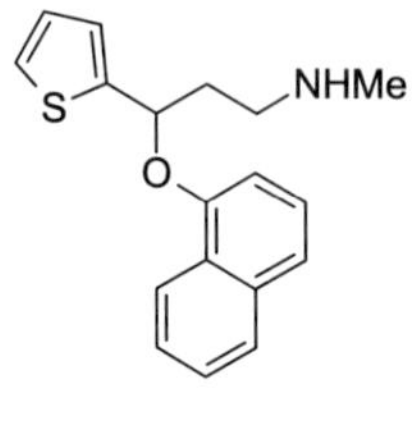 (+/-)   **59** (+/-)
                    60 (R) enantiomer
                    11 (S) enantiomer - Duloxetine

Fig. 21 Structures of aryloxy propanamines **11** and **58–60**

Table 8 Binding affinities of compounds **11** and **58–60**

Compound	SERT K_i (nM)	NET K_i (nM)	DAT K_i (nM)
58	2.4	20	NT
59	1.4	20	NT
60	8.8	16	660
11	4.6	16	370

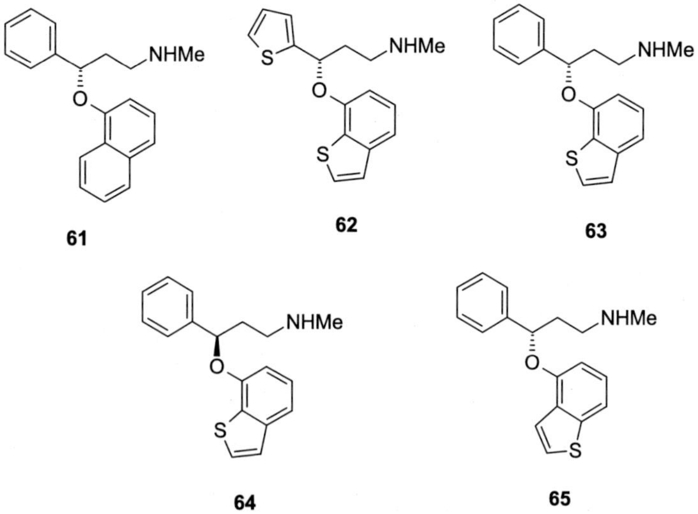

Fig. 22 Structures of duloxetine analogues 61–65

have been recently reviewed elsewhere [54, 55] and therefore will not be covered in this chapter.

Acid instability studies with duloxetine then prompted Gallagher et al. to investigate the acid stability of new SNRIs from the thienyl series [56]. Analysis of compounds 11 and 61–63 (Fig. 22) showed that removal of the thienyl group significantly improved acid stability (Table 9), and compound 63 proved to be the most attractive lead which combined stability and good SNRI activity. Further benzothiophenes were then synthesized as single enantiomers (64 and 65, Fig. 22). From the resulting SAR, 7-isomer 65 was progressed further; however, in vitro metabolic studies showed 65 formed diol metabolites which were postulated to arise from a reactive epoxide interme-

Table 9 Binding affinities and acid stability of compounds 11 and 61–65

Compound	SERT K_i (nM)	NET K_i (nM)	DAT K_i (nM)	Acid stability[a]
11	4.6	16	370	18
61	1	6.1	190	99
62	1.5	1.4	120	40
63	8.2	2.2	220	99
64	1.9	1.8	220	NT
65	0.5	4.4	440	NT

[a] % Parent compound remaining at 2 h and 37 °C

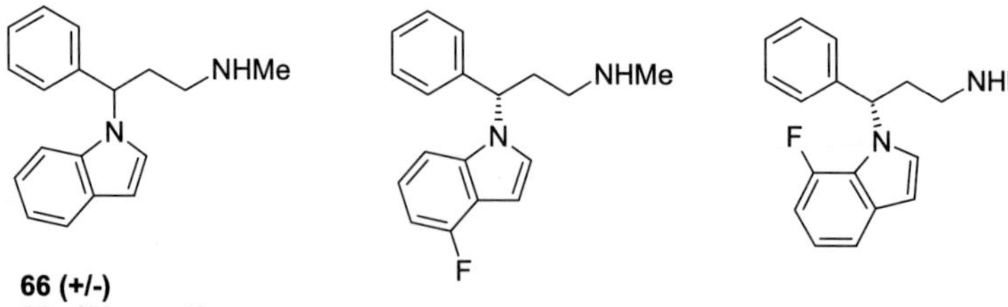

66 (+/-)
66a (R) enantiomer
66b (S) enantiomer
67 (S) enantiomer
68 (S) enantiomer

Fig. 23 Structures of indole compounds **66–68**

Table 10 Binding affinities of compounds **66–68**

Compound	SERT K_i (nM)	NET K_i (nM)	DAT K_i (nM)
66	32	47	1243
66a	35	34	3200
66b	41	33	> 10 000
67	12	12	680
68	24	18	16% at 2 μM

diate. Due to the potential for such epoxide metabolites to lead to in vivo toxicity, this compound was halted. The 4-isomer **64** was shown not to form diol metabolites, and was then progressed further. Following po dosing of **64** at 3 mg/kg in a rat microdialysis study, increases over basal levels of 5-HT and NA of $222 \pm 14\%$ and $215 \pm 9\%$, respectively, were observed.

Mahaney et al. at Wyeth have recently disclosed a series of indolylpropanamines [57] and investigated their monoamine reuptake activity (Fig. 23). The N-linked indole group was introduced as a replacement for the aryloxy group in the above aryloxy propylamine template. The unsubstituted indole lead **66** showed balanced SNRI activity with encouraging selectivity over dopamine reuptake activity (Table 10). The single enantiomers **66a** and **66b** had very similar reuptake pharmacology. Introduction of a fluoro substituent onto the indole ring gave an increase in 5-HT and NA activity, with both **67** and **68** having balanced SNRI activity with good dopamine (DA) selectivity. No in vitro ADME or in vivo studies have yet been reported on this series of compounds.

3.4 Piperazines

Fray et al. at Pfizer have reported the discovery of a novel series of piperazine derivatives as dual SNRIs [58, 59]. Lead compound **69** originated from targeted file screening looking for compounds with similar structural features to known monoamine reuptake inhibitors (Fig. 24). It was then established that mono-substitution of the A-ring was preferred for dual pharmacology, with *ortho* substitution generally providing the most potent and balanced SNRIs. Although several substituents were tolerated, it was also found that small substitution changes could result in subtle differences in the 5-HT to NA ratio of activities. Finally, it was found that substitution of the phenyl B-ring or replacement by heterocycles generally resulted in a loss of potency against both the 5-HT and NA transporters. Optimization of the A-ring substituent led to the discovery of racemic compounds **70** and **71**, which were subsequently separated into single enantiomers **72–75** (Fig. 24).

Interestingly, racemates and single enantiomers exhibited potent SNRI activity and good selectivity over the uptake of DA (Table 11). Single enantiomers **72–75** were further profiled in ADME assays. Although all compounds exhibited good metabolic stability in HLMs, the 2-Cl-containing analogues **74** and **75** were found to be more stable. This could indicate O dealkylation of the 2-alkoxy group as a potential metabolic pathway resulting in more rapid turnover. However, 2-OEt containing compounds **72** and **73** were found to be significantly weaker CYP2D6 inhibitors. Compound **73** was also

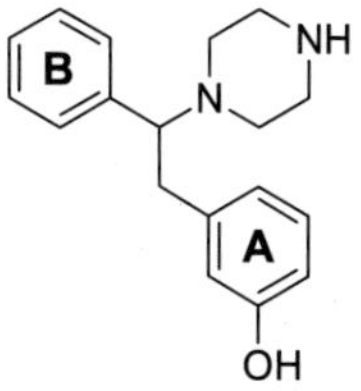

69

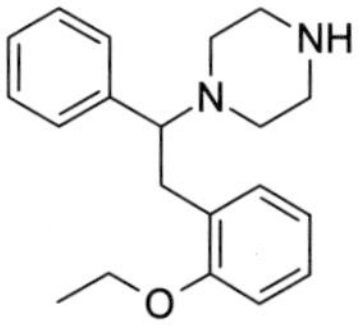

70 racemic
72 (S)(+)- isomer
73 (R)(-)- isomer

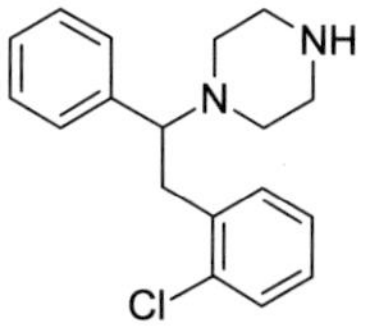

71 racemic
74 (+)- isomer
75 (-)- isomer

Fig. 24 Structures of piperazines **69–75**

Table 11 Uptake inhibition and ADME profile of compounds **69–75**

Compound	[^{3}H]5-HT IC$_{50}$ (nM)	[^{3}H]NE IC$_{50}$ (nM)	[^{3}H]DA IC$_{50}$ (nM)	$c\log P/$ $\log D$	HLM $t_{1/2}$ (min)	CYP2D6inh IC$_{50}$ (μM)
Duloxetine	6	19	870	4.3/–	–	–
69	9.5	14	1400	3/–	–	–
70	13	16	> 4000	4.2/–	–	–
71	5.4	22	1300	4.5/–	–	–
72	9.8	19	> 4000	4.2/1.5	76	10
73	13	16	> 4000	4.2/1.5	96	30
74	13	50	1600	4.5/1.8	> 120	0.6
75	3.9	19	2900	4.5/1.8	> 120	0.5

found to be a weak inhibitor of CYPs 3A4, 2C9, 2C19 and 1A2 (IC_{50} values all $\geq 30\,\mu M$) and demonstrated good membrane permeability (Caco-2 A-B 33×10^{-6} cm/s/B-A 36×10^{-6} cm/s), which may be predictive of good oral absorption.

Compound **73** was selected to further evaluate the series. Compound **73** demonstrated > 100-fold selectivity for serotonin and noradrenaline reuptake inhibition over a variety of G-protein coupled receptors including adrenergic, dopaminergic, muscarinic, nicotinic and opioate receptors. However, it did exhibit some affinity for sodium channels (site 2, IC_{50} $0.41\,\mu M$), calcium channels (L-type, diltiazem site, IC_{50} $0.73\,\mu M$) and the hERG potassium channel (dofetilide binding IC_{50} $16\,\mu M$). The impact of these findings on potential cardiovascular effects, as well as any in vivo pharmacology, pharmacokinetics or efficacy data in appropriate disease models have yet to be reported.

3.5
Pyrrolidines

Whitlock et al. at Pfizer have reported the discovery of a potent series of substituted 3-aminopyrrolidines as dual SNRIs, derived from the piperazine series described above by scaffold-hopping (Fig. 25) [60]. To rapidly optimize the drug-like properties of the series the authors determined potency, selectivity, CYP2D6 inhibition and metabolic stability in parallel. Initial SAR study was carried out with 2,3-di-Me and 2,3-di-Cl as the aryl A-ring R' substituents. This demonstrated that a branched 4-tetrahydropyranyl was preferred to combine SNRI activity with reduced CYP2D6 inhibition. It also found that both pyrrolidine enantiomers did exhibit SNRI activity but that the (S) enantiomer generally had better potency. Subsequent SAR studies with 4-tetrahydropyranyl as the R group established that di-substitution of phenyl ring-A was generally required for dual 5-HT and NA reuptake inhibition, and that small modifications of the substituents could result in subtle differences in the 5-HT to NA ratio of activities. This analysis led to the identification of several potent and balanced SNRIs, including compounds **76–79**

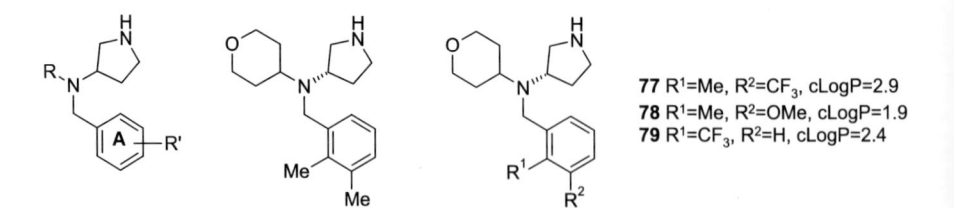

77 R^1=Me, R^2=CF_3, cLogP=2.9
78 R^1=Me, R^2=OMe, cLogP=1.9
79 R^1=CF_3, R^2=H, cLogP=2.4

76 cLogP 2.4

Fig. 25 Structures and $c \log P$ of pyrrolidines **76–79**

Table 12 In vitro inhibition of monoamine reuptake, ADME profiles and hERG channel activity of compounds 76–79

Compound	$[^3H]$5-HT IC_{50} (nM)	$[^3H]$NE IC_{50} (nM)	$[^3H]$DA IC_{50} (nM)	HLM/hheps $t_{1/2}$ (min)	CYP2D6inh IC_{50} (μM)	hERG IC_{50} (μM)
Duloxetine	6	19	870	–	–	–
76	9	7	727	> 120/144	5.9	14.4
77	13	20	> 4000	> 120/> 240	5.5	2.9
78	11	11	968	94/–	8.2	–
79	21	9	> 4000	> 120/–	30	1.1

(Fig. 25), which were further differentiated based on their human hepatocyte stability and potential to block the hERG channel. It is interesting to notice that in this series hERG activity was found to be more sensitive to structural changes than lipophilicity, as can be seen when comparing compounds 76, 77 and 79 (Table 12).

From their analysis, compound 76 emerged as optimal, combining potent dual SNRI activity with good selectivity over dopamine reuptake, good in vitro metabolic stability, weak CYP2D6 inhibition and weak affinity for the hERG channel (Table 12). Compound 76 also exhibited more than 200-fold selectivity for serotonin and noradrenaline reuptake inhibition when assessed over a panel of 150 receptors, enzymes and ion channels.

More recently, Wakenhut et al. at Pfizer have reported further advances in the 3-amino pyrrolidine series [61]. Lead compound 76 was found to be predominantly metabolized by the polymorphic enzyme CYP2D6, via hydroxylation of the 2,3-di-Me aryl ring (Fig. 26). This metabolic pathway had the potential for significant overexposure in CYP2D6 poor metabolizers and so the progression of 76 was halted. A strategy of metabolic blocking was initially employed, leading to compound 80. SNRI activity was retained in 80, and in vitro studies showed that the primary route of metabolism was now N dealkylation of the tetrahydropyranyl group. However, the metabolizing enzyme was still found to be CYP2D6, resulting in no mitigation of the drug–drug interaction risk.

The N-dealkylation pathway was then blocked by introduction of an amide substituent (81, Fig. 27). Compound 81 combined SNRI activity, good selectivity, excellent metabolic stability and good membrane permeability when determined in Caco-2 cells. Compound 81 exhibited no significant inhibition of CYP450 enzymes and was > 200-fold selective when screened against a panel of receptors, enzymes and ion channels. Compound 81 was found to be so metabolically stable that no measurable turnover was detected. This extreme metabolic stability resulted in no assessment of CYP2D6 contribution to metabolism being possible, and the potential for increased exposure in CYP2D6 poor metabolizers would need to be monitored in clinical studies.

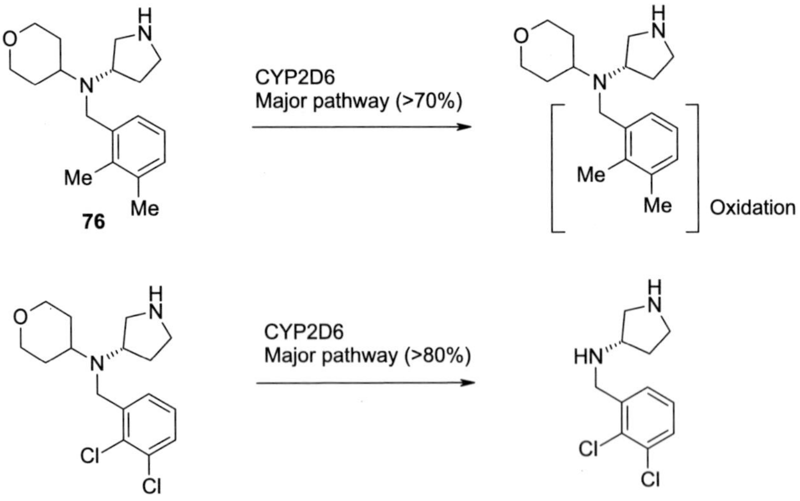

Fig. 26 Metabolic pathways of compounds 76 and 80

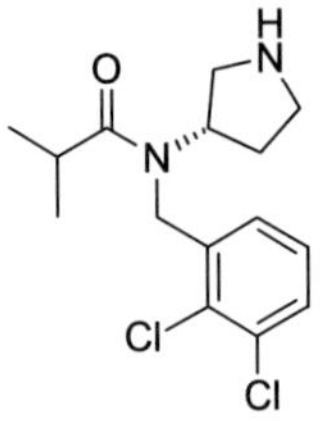

[³H]5-HT IC$_{50}$ 12nM
[³H]NE IC$_{50}$ 23nM
[³H]DA IC$_{50}$ 1270nM
HLM T$_{1/2}$ >120min
h.heps Cl$_{int}$ 2.7 µl/min/million cells
CaCO-2 A-B/B-A 19/28
hERG IC$_{50}$ 24.7µM
CYP2D6 IC$_{50}$ 27.1µM

Dog pharmacokinetics:
iv Cl 1.1 ml/min/kg
iv Vd 0.5 l/kg
iv T$_{1/2}$ 5.7 h

po T$_{1/2}$ 5.7 h
po F 91%

Fig. 27 Structure, pharmacology and ADME properties of compound 81

Dog pharmacokinetic data on 81 were disclosed, and showed that 81 had low clearance, low volume, long half-life and high bioavailability (Fig. 27).

3.6
Piperidines

Gallagher et al. at Eli Lilly have reported the discovery of a novel series of N-alkyl-N-arylmethylpiperidine-4 amines as dual SNRIs [62]. Initial investigations with 2-CF$_3$ as the R' substituent established that dual SNRI potency could be achieved with several alkyl R substituents (compounds 82–85, Fig. 28, Table 13). Subsequently the R substituent was fixed as isobutyl, and then SNRI activity was investigated as a function of the aryl ring R' substituent. Mono-substitution in the *ortho* position delivered several potent SNRIs, with Me, CF$_3$ and Cl amongst the preferred groups (compounds 83, 86, 87), whereas *meta* or *para* substitution resulted in a drop in NRI ac-

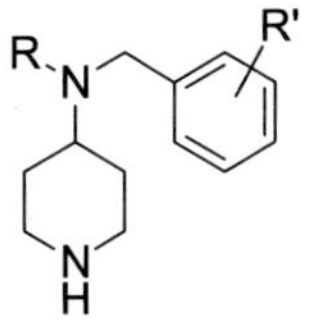

82 R=n-Pr, R'=2-CF$_3$
83 R=i-Bu, R'=2-CF$_3$
84 R=c-C$_5$H$_9$, R'=2-CF$_3$
85 R=c-C$_3$H$_5$CH$_2$, R'=2-CF$_3$
86 R=i-Bu, R'=2-Me
87 R=i-Bu, R'=2-Cl
88 R=i-Bu, R'=2-Cl 4-Cl
89 R=i-Bu, R'=2-CF$_3$ 4-F
90 R=i-Bu, R'=2-CF$_3$ 5F
91 R=i-Bu, R'=2-CF$_3$ 6F

Fig. 28 Structures of piperidines **82–91**

Table 13 Binding affinities of compounds **82–91**

Compound	SERT K_i (nM)	NET K_i (nM)	DAT K_i (nM) or %inh at 1 μM
Duloxetine	0.8 ± 0.04	7.5 ± 0.3	240 ± 23
82	2.2 ± 0.2	6.5 ± 0.4	− 12 ± 1%
83	1.4 ± 0.1	3.2 ± 0.3	210 ± 30
84	3.6 ± 0.1	4.1 ± 0.1	− 45 ± 0.7%
85	1.5 ± 0.1	6 ± 1	− 27 ± 0.1%
86	1.5 ± 0.4	3.3 ± 0.2	240 ± 19
87	1.6 ± 0.1	1.4 ± 0.1	83 ± 3
88	0.96 ± 0.1	2.0 ± 0.2	230 ± 20
89	0.30 ± 0.1	1.8 ± 0.3	190 ± 20
90	7.9 ± 0.9	3.4 ± 0.1	− 19 ± 2%
91	4.1 ± 0.40	6.0 ± 0.8	− 18 ± 2%

tivity. Di-substitution with preferred groups in the *ortho* position delivered additional potent SNRIs with several substitution patterns tolerated (compounds **88–91**).

Compound **83** was selected for evaluation in in vivo microdialysis experiments, and elevated both synaptic 5-HT and NA levels by 202 and 249%, respectively (3 mg/kg po). No pharmacokinetics, off-target pharmacology or efficacy data in disease models have been reported yet.

3.7
Constrained Tropanes

Kozikowski et al. at Georgetown University Medical Center have reported a series of (1S,3S,6R,10S)-(Z)-9-(thienylmethylene- or substituted thienylmethylene)-7-azatricyclo[4.3.1.0^{3,7}]decane derivatives as monoamine reuptake inhibitors [16, 63]. Among the structures shown, compound **92** was the most potent SNRI (SERT K_i 29.0 ± 1.6 nM, NET K_i 5.0 ± 1.3 nM) and exhibited good selectivity over dopamine reuptake (DAT K_i 368 ± 1.6 nM) (Fig. 29).

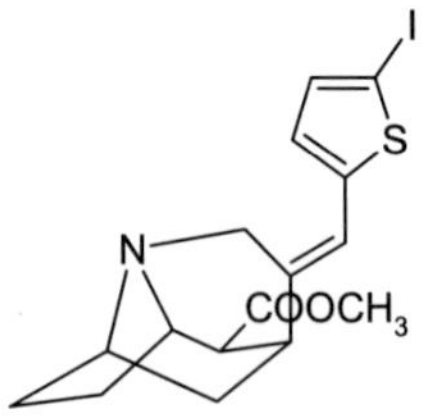

92
SERT K_i = 29.0+/-1.6nM
NET K_i = 5.0+/-1.3nM
DAT K_i = -368+/-1.6nM

Fig. 29 Structure of constrained tropane **92**

4
Noradrenaline Reuptake Inhibitors

Selective inhibition of noradrenaline reuptake has been shown to be an attractive approach for the treatment of a number of diseases. For example, reboxetine **14** (Sect. 4.1.1) is used clinically for the treatment of depression and atomoxetine **13** (Sect. 4.2.1) is a new therapy for the treatment of ADHD (Fig. 30). Furthermore, NRIs are under investigation as potential therapies for chronic pain, urinary incontinence, fibromyalgia and a variety of other indications [64, 65]. A number of reviews have been published that summarize discoveries of NRIs [66–69].

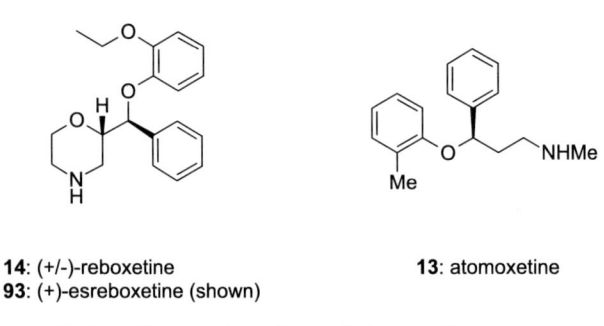

14: (+/-)-reboxetine
93: (+)-esreboxetine (shown)

13: atomoxetine

Fig. 30 Structures of reboxetine, esreboxetine and atomoxetine

4.1
Morpholine Analogues

4.1.1
Reboxetine

(±)-Reboxetine **14** [RBX; racemic mixture of (S,S) and (R,R) enantiomers] (Fig. 30) is an orally active, selective NRI developed and launched by Pharmacia & Upjohn (now Pfizer) for the treatment of depression. Reboxetine was launched in Europe in 1997 and since that time a number of primary papers

Table 14 Binding affinities of (±)-racemic, (+)-(S,S) and (−)-(R,R) reboxetine

Compound	NET K_i (nM)	SERT K_i (nM)	Selectivity of K_i of SERT/NET
Reboxetine	1.1	129	124
(+)-(S,S)-Reboxetine	0.2	2900	14 500
(−)-(R,R)-Reboxetine	7.0	104	15

Table 15 Binding affinities of (±)-racemic and (+)-(S,S)-reboxetine for additional receptor systems [K_i (nM)]

Compound	Histamine H_1	Serotonin 5-HT$_{2C}$	Muscarinic	Adrenergic α_1
Reboxetine	1400	1500	3900	10 000
(+)-(S,S)-Reboxetine	> 30 000	3000	3000	> 10 000

Other receptors and enzymes tested with $K_i > 10\,000$ nM include: dopamine reuptake (DRI), muscarinic (M1, M2, M3, M4, M5), adrenergic (α2, β1, β2), dopaminergic (D1, D2, D3, D4), serotinergic (5-HT1A, 5-HT2A, 5-HT3, 5-HT4, 5-HT6, 5-HT7), adenosine (A1, A2), benzodiazepine (peripheral and central), L-type calcium channels, histaminergic (H2), imidazoline (I2), melatonin (mt1), N-methyl-D-aspartate (glutamate and channel sites), neurokinin (NK1), nicotinic (α3, α4, α7), sigma, MAO-A, MAO-B, NOS, tyrosine hydroxylase and xanthine oxidase

and comprehensive reviews have been published describing reboxetine's pharmacology, clinical efficacy, pharmacokinetics, safety and toleration [70–76].

The (+) enantiomer of reboxetine is being developed by Pfizer as a potential oral treatment of neuropathic pain and fibromyalgia [77]. Esreboxetine [(+)-(S,S)-reboxetine; SS-RBX] **93** is more potent than reboxetine racemate with respect to inhibiting the uptake of NA and much weaker at inhibiting uptake of 5-HT; the combination of these two pharmacologies makes esreboxetine highly selective for NRI over SRI [76] (Table 14). In addition, esreboxetine also demonstrates good selectivity over a range of aminergic receptors and other targets [78] (Table 15).

4.1.2
Arylthiomethyl Morpholines

Walter et al. at Eli Lilly have reported novel arylthiomethyl morpholines **94** (Fig. 31), including an example of a selective NRI [79]. All target compounds were initially tested as racemic (S,S)/(R,R) diastereoisomers and the SAR showed that replacement of the aryloxy ether in reboxetine by an arylthio ether retained good levels of NRI activity. Furthermore, significant SRI activity had been introduced to yield dual SNRIs **95**.

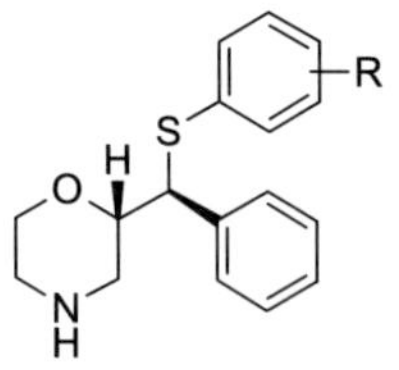

Fig. 31 Structures of compounds 94 and 95

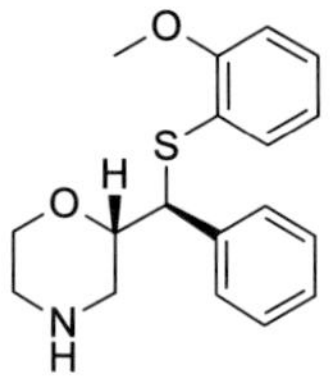

Table 16 Binding affinities of compound 95 and its enantiomers

Compound	SERT K_i (nM)	NET K_i (nM)	DAT K_i (nM)
95	10.7 ± 2.2	1.2 ± 2.2	> 200 $(9.7 \pm 3.0\%)$ [a]
95a	66.2 ± 3.0	1.7 ± 2.2	> 200 $(2.8 \pm 1.8\%)$ [a]
95b	1.5 ± 0.2	24.6 ± 2.3	> 200 $(19.0 \pm 3.4\%)$ [a]

[a] % displacement at 1000 nM

When racemic 2-methoxy analogue **95** was separated into its enantiomers **95a** and **95b** by HPLC (no assigned absolute stereochemistry), there was a degree of disconnection of NRI and SRI activities to give **95a** as a selective NRI and **95b** as a dual SNRI (Table 16).

Caution must be exercised when evaluating SAR from the biological activities of racemic compounds as the contribution to the observed effects may not be equal for each enantiomer; this is further complicated when seeking dual activities across two biological targets.

4.1.3
Hydroxy-Reboxetine Analogues

The synthesis and biological activity of a series of tertiary alcohol containing benzyl morpholines **96** have been reported by Cases et al. at Eli Lilly [80, 81] (Fig. 32). Within this select set of compounds, SAR established a preference for the (2S, 2'R) stereochemistry, substitution at the 2 position of the benzyl ring and that a number of different groups at the 2 position would give potent and selective NRIs (NET, $K_i < 10$ nM when R = OMe, OEt, OiPr, Ph).

The NRI activity of 2-OMe compound **97** (K_i 3.2 nM) was comparable to that of atomoxetine and reboxetine and **97** had no significant affinity for either the 5-HT or DA transporters. Additional characterization in vitro found **97** to be extensively metabolized in RLMs but much more stable in dog liver microsome (DLM) and HLM preparations. Consistent with the microsomal

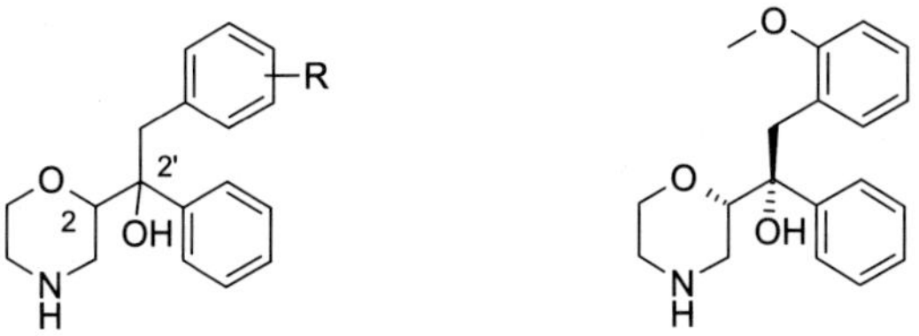

Fig. 32 Hydroxy-reboxetine analogues **96** and **97**

data, **97** had significantly improved oral bioavailability in the dog ($F_o \sim 61\%$; beagle dogs, 1 mg/kg) compared to the rat ($F_o \sim 5\%$; F344 rats, 10 mg/kg). Compound **97** was both chemically and configurationally stable under the aqueous acidic conditions disclosed.

Despite this extensive metabolism in RLMs, **97** was found to have potent oral in vivo activity in the α-methyl-m-tyrosine (α-MMT)-induced cortical NA depletion model in rats. In microdialysis experiments orally administered **97** (10 mg/kg) increased NA levels in interstitial fluid of the prefrontal cortex of rats by 250–300% above pre-drug baseline levels [81].

4.1.4
New Reboxetine Analogues

New derivatives of 2-[(phenoxy)(phenyl)-methyl]morpholine **98** have been disclosed as inhibitors of monoamine reuptake by Fish et al. at Pfizer [82]. SARs established that serotonin and noradrenaline reuptake inhibition were functions of stereochemistry and aryl/aryloxy ring substitution. Consequently, selective NRIs, SSRIs and dual SNRIs were all identified (Fig. 33, Table 17).

The absolute stereochemistry of **98** was found to have a significant effect on activity. For example, (*S,S*)-**99a** was a potent and selective NRI whereas (*R,R*)-**99b** was a potent SRI with NRI activity. This split in NRI vs SRI activity was also observed with five additional pairs of enantiomers. Overall,

Table 17 Binding affinities of compounds **99–101**

Compound	SERT K_i (nM)	NET K_i (nM)	DAT K_i (nM)
99a	3390	10	2600
99b	12	130	380
100	740	6	3080
101	1800	12	NT

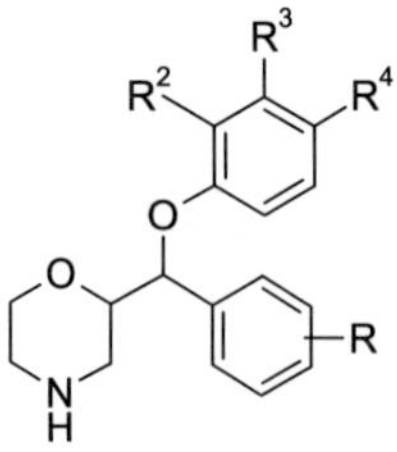

98

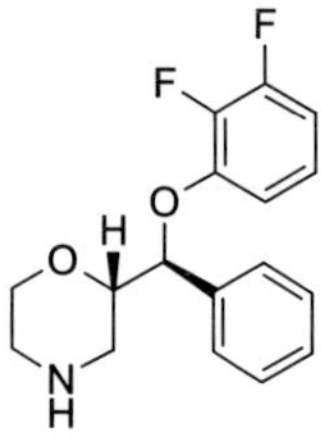

99a: (S,S)

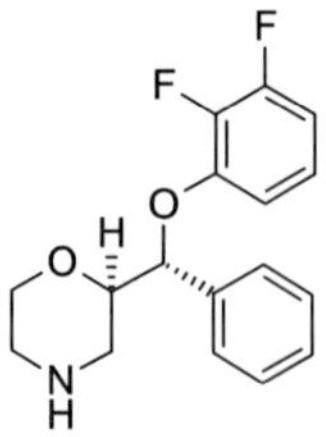

99b: (R,R)

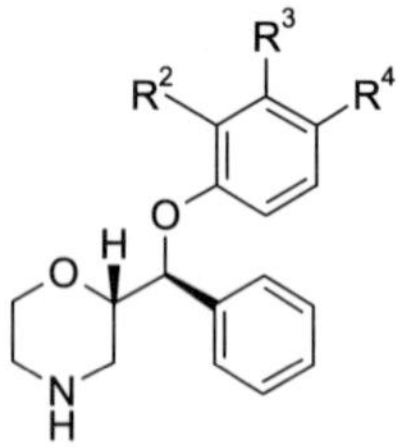

100: 2-OMe, 4-F
101: 2,3-OCH$_2$CH$_2$

Fig. 33 Structures of compounds 98–101

there was a broad range in both SRI and NRI activities and ratios of activity (SRI : NRI, 350 : 1 to 1 : 40). Additional examples of selective NRIs were **100** and **101**.

4.1.5
Viloxazine Analogues

The synthesis and pharmacological evaluation of thiophene analogues of viloxazine have been reported by Lissavetzky et al. at the Instituto de Quimica Medica (CSIC) [83]. Viloxazine **102** is a selective NRI used in the treatment of depression and the pharmacological properties and human pharmacokinetics have been reviewed [84, 85]. Preliminary pharmacological assessment showed eight thiophenes (e.g. **103**) to be generally active in a series of mouse

102: viloxazine   **103**

Fig. 34 Structures of viloxazine 102 and compound 103

models predictive of antidepressant properties. However, no primary pharmacology was reported and so it is yet to be established by which mechanism(s) these compounds are having their effects (Fig. 34).

4.2
Aryloxypropanamine Analogues

4.2.1
Atomoxetine

Atomoxetine **13** (Fig. 2, Table 18) [ATX; (*R*) enantiomer of tomoxetine; Strattera] is an orally active, selective NRI developed and launched by Eli Lilly as a new treatment of child, adolescent and adult ADHD. Atomoxetine hydrochloride was first launched in the USA in 2003 and is the first non-stimulant marketed for the treatment of ADHD. The pre-clinical, clinical and pharmacokinetic profile of ATX has been extensively reviewed [86–91].

Table 18 Inhibition of monoamine reuptake by atomoxetine in human recombinant assays

Compound	NET K_i (nM)	SERT K_i (nM)	DAT K_i (nM)	SERT/NET ratio
Atomoxetine	3.0	64	2440	21

4.2.2
Atomoxetine Analogues

The design, synthesis and SAR of a series of ring-constrained atomoxetine analogues were described by Wade et al. at Neurocrine Biosciences with templates derived from 1-aminoindanes **104**, 1-aminotetralins **105** and 2-aminotetralins **106** (Fig. 35) [92]. (2*S*,4*R*)-*trans*-Aminotetralin **107**

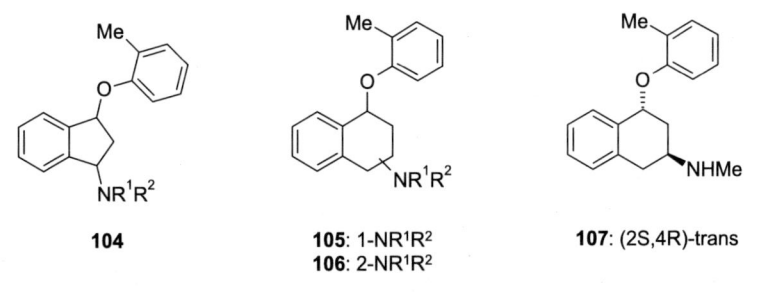

104	**105**: 1-NR¹R²	**107**: (2S,4R)-trans
	106: 2-NR¹R²	

Fig. 35 Structures of aminoindanes and aminotetralins **104–107**

(NBI 80532) proved to be the most potent compound disclosed with NRI activity (IC_{50} 8 nM) similar to that of atomoxetine. Furthermore, **107** had reduced affinity for the 5-HT transporter (IC_{50} 590 nM) giving moderately improved levels of selectivity over SRI (73-fold) compared to atomoxetine.

Assessment of compound **107** for selectivity against a panel of amine neurotransmitters revealed activity for the 5-HT$_{2B}$ receptor (IC_{50} 250 nM). It is important to note that no functional 5-HT$_{2B}$ was reported, but should **107** prove to be a potent agonist at the 5-HT$_{2B}$ receptor then it is likely to possess an unacceptable safety risk for cardiac toxicity [93]. Additional screening of **107** in in vitro ADME assays showed **107** to be a CYP2D6 inhibitor (IC_{50} 250 nM) and it was concluded that this level of inhibition could lead to potential drug–drug interactions.

4.3
Biphenyl Pyrrolidine Analogues

Researchers at Pfizer have disclosed derivatives of (3*S*)-*N*-(biphenyl-2-ylmethyl)pyrrolidin-3-amine as a new series of NRIs (Fig. 36). *N*-(Benzyl)pyrrolidin-3-amines **76** were first reported by Whitlock et al. as dual SNRIs [60]. As part of this study, during the exploration of the SAR of the aryl ring, biphenyl compound **108** was identified as a potent NRI (IC_{50} 12 nM) with selectivity over SRI and DRI (> 30-fold). However, in contrast to **76**, biphenyl **108** also possessed an unacceptable level of CYP2D6 inhibition (IC_{50} 100 nM) and so could potentially inhibit the metabolism of CYP2D6 substrates.

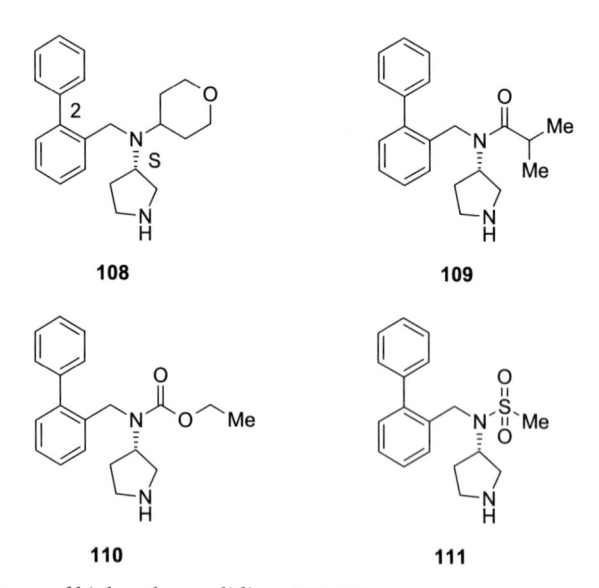

Fig. 36 Structures of biphenyl pyrrolidines **108–111**

Wakenhut et al. discovered that further modification of this template by acylation of the (3S)-amino group gave amide **109** which retained NRI activity (IC$_{50}$ 14 nM) whilst significantly reducing CYP2D6 inhibition to an acceptable level (IC$_{50}$ 1860 nM) [61]. Further exploration of the SAR within this template by Fish et al. identified carbamate **110** and sulfonamide **111** as two additional examples with a superior combination of NRI activity (K_i < 10 nM) combined with selectivity over SRI and DRI (> 100-fold) (Table 19) [94]. Additional screening of **109–111** in high throughput in vitro ADME and safety screens showed all three compounds to have excellent metabolic stability in HLMs and human hepatocytes consistent with low predicted clearance, weak CYP450 enzyme inhibition and modest ion channel activity as measured by binding to representative potassium, sodium and calcium channels (Table 19).

An in vitro screen that clearly differentiated **109**, **110** and **111** was transit performance in the MDCK-mdr1 cell line, which is commonly used as a model to estimate CNS penetration. All three compounds have good passive permeability but amide **109** and sulfonamide **111** demonstrate efflux ratios (ER > 5) which were consistent with recognition and efflux by the P-gp transporter. Compounds with significant efflux by P-gp tend to have poorer CNS penetration than compounds that are not. Performance in this assay proved to be decisive in selecting carbamate **110** for in vivo assessments of CNS penetration and functional activity. Pharmacological evaluation in vivo, in microdialysis experiments, showed **110** increased NA levels in interstitial fluid of the prefrontal cortex of conscious rats by 300–400% above pre-drug

Table 19 In vitro inhibition of monoamine reuptake, ADME profiles and ion channel affinities of **109–111**

	109	110	111
NET, K_i (nM)	6	8	5
SERT, K_i (nM)	960	1110	4120
DAT, K_i (nM)	3740	3030	2670
HLM, Cl$_i$ μl/min/mg	< 7	< 7	7
h.heps, Cl$_i$ μl/min/mg	NT	< 5	< 5
CYP1A2 inhib., IC$_{50}$ nM	> 3000	> 3000	> 3000
CYP2C9 inhib., IC$_{50}$ nM	> 3000	> 3000	> 3000
CYP2D6 inhib., IC$_{50}$ nM	1860	1210	3100
CYP3A4 inhib., IC$_{50}$ nM	> 3000	> 3000	> 3000
Caco-2, A-B/B-A	NT	27/28	NT
MDCK-mdr1, A-B/B-A	11/68	22/44	10/51
MDCK-mdr1, ER	6.1	2.0	5.1
K$^+$, hERG, K_i nM	2840	2830	> 2700
Ca^{2+}, L-type, K_i nM	> 10 000	1500	1900
Na$^+$, site 2, K$_i$ nM	1200	1100	1900

baseline levels (0.3–3 mg/kg administered s.c., $n = 5$–6). The magnitude and duration of the responses observed with **110** were equivalent (by dose) to those observed with ATX which was used as a control. Based on this profile, **110** was selected as a candidate for further evaluation in pre-clinical disease models.

4.4
Miscellaneous Series

4.4.1
1-Aryl-3,4-dihydro-1*H*-quinolin-2-ones

A novel series of dihydro-1*H*-quinolin-2-ones **112** have been discovered as potent and selective NRIs by Camp et al. at Eli Lilly [95]. The series was designed based on results from an in-house HTS combined with a ligand based pharmacophore model derived from the 3-aryloxypropylamine structural motif (Fig. 37).

An initial lead **113** (single enantiomer of unknown absolute stereochemistry) had good NET potency (K_i 8 nM) and excellent selectivity over SERT and DAT but measurement of physicochemical properties showed **113** to be hydrophilic ($\log D$ – 0.48), predicting for poor brain penetration. The strategy was to increase the overall lipophilic nature of the series by the systematic introduction of lipophilic groups at the western and southern N-aromatic rings (X and Y, respectively) and small alkyl chains at the C-3 position (R = Me to *n*-Bu). The most successful modifications were the introduction of a 4-Me on the southern N-aromatic ring **114-rac** (K_i 3 nM) and the incorporation of an *n*-Pr at C-3 **115-rac** (K_i 7 nM). However, the combination of these two modifications did not give an additive boost but a dual NSRI **116-rac** (K_i 11 and 160 nM). Camp's conclusions were: "there were limitations within this SAR and often increasing lipophilicity within this series resulted in a reduction in NRI activity".

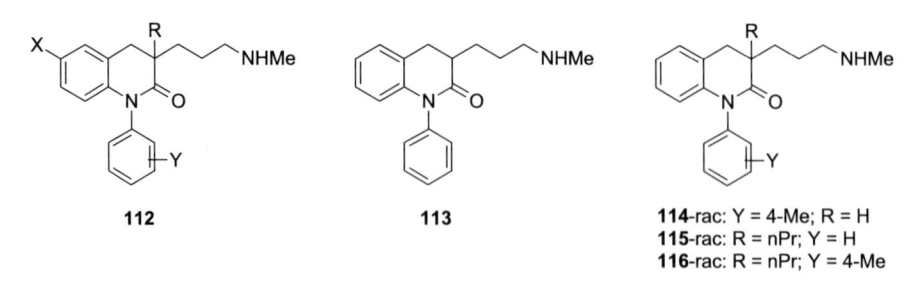

112   **113**   **114-rac:** Y = 4-Me; R = H
                                                       115-rac: R = nPr; Y = H
                                                       116-rac: R = nPr; Y = 4-Me

Fig. 37 Structures of dihydro-1*H*-quinolin-2-ones 112–116

4.4.2
4-(4-Chlorophenyl)piperidines

Kozikowski et al. at the University of Illinois at Chicago have reported further SARs in a series of 4-(4-chlorophenyl)piperidine analogues each bearing a thioacetamide group at C-3 (117) [96]. The paper describes a comprehensive analysis of the influence of relative and absolute stereochemistry at C-3 and C-4 of the piperidine ring on inhibition of the NA, 5-HT and DA transporters (Fig. 38).

Among the analogues reported in this paper, (+)-*cis*-118 exhibited good activity for NET (K_i 5.5 nM) with some degree of selectivity over DAT and SERT (39- and 321-fold, respectively).

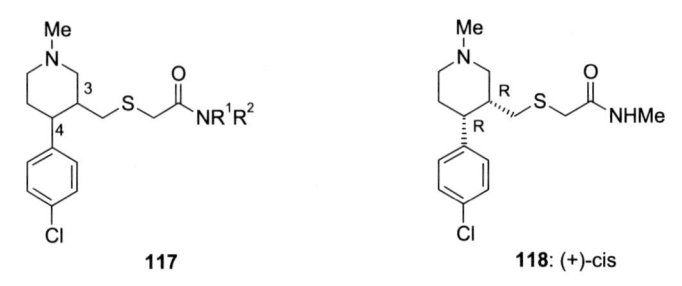

Fig. 38 Structures of 4-(4-chlorophenyl)piperidines **117** and **118**

4.4.3
Pyridinyl Phenyl Ethers

[4-(Phenoxy)pyridine-3-yl]methylamines **121** have been disclosed as a new series of selective NRIs (Fig. 39) [97]. Pyridinyl phenyl ethers **119** have been reported in the literature as SSRIs [98] and further modification of this template afforded dual SNRIs **120** [41]. An emerging understanding of the SAR showed that the aryloxy ring played an important role in modulating SRI and NRI activity, i.e. appropriate substitution at the 2 position conferred NRI activity whereas substitution at the 4 position gave SRI activity. Hence, as an initial venture, Fish et al. at Pfizer elected to delete the 4-Cl substituent from the aryloxy ring of **120** with the aim of reducing the SRI activity to furnish selective NRIs.

Compound SAR was then directed at improving NRI activity by exploring a broad set of 2-substituents on the aryloxy ring in both the *sec*- and *tert*-amine series (**121**: R^3 = Et, OMe, OEt, Me, SMe, Cl, *i*-Pr, *n*-Pr, OPh). NRI activity was clearly dependent on the lipophilicity of the 2-substituent with potent NRI activity tracking with increased lipophilicity. In addition, clear trends in the SAR of the *sec*- and *tert*-amines also emerged: *tert*-amines invariably showed a significant advantage in NRI ac-

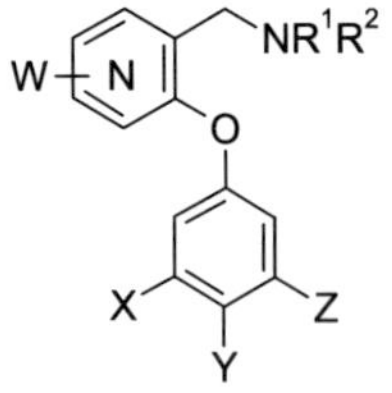

Fig. 39 Structures of pyridinyl phenyl ethers 119–122

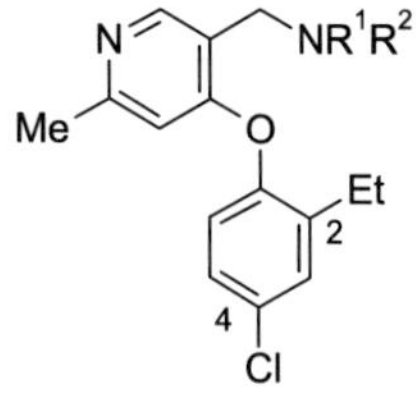

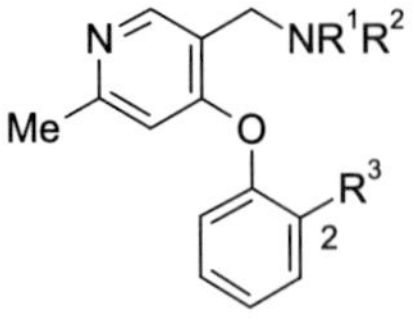

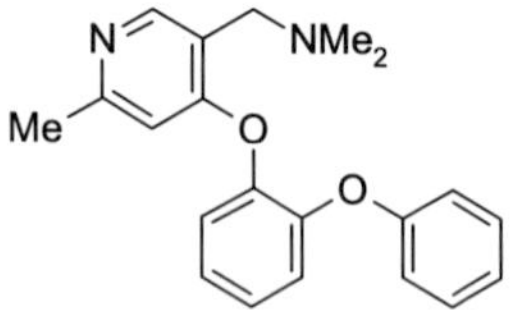

tivity compared to the corresponding *sec*-amines, and *tert*-amines were generally weaker SRIs compared to *sec*-amines. No compound demonstrated any significant DRI activity. The combination of these SARs resulted in *tert*-amines yielding the most potent NRIs with better selectivity for SRI and DRI.

From these experiments 122 emerged as having a superior combination of NET activity (K_i 10 nM) combined with selectivity over SERT and DAT (> 80-fold). Additional screening in vitro showed 122 to have good membrane permeability (Caco-2 11/19) with low affinity for P-gp efflux transporters (MDCK-mdr1 25/41) suggesting the potential for good oral absorption and CNS penetration. Compound 122 had good metabolic stability in HLMs ($Cl_{int} < 7 \mu l/min/mg$) and human hepatocytes (Cl_{int} $8 \mu l/min/million$ cells) consistent with low predicted clearance. Compound 122 had no significant inhibition of CYP450 enzymes (1A2, 2D6, 3A4; IC_{50} values $> 3 \mu M$), modest ion channel activity as measured by binding to potassium hERG ($[^3H]$-dofetilide, K_i 5.8 μM), sodium (site 2, K_i 3.4 μM) and calcium (L-type diltiazem site, K_i 1.2 μM) channels and no significant off-target pharmacology. Further pharmacological evaluation in vivo, in microdialysis experiments, showed 122 increased NA levels in interstitial fluid of the prefrontal cortex of conscious rats by 400% above pre-drug baseline levels (0.3 mg/kg administered s.c., $n = 4$). Based on this profile, 122 was selected as a candidate for further evaluation in pre-clinical disease models.

5
Conclusions

Selective monoamine reuptake inhibitors (SSRIs and NRIs) have had a significant impact on the treatment of several serious diseases, including depression, ADHD, neuropathic pain and fibromyalgia. The discovery and launch of dual pharmacology SNRIs such as duloxetine has further expanded the treatment options for these diseases. This review has highlighted the discovery of many new compounds which combine selective inhibition of SERT and/or NET activity with a large degree of both structural and physicochemical diversity. The coming years will determine whether any of these recently discovered compounds will advance to clinical development and ultimately to regulatory approval as new medicines.

References

1. Hertting G, Axelrod J (1961) Nature 192:172
2. Iversen LL (1971) Br J Pharmacol 41:571
3. Ritz MC, Cone EJ, Cone MJ (1990) Life Sci 46:635
4. Tatsumi M, Groshan K, Blakely RD, Richelson E (1997) Eur J Pharmacol 340:249
5. Moltzen EK, Bang-Andersen B (2006) Curr Top Med Chem 6:1801
6. Torres GE, Gainetdinov RR, Caron MG (2003) Nat Rev Neurosci 4:13
7. Steffens DC, Krishnan KR, Helms MJ (1997) Depress Anxiety 6:10
8. Butler SG, Meegan MJ (2008) Curr Med Chem 15:1737
9. Liu S, Molino BF (2007) Annu Rep Med Chem 42:13
10. Patent activity based on keyword search using CAS Scifinder search engine. Search carried out June 2008
11. Singh S (2000) Chem Rev 100:925
12. Ravna AW, Sylte I, Kristiansen K, Dahl SG (2006) Bioorg Med Chem 14:666
13. Hanna MM, Eid NM, George RF, Safwat HM (2007) Bioorg Med Chem 15:7765
14. Emond P, Helfenbein J, Chalon S, Garreau L, Vercouille J, Frangin Y, Besnard JC, Guilloteau D (2001) Bioorg Med Chem 9:1849
15. Tamagnan G, Alagille D, Fu X, Kula NS, Baldessarini RJ, Innis RB, Baldwin RM (2005) Bioorg Med Chem Lett 15:1131
16. Hoepping A, Johnson KM, George C, Flippen-Anderson J, Kozikowski AP (2000) J Med Chem 43:2064
17. Zhang A, Zhou G, Hoepping A, Mukhopadhyaya J, Johnson KM, Zhang M, Kozikowski AP (2002) J Med Chem 45:1930
18. Schmitz WD, Denhart DJ, Brenner AB, Ditta JL, Mattson RJ, Mattson GK, Molski TF, Macor JE (2005) Bioorg Med Chem Lett 15:1619
19. Mattson RJ, Catt JD, Denhart DJ, Deskus JA, Ditta JL, Higgins MA, Marcin LR, Sloan CP, Beno BR, Gao Q, Cunningham MA, Mattson GK, Molski TF, Taber MT, Lodge NJ (2005) J Med Chem 48:6023
20. Deskus JA, Epperson JR, Sloan CP, Cipollina JA, Dextraze P, Qian-Cutrone J, Gao O, Ma B, Beno BR, Mattson GK, Molski TF, Krause RG, Taber MT, Lodge NJ, Mattson RJ (2007) Bioorg Med Chem Lett 17:3099

21. King HD, Denhart DJ, Deskus JA, Ditta JL, Epperson JR, Higgins MA, Kung JE, Marcin LR, Sloan CP, Mattson GK, Molski TF, Krause RG, Bertekap LR Jr, Lodge NJ, Mattson RJ, Macor JE (2007) Bioorg Med Chem Lett 17:5647
22. Meagher KL, Mewshaw RE, Evrard DA, Zhou P, Smith DL, Scerni R, Spangler T, Abulhawa S, Shi X, Schechter LE, Andree TH (2001) Bioorg Med Chem Lett 11:1885
23. McMahon CG, McMahon CN, Leow LJ (2006) Neuropsychiatr Dis Treat 2:489
24. Pryor JL, Althof SE, Steidle C, Rosen RC, Hellstrom WJG, Shabsigh R, Miloslavsky M, Kell S (2006) Lancet 368:929
25. Wyllie MG (2006) BJU Int 98:227
26. Waldinger MD, Schweitzer DH (2008) J Sex Med 5:966
27. Middleton DS, Andrews M, Glossop P, Gymer G, Jessiman A, Johnson PS, MacKenny M, Pitcher MJ, Rooker T, Stobie A, Tang K, Morgan P (2006) Bioorg Med Chem Lett 16:1434
28. Middleton DS, Andrews M, Glossop P, Gymer G, Hepworth D, Jessiman A, Johnson PS, MacKenny M, Pitcher MJ, Rooker T, Stobie A, Tang K, Morgan P (2008) Bioorg Med Chem Lett 18:4018
29. Goodman MM, Chen P, Plisson C, Martarello L, Galt J, Votaw JR, Kilts CD, Malveaux G, Camp VM, Shi B, Ely TD, Howell L, McConathy J, Nemeroff CB (2003) J Med Chem 46:925
30. Wilson AA, Houle S (1990) J Labelled Comp Radiopharm 42:1277
31. Stehouwer JS, Jarkas N, Zeng F, Voll RJ, Williams L, Owens MJ, Votaw JR, Goodman MM (2006) J Med Chem 49:6760
32. Plisson C, Stehouwer JS, Voll RJ, Howell L, Votaw JR, Owens MJ, Goodman MM (2007) J Med Chem 50:4553
33. Plisson C, McConathy J, Martarello L, Malveaux EJ, Camp VM, Williams L, Votaw JR, Goodman MM (2004) J Med Chem 47:1122
34. Parhi AK, Wang JL, Oya S, Choi S, Kung M, Kung HF (2007) J Med Chem 50:6673
35. Oya S, Choi SR, Coenen H, Kung HF (2002) J Med Chem 45:476
36. Wilson AA, Ginovart N, Schmidt M, Meyer JH, Threlkend PG, Houle S (2000) J Med Chem 43:3103
37. Jarkas N, McConathy J, Voll RJ, Goodman MM (2005) J Med Chem 48:4254
38. Emond P, Vercouillie J, Innis R, Chalon S, Mavel S, Frangin Y, Halldin C, Besnard JC, Guilloteau D (2002) J Med Chem 45:1253
39. Huang Y, Bae SA, Zhu Z, Guo N, Roth BL, Laruelle M (2005) J Med Chem 48:2559
40. Whitlock GA, Blagg J, Fish PV (2008) Bioorg Med Chem Lett 18:596
41. Whitlock GA, Fish PV, Fray MJ, Stobie A, Wakenhut F (2008) Bioorg Med Chem Lett 18:2896
42. Bisserbe JC (2002) Int Clin Psychopharmacol 17:S43
43. Leo RJ, Brooks VL (2006) Curr Opin Investig Drugs 7:637
44. Puozzo C, Panconi E, Deprez D (2002) Int Clin Psychopharmacol 17:S25
45. Roggen H, Kehler J, Stensbol TB, Hansen T (2007) Bioorg Med Chem Lett 17:2834
46. Chen C, Dyck B, Fleck BA, Foster AC, Grey J, Jovic F, Mesleh M, Phan K, Tamiya J, Vickers T, Zhang M (2008) Bioorg Med Chem Lett 18:1346
47. Vickers T, Dyck B, Tamiya J, Zhang M, Jovic F, Grey J, Fleck BA, Aparico A, Johns M, Jin L, Tang H, Foster AC, Chen C (2008) Bioorg Med Chem Lett 18:3230
48. Tamiya J, Dyck B, Zhang M, Phan K, Fleck BA, Aparico A, Jovic F, Tran JA, Vickers T, Grey J, Foster AC, Chen C (2008) Bioorg Med Chem Lett 18:3328
49. Goldstein DJ (2007) Neuropsychiatr Dis Treat 3:193
50. Smith T, Nicholson RA (2007) Vasc Health Risk Manag 3:833

51. Russell IJ, Mease PJ, Smith TR, Kajdasz DK, Wohlreich MM, Detke MJ, Walker DJ, Chappell AS, Arnold LM (2008) Pain 136:432
52. Agur W, Abrams P (2007) Exp Rev Obstet Gynecol 2:133
53. Bymaster FP, Beedle EE, Findlay J, Gallagher PT, Krushinski JH, Mitchell S, Robertson DW, Thompson DC, Wallace L, Wong DT (2003) Bioorg Med Chem Lett 13:4477
54. Bymaster FP, Lee TC, Knadler MP, Detke MJ, Iyengar S (2005) Curr Pharm Des 11:1475
55. Karpa KD, Cavanaugh JE, Lakoski JM (2002) CNS Drug Rev 8:361
56. Boot JR, Brace G, Delatour CL, Dezutter N, Fairhurst J, Findlay J, Gallagher PT, Hoes I, Mahadevan S, Mitchell SN, Rathmell RE, Richards SJ, Simmonds RG, Wallace L, Whatton MA (2004) Bioorg Med Chem Lett 14:5395
57. Mahaney PE, Vu AT, McComas CC, Zhang P, Nogle LM, Watts WL, Sarkahian A, Leventhal L, Sullivan NR, Uveges AJ, Trybulski EJ (2006) Bioorg Med Chem Lett 16:8455
58. Fray MJ, Bish G, Brown AD, Fish PV, Stobie A, Wakenhut F, Whitlock GA (2006) Bioorg Med Chem Lett 16:4345
59. Fray MJ, Bish G, Fish PV, Stobie A, Wakenhut F, Whitlock GA (2006) Bioorg Med Chem Lett 16:4349
60. Whitlock GA, Fish PV, Fray MJ, Stobie A, Wakenhut F (2007) Bioorg Med Chem Lett 17:2022
61. Wakenhut F, Fish PV, Fray MJ, Gurrell I, Mills JEJ, Stobie A, Whitlock GA (2008) Bioorg Med Chem Lett 18:4308
62. Gallagher PT, Boot JR, Boulet SL, Clark BP, Cases-Thomas MJ, Delhaye L, Diker K, Fairhurst J, Findlay J, Gilmore J, Harris JR, Masters JJ, Mitchell SN, Naik M, Simmonds RG, Smith SM, Richards SJ, Timms GH, Whatton MA, Wolfe CN, Wood VA (2006) Bioorg Med Chem Lett 16:2714
63. Kozikowski AP, Johnson KM, Rong SB, Zhang A, Zhang M, Zhou G (2002) Bioorg Med Chem Lett 12:993
64. Krell HV, Leuchter AF, Cook IA, Abrams M (2005) Psychsomatics 45:379
65. Zhou J (2004) Drugs Future 29:1235
66. Walter MW (2005) Drug Dev Res 65:97
67. Kamalesh Babu RP, Maiti SN (2006) Heterocycles 69:539
68. Huang Y, Williams WA (2007) Expert Opin Ther Pat 17:889
69. Liu S, Molino BF (2007) Ann Rep Med Chem 42:413
70. Wong EHF, Sonders MS, Amara SG, Tinbolt PM, Piercey MFP, Hoffmann WP, Hyslop DK, Franklin S, Porsolt RD, Bonsignori A, Carfagna N, McArthur RA (2000) Biol Psychiatry 47:818
71. De Maio D, Johnson FN (2000) Rev Contemp Pharmacother 11:303
72. Szabadi E, Bradshaw CM (2000) Rev Contemp Pharmacother 11:267
73. Fleishaker JC (2000) Clin Pharmacokinet 39:413
74. Fleishaker JC (2000) Rev Contemp Pharmacother 11:283
75. Frigerio E, Benecchi A, Brianceschi G, Pellizzoni C, Poggesi I, Strolin Benedetti M, Dostert P (1997) Chirality 9:303
76. Baldwin DS, Buis C, Carabal E (2000) Rev Contemp Pharmacother 11:321
77. http://www.pfizer.com. Information accurate as of website access in October 2008
78. Hajos M, Fleishaker JC, Filipiak-Reisner JK, Brown MT, Wong EHF (2004) CNS Drug Rev 10:23
79. Boot J, Cases M, Clark BP, Findlay J, Gallagher PT, Hayhurst L, Man T, Montalbetti C, Rathmell RE, Rudyk H, Walter MW, Whatton M, Wood V (2005) Bioorg Med Chem Lett 15:699

80. Cases MJ, Masters JJ, Walter MW, Campbell G, Haughton L, Gallagher PT, Dobson DR, Mancuso V, Bonnier B, Giard T, Defrance T, Vanmarsenille M, Ledgard A, White C, Ouwerkerk-Mahadevan S, Brunelle FJ, Dezutter NA, Herbots CA, Lienard JY, Findlay J, Hayhurst L, Boot J, Thompson LK, Hemrick-Luecke S (2006) Bioorg Med Chem Lett 16:2022

81. Cases M, Campbell G, Haughton L, Masters JJ, Walter MW, Gallagher PT, Dobson DR, Finn T, Bonnier B, White C, Findlay JD, Hayhurst L, Kluge AH, Mahadevan S, Brunelle FJ, Delatour CL, Lavis AA, Dezutter NA, Vervaeke VN, Lienard JY, Boot R (2005) Abstracts of papers, 229th ACS National Meeting, San Diego, 13–17 March 2005, MEDI-058-059

82. Fish PV, Deur C, Gan X, Greene K, Hoople D, MacKenny M, Para KS, Reeves K, Ryckmans T, Stiff C, Stobie A, Wakenhut F, Whitlock GA (2008) Bioorg Med Chem Lett 18:2562

83. Corral C, Lissavetzky J, Manzanares I, Darias V, Exposito-Orta MA, Martin Conde JA, Sanches-Mateo C (1999) Bioorg Med Chem 7:1349

84. Pinder RM, Brogden RN, Speight TM, Avery GS (1977) Drugs 13:401

85. Case DE, Reeves PR (1975) Xenobiotica 5:113

86. Ledbetter M (2006) Neuropsychiatr Dis Treat 2:455

87. Preti A (2002) Curr Opin Investig Drugs 3:272

88. Thomason C, Michelson D (2004) Drugs Today 40:465

89. Turgay A (2006) Therapy 3:19

90. Bymaster FP, Katner JS, Nelson DL, Hemrick-Luecke SK, Threlkeld PG, Heiligenstein JH, Morin SM, Gehlert DR, Perry KW (2002) Neuropsychopharmacology 27:699

91. Sauer JM, Ring BJ, Witcher JW (2005) Clin Pharmacokinet 44:571

92. Wu D, Pontillo J, Ching B, Hudson S, Gao Y, Fleck BA, Gogas K, Wade WS (2008) Bioorg Med Chem Lett 18:4224

93. Roth BL (2007) N Engl J Med 356:6

94. Fish PV, Barta NS, Gray DLF, Ryckmans T, Stobie A, Wakenhut F, Whitlock GA (2008) Bioorg Med Chem Lett 18:4355

95. Beadle CD, Boot J, Camp NP, Dezutter N, Findlay J, Hayhurst L, Masters JJ, Penariol R, Walter MW (2005) Bioorg Med Chem Lett 15:4432

96. He R, Kurome T, Giberson KM, Johnson KM, Kozikowski AP (2005) J Med Chem 48:7970

97. Fish PV, Ryckmans T, Stobie A, Wakenhut F (2008) Bioorg Med Chem Lett 18:1795

98. Adam MD, Andrews MD, Gymer GE, Hepworth D, Howard HR, Middleton DS, Stobie A (2002) WO Patent 083643

Top Med Chem (2009) 4: 95–129
DOI 10.1007/7355_2008_027
© Springer-Verlag Berlin Heidelberg
Published online: 18 October 2008

Atypical Dopamine Uptake Inhibitors that Provide Clues About Cocaine's Mechanism at the Dopamine Transporter

Amy Hauck Newman[1,2] (✉) · Jonathan L. Katz[1,3]

[1]Medications Discovery Research Branch, National Institute on Drug Abuse
Intramural Research Program, National Institutes of Health, Baltimore, MD 21224, USA
anewman@intra.nida.nih.gov

[2]Medicinal Chemistry Section, NIDA-IRP, NIH,
333 Cassell Dr., Baltimore, MD 21224, USA

[3]Psychobiology Section, NIDA-IRP, NIH,
5500 Nathan Shock Dr., Baltimore, MD 21224, USA

Abstract The dopamine transporter (DAT) has been a primary target for cocaine abuse/addiction medication discovery. However predicted addiction liability and limited clinical evaluation has provided a formidable challenge for development of these agents for human use. The unique and atypical pharmacological profile of the benztropine (BZT) class of dopamine uptake inhibitors, in preclinical models of cocaine effects and abuse, has encouraged further development of these agents. Moreover, in vivo studies have challenged the original DAT hypothesis and demonstrated that DAT occupancy and subsequent increases in dopamine produced by BZT analogues are significantly delayed and long lasting, as compared to cocaine. These important and distinctive elements are critical to the lack of abuse liability among BZT analogues, and improve their potential for development as treatments for cocaine abuse and possibly other neuropsychiatric disorders.

Keywords Addiction medication discovery · Benztropine · Cocaine · Dopamine transporter · Dopamine uptake inhibitor

Abbreviations

5-HT	Serotonin
ACE-Cl	α-Chloroethylchloroformate
AlH$_3$	Alane
BZT	Benztropine
DA	Dopamine
DAT	Dopamine transporter
FR	Fixed ratio
LiAlH$_4$	Lithium aluminum hydride
MTSET	Methanethiosulfonate ethyltrimethyl-ammonium
NET	Norepinephrine transporter
SAR	Structure–activity relationships
SERT	Serotonin transporter
TM	Transmembrane

1
Introduction –
Chemical and Pharmacological Diversity of Dopamine Uptake Inhibitors

Over the past 2 decades, a significant number of studies have supported the dopamine transporter (DAT) hypothesis of cocaine's behavioral effects first described by Ritz et al. [1]. In that study, a significant and positive correlation of binding affinities at the DAT and the potency for self-administration of a variety of monoamine uptake inhibitors was reported. That correlation was greater than the correlations for these compounds among self-administration potencies and affinities for either the norepinephrine (NET) or serotonin transporters (SERT). Thus, the DAT was considered the primary biological target relevant to the effects of cocaine contributing to its abuse liability, and became the mechanistic target for medication development.

One decade ago, a special issue of *Medicinal Chemistry Research*, was dedicated to the design and synthesis of novel dopamine uptake inhibitors [2]. Structure–activity relationships (SAR) and combinations with the molecular biology techniques of the time were described in detail for each chemical class and many of these compounds have been further developed preclinically. Since that time, numerous primary papers and several excellent and comprehensive reviews (e.g., [3–5]) have been published. Therefore a review of all structural classes and SAR within each will not be included in this chapter. However, it should be noted that the chemical classes based on the 3-aryltropanes (e.g., cocaine analogues), the 1,4-dialkylpiperazines (e.g., GBR 12909 and its analogues), and analogues of BZT have all been synthetically "mined", evaluated in vitro for binding and function at all three monoamine transporters as well as other potentially relevant receptors (e.g., muscarinic, histaminic, sigma). In the process, modifications to these structures have improved DAT affinity and selectivity. Further, several candidates in each class have been extensively evaluated in a variety of animal models of psychostimulant actions related to drug abuse, as well as other models of neuropsychiatric disorders. Several of these agents show potential for further clinical investigation [5, 6].

With ongoing progress came several lines of research that revealed some limitations to the hypothesis of the DAT as the mechanism for cocaine's effects and its abuse liability. For example, studies on DAT knockout mice showed place preference and self-administration of cocaine [7, 8]. While the mechanisms underlying these effects of cocaine in DAT knockout mice are not fully understood at this time, it is clear that the reinforcing effects of cocaine can be obtained in animals lacking what is thought to be the primary biological substrate for cocaine's actions. Another limitation, which will be elaborated upon in this review, was initially suggested in papers by Rothman et al. [9] and Vaugeois et al. [10]. In the latter study, the ED_{50} doses of various DA uptake inhibitors for in vivo displacement of a radiolabel for the DAT ([^{3}H]GBR 12783), produced quite different amounts of behavioral stimulation as evidenced by elevations in locomotor activity, ranging from only a slight effect by mazindol ($\sim$ 106% of control) and pyrovalerone (123%) to an almost three-fold increase produced by the phenylpiperazine, GBR 13069. Thus, there are some DA uptake inhibitors that have behavioral effects that differ from those of cocaine in animal models. In addition, there are other DA uptake inhibitors that are used clinically and are not subject to substantial abuse; one of these is benztropine.

2
Case Study: Benztropine Analogues

As can be seen in Fig. 1, benztropine (3α-diphenylmethoxytropane, abbreviated herein as BZT) shares structural features with cocaine (tropane ring) and the diphenyl ether of the phenylpiperazines. A prototype of this latter

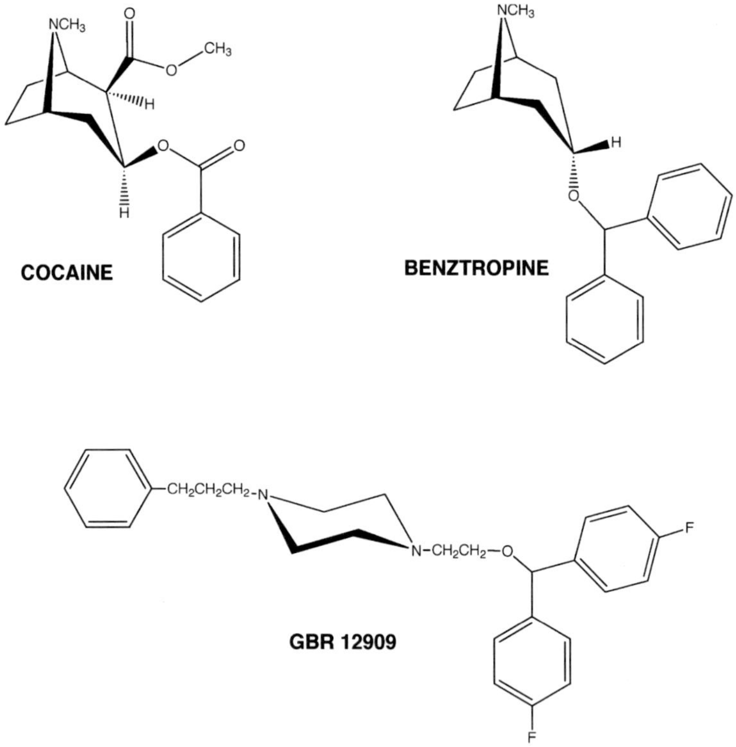

Fig. 1 Chemical structures of cocaine, benztropine (BZT) and GBR 129009

class is GBR 12909, which is a selective DA uptake inhibitor that was selected for clinical evaluation as a cocaine-abuse medication. Therefore, from an initial structural perspective BZT and its analogues were of interest. Further piquing our interest was that BZT is in clinical use for treatment of symptoms associated with Parkinson's Disease and is not subject to any significant abuse. Moreover, though not a focus of the study, Colpaert et al. [11] showed that BZT did not fully substitute in rats trained to discriminate cocaine from saline injections. These additional considerations suggested that BZT analogues may also be of some interest for a potentially different mode of action with the DAT, compared to traditional dopamine uptake inhibitors. However, it is also possible that BZT is a typical DA uptake inhibitor that has other actions that interfere with its cocaine-like effects. In order to pursue these questions further, we embarked on a medicinal chemistry approach to the design, synthesis and evaluation of BZT analogues.

It is important to emphasize that while BZT analogues have pharmacological effects that differ from those of cocaine, this group of drugs may not be the only chemical class of DAT inhibitors for which this disparity exists. For example, some analogues of the sigma receptor ligand, rimcazole, bind to the DAT but do not produce cocaine-like effects [12–14]. Further, there has been a preliminary report [15] of a cocaine analogue with high affinity for the DAT, low efficacy in stimulating locomotor activity, and that antagonizes the stimulation of locomotor activity produced by cocaine. Taken together, these studies are consistent with the idea that binding to the DAT and its resultant inhibition of DA uptake does not uniformly produce cocaine-like effects.

3
Design, Synthesis and SAR of the Benztropine Analogues

As mentioned above, BZT shares structural features of cocaine and the phenylpiperazines. We viewed the tropane ring as a structurally rigid isostere of the piperazine ring system of the phenylpiperazines, and were particularly intrigued with the lack of a 2-position substitution for BZT, which in the cocaine-class of DAT inhibitors renders these molecules inactive (see [16] for a review). In Fig. 2, the chemical modifications that have been explored in our laboratory on the parent molecule BZT are illustrated. Initially, we noted that the diphenyl ether at the 3-position of the tropane ring was in the α-configuration and that this opposed most of the cocaine analogues that had been reported to have high affinity for the DAT [16]. Further, replacement of the β-benzoyl group of cocaine with various substituted phenyl rings, significantly affected binding affinities at the DAT and generally improved potency for inhibition of DA uptake. We therefore initially explored both stereochemistry and optimal substitution on these phenyl rings.

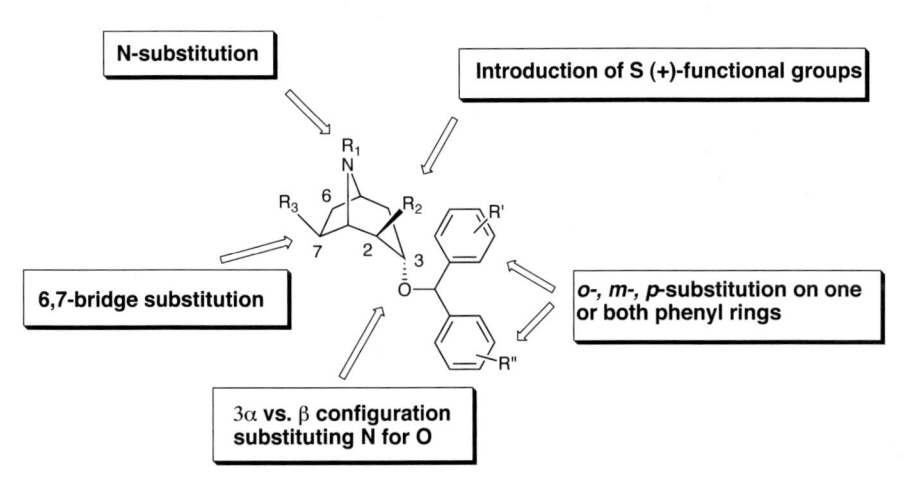

Fig. 2 Design strategy for BZT analogues

3.1
3-Substituted Benztropine Analogues

More than 40 3α-diphenyl ether analogues of BZT were prepared and evaluated for binding at transporters for DA, serotonin (5-HT), and norepinephrine (NE), as well as at muscarinic M_1 and histamine H_1 receptors
in membrane preparations. Effects of selected analogues are displayed in
Table 1, as well as data for inhibition of DA uptake in synaptosomes or
a chopped tissue preparation [17–19]. We discovered that the 3α-stereo-
chemistry provided optimally active compounds as did small substituents
such as F or Cl in the *para*- and or *meta*-positions, with 4′,4″-diF giving
the highest affinity analogue in this series (AHN 1-055). However, it must
be noted that small halogens substituted in these positions uniformly gave
high affinity analogues (K_i values from 11 to 30 nM). Whereas, increasing
steric bulk or substitution in the 2′-position decreased DAT affinity [19]. Note
that compared to BZT and cocaine, most of these halogenated analogues had
higher affinity at the DAT (Table 1). Only one β-stereoisomer (AHN 1-063 in
Table 1) was prepared and it had lower affinity for DAT than its 3α-epimer
demonstrating opposite stereoselectivity of the effects of these BZT analogues
as compared to cocaine.

In addition muscarinic receptor affinities were generally lower than that
for BZT. Thus, these analogues showed relatively similar affinities at DAT
and M_1 receptors, rather than the 56- to 200-fold selectivity for muscarinic
receptors exhibited by the parent compound, BZT. Although none of these
analogues demonstrated high affinity for SERT or NET, the high to moderate muscarinic receptor binding could contribute to their CNS effects. Like
the parent compound, it should also be noted that several of these analogues
demonstrated high to moderate affinity for histamine H_1 receptors [20, 21].
Because these additional actions may significantly impact the behavioral
profile of these agents their contributions were investigated (see below). In
addition, it was desirable to find structural features that would discriminate
between the binding at these sites, and we thus initiated studies to alter muscarinic M_1 receptor affinity.

3.2
N-Substituted Benztropine Analogues

As noted above, the BZT analogues share structural features with GBR 12909,
but unlike this compound, the BZT analogues synthesized to date were all *N*-
methyl tropanes. GBR 12909 has a propylphenyl substituent appended to its
piperazine *N*-terminus, and does not bind appreciably to muscarinic M_1 receptors. As BZT can be considered a rigid analogue of GBR 12909, a series of
N-substituted analogues of AHN 1-055 and several other small halogenated
diphenyl ether analogues were designed. *N*-substitution turned out to be

Table 1 Binding data for *N*-substituted-3-substituted benztropine analogues at DAT, SERT, NET, M_1 and H_1 receptors and inhibition of dopamine uptake[1]

Compound	R	R', R''	DAT	SERT[2]	NET[2]	M_1	H_1	DAUI[2,3]
GA 103	4''-Phenyl-n-butyl	4',4''-diF	8.51±1.2[c]	376±51.8[m]	2210±240[m] 1 Hr	576±10.7[g]	141±6.72[n]	
AHN 2-003	H	4',4''-diF	11.2±1.2[c]	922±87.2[o]	902±60.9 1 Hr; 457±69.8[o] 3 Hr	203±16.5[g]	65.4±7.28[n]	15.1±1.27
AHN 1-055	CH_3	4',4''-diF	11.8±1.3[a]	3260±110[k]	610±80.5[k] 1 Hr; 844±57 3 Hr	11.6±0.930[g]	19.7±1.32[n]	71±12[a]; 13.8±1.71[m]
GA 2-99	2''-Aminoethyl	4',4''-diF	13.9±1.7[g]	4600±680[m]	1420±130[m] 1 Hr	1250±138[g]	240±32.6[n]	26.0±2.88[m]
4',4''-DiCl-BZT	CH_3	4',4''-diCl	20.0±2.8[a]	1640±236[f]	2980±182[f] 1 Hr	47.9±5.18[f]	122±4.55[n]	75±24[a]; 23.4±3.00[f]

Table 1 (continued)

Compound	R	R′, R″	DAT	SERT[2]	NET[2]	M_1	H_1	DAUI[2,3]
JHW 007	*n*-Butyl	4′,4″-diF	24.6 ± 2.0[c]	1350 ± 151[o]	1490 ± 190 3 Hr 1670 ± 232[o] 1 Hr	251 ± 12.5[c] 399 ± 28.3[o]		24.6 ± 1.97[l]
SSK 2-046	$CH_2CH_2CONHPh$	4′,4″-diF	27.6 ± 1.92[m]	1490 ± 81.7[m]	1420 ± 165[m] 1 Hr	3280 ± 420[m]	140 ± 17.3[n]	4.99 ± 0.011[m]
GA 2-95	*S*-2″-Amino-3″-Methyl-*n*-butyl	4′,4″-diF	29.5 ± 3.5[g]			2180 ± 85.8		
AHN 2-005	Allyl	4′,4″-diF	29.9 ± 3.0[c]	2850 ± 62.5[o]	1570 ± 162 1 Hr 1740 ± 242[o] 3 Hr	126 ± 7.6[c] 177 ± 21[g]	24.9 ± 1.16[n]	19.7 ± 0.568
4′-Cl-BZT	CH_3	4′-Cl	30.0 ± 3.6[a]	5120 ± 395[i]	1470 ± 180[i] 1 Hr	1.48 ± 0.02[b] 7.90 ± 0.85[j]	39.9 ± 1.57[n]	115 ± 27.6[a] 23.1 ± 1.80[i]

Table 1 (continued)

Compound	R	R', R''	DAT	SERT[2]	NET[2]	M_1	H_1	DAUI[2,3]
JHW 013	Cyclopropylmethyl	4',4''-diF	32.4 ± 2.9[c]	1420 ± 116	1640 ± 153 1 Hr	136 ± 11[c] 257 ± 28.9[g]		
GA 1-89	CH_3	3',4''-diCl	32.5 ± 4.88[i]	3870 ± 303[i]	1660 ± 239[i]	21.5 ± 2.63[i]		12.3 ± 1.23[i]
GA 1-69	Indol-3''-ethyl	4',4''-diF	44.6 ± 4.9[c]	490 ± 56.4	7350 ± 934	3280 ± 220[g]	333 ± 22.6[n]	
GA 2-50	R-2''-Amino-3''-methyl-n-butyl	4',4''-diF	56.4 ± 9.6[g]	3870 ± 135	2130 ± 160 1 Hr	4020 ± 592	218 ± 15.5[n]	20.7 ± 2.85
BZT	CH_3	H, H	118 ± 10.6[a]	$> 10\,000$[a]	1390 ± 134[o] 1 Hr	0.59 ± 0.01[b] 2.1 ± 0.29[i]	15.7 ± 2.13[n]	403 ± 115[a,i]
AHN 1-063	CH_3	3β-4'-Cl	854 ± 59.8[a]	$> 10\,000$[a]	2120 ± 276 1 Hr	18.9 ± 1.7[h]	2200 ± 373[n]	

Table 1 (continued)

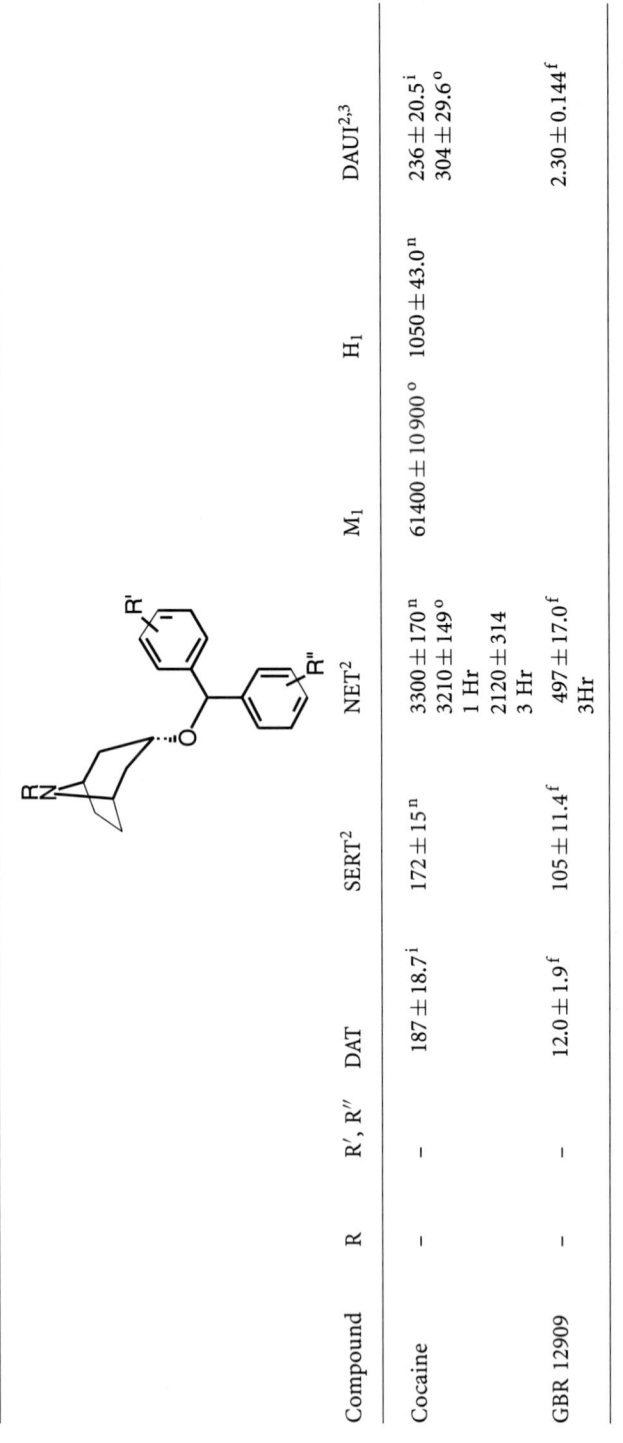

Compound	R	R', R''	DAT	SERT[2]	NET[2]	M$_1$	H$_1$	DAUI[2,3]
Cocaine	–	–	187 ± 18.7[i]	172 ± 15[n]	3300 ± 170[n] 3210 ± 149[o] 1 Hr 2120 ± 314 3 Hr	61400 ± 10900[o]	1050 ± 43.0[n]	236 ± 20.5[i] 304 ± 29.6[o]
GBR 12909	–	–	12.0 ± 1.9[f]	105 ± 11.4[f]	497 ± 17.0[f] 3Hr			2.30 ± 0.144[f]

[1] All binding data are recorded in K_i ± SEM; these data and the methods used to obtain them are published in the cited references: [a] [18]; [b] [19]; [c] [26]; [f] [12]; [g] [23]; [h] [42]; [i] [3]; [j] [65]; [k] [33]; [l] [69]; [m] [25]; [n] [21]; [o] [61]. If there is no reference cited, the data are previously unpublished but the methods used are identical to those used for the referenced values in this column.
[2] Where there are two values, different radiolabeled ligands or analysis programs or incubation times were used, the details of which can be found in the primary references.
[3] The inhibition of dopamine uptake (DAUI) data is recorded as IC_{50} ± SEM and was obtained in rat brain chopped tissue or brain synaptosomes according to methods detailed in the primary references.

a fruitful area of investigation and more than 60 analogues have been prepared and evaluated in vitro [22–25] (selected N-substituted analogues are displayed in Table 1).

In general, N-substituted analogues with the 4′,4″-diF substitution on the 3α-diphenyl ether resulted in high affinity binding at the DAT, with several extended alkyl and alkylaryl substituents being well tolerated. However, there was an optimal length of the N-substituents, which if exceeded, resulted in compounds with low affinity at the DAT [25]. Furthermore, the tropane N must be a secondary or tertiary amine, as the amides were inactive at the DAT [22]. When the diphenyl ether substituents increased in steric bulk (e.g., 4′,4″-diCl) N-substitution resulted in further decreases in DAT affinity [24]. However, a notable separation of DAT from muscarinic M_1 receptor binding was achieved with these N-substitutions, with several analogues in this series having greater than 100-fold selectivity for the DAT. This is in remarkable contrast to the parent compound, which as mentioned above has significant selectivity for muscarinic M_1 receptors over the DAT. Stereoselectivity was originally reported for DAT binding with GA 2-50 and GA 2-95, however, subsequent testing (unpublished results) of these analogues showed no stereoselectivity at DAT and only a 2-fold difference between enantiomers in affinity at muscarinic M_1 receptors. Several of the analogues in this series have been evaluated in vivo and will be described in detail below. Further, because many N-substituents were well tolerated at the DAT, this position has been used to design and synthesize molecular tools such as the photoaffinity ligand GA 2-34 for structure function studies of the DAT [26, 27].

3.3
6/7-Substituted Benztropine Analogues

Once identification of the optimal substitutions on the 3α-diphenyl ether and the tropane N8 were identified, additional modifications at the 6/7 bridge were investigated. In addition to the 6β-analogues, the 6-β-OH BZT was oxidized and then stereoselectively reduced to the 6α-alcohol to compare the effect of stereochemistry at this position [28]. Upon testing, it was discovered that most of these compounds had minimal activity at the DAT, with the exception of those analogues with very small substituents in the 6/7-bridge. Further, none of the analogues showed high affinity for the SERT or the NET [28]. Similar results were reported by Simoni and colleagues [29] and thus, studies on compounds in this series were not pursued further.

3.4
Benztropinamines

While preparing the BZT analogues, it was noted that under acidic conditions, the diphenyl ethers would sometimes cleave from the tropane ring. As such, we decided to explore a series of compounds wherein the ether linkage was replaced by a secondary or tertiary amine, which we refer to as benztropinamines. This substitution provided some additional flexibility for SAR and also had potential to improve water solubility of these compounds, due to the possibility of forming a more water soluble salt. A number of *N*-alkyl and *N*-arylalkyl substituted benztropinamine analogues, with varying substituents on the diphenyl moiety were prepared. These substituents were selected from those that gave optimal binding in the BZT series for direct comparison of SAR [30].

In this series, the 3-amino substitution readily replaced the ether linkage and gave a series of highly active compounds at the DAT that were comparable to the similarly substituted BZT analogues. As with the 3β-BZT, AHN 1-063, the least active analogue in the benztropinamine series was the similarly substituted 3β-analogue, PG 02048, showing that stereochemistry is important within this series as well [30]. Once again, none of the compounds showed high affinity for the SERT or the NET. Most of the compounds were similarly selective for DAT over muscarinic M_1 receptors as their ether counterparts. These compounds appeared to be more stable to acidic conditions and in vivo evaluation of selected analogues is underway.

3.5
N-Substituted-2-Substituted Benztropine Analogues

The 2-substituted BZT analogues were first prepared by Meltzer and colleagues, who made all 8 stereoisomers and found that only the S-(+)-2-COOCH$_3$-substituted analogue of $4',4''$-diF BZT (difluoropine, MFZ 1-76 in Table 2) showed any activity at the DAT [31] This observation was notable from the standpoint that although cocaine and its 3-phenyl analogues all need a substituent at the 2-position, it must also be of R-(–)-stereochemistry. The BZT analogues clearly do not need a substituent in this position, as analogues from our laboratory and others have shown. However, if this substituent is on the tropane ring, it must be the opposite epimer to cocaine [31–33]. These findings supported our original hypothesis that the cocaine-like and BZT analogues demonstrated very different SAR (see [3] for a review). Thus we devised a stereoselective synthesis for the S-(+)-substituted BZT analogues using chiral amine technology and published the first series of 2-substituted compounds using this method [33]. This strategy has allowed variation at all three positions (N, 2 and 3) on the tropane ring, which has allowed for rich deduction of SAR in this series and direct comparison to cocaine analogues that are substituted in these three positions [34, 35]. Interestingly, Meltzer and colleagues showed that the tropane-N of a series of 3β-aryl tropane analogues could be replaced with either O- or CH$_2$, without significantly altering DAT binding. However, replacing the tropane-N in their series of S-(+)-2-carbomethoxy BZT analogues had significant and deleterious effects on DAT

Table 2 Binding data for the 2-substituted benztropines at DAT, SERT, NET, M_1 and inhibition of dopamine uptake[1]

Compound	R_1, R_2	X	DAT	SERT	NET	M_1	DAUI
MFZ 6-35	CH=CH$_2$, CH$_3$	F	1.81 ± 0.21 [a]	1790 ± 115 [a]	473 ± 64 [a]	163 ± 23.1 [a]	3.99 ± 0.520
MFZ 1-76	COOCH$_3$, CH$_3$	F	12.9 ± 1.80 [b] 2.94 ± 0.360 [a]	690 ± 58.4 [b]	269 ± 38.9 [b]	133 ± 4.16 [b]	1.50 ± 0.020 [b]
MFZ 6-21	CH$_2$OH, CH$_3$	F	3.40 ± 0.321 [a]	910 ± 64.5 [a]	983 ± 97.0 [a]	109 ± 15.1 [a]	13.8 ± 1.55
MFZ 6-30	CH$_2$CH$_2$CO$_2$ CH$_3$, CH$_3$	F	3.74 ± 0.070 [a]	1070 ± 118 [a]	454 ± 43.2 [a]	3110 ± 435 [a]	3.66 ± 0.526
MFZ 6-26	CH=CHCO_2, CH$_3$, CH$_3$	F	4.69 ± 0.601 [a]	572 ± 51.0 [a]	269 ± 34.8 [a]	1380 ± 81 [a]	1.97 ± 0.090
MFZ 6-95	CO$_2$CH$_2$CH$_3$, butyl	F	8.18 ± 0.171 [a]	2500 ± 327 [a]	580 ± 15.9 [a]	3200 ± 298 [a]	9.96 ± 1.03
MFZ 6-93	CO$_2$, CH$_2$CH$_3$, H	F	8.87 ± 1.02 [a]	2150 ± 221 [a]	563 ± 73.9 [a]	17 100 ± 2480 [a]	4.71 ± 0.667
MFZ 6-96	CO$_2$ CH$_2$CH$_3$, allyl	F	11.0 ± 0.941 [a]	2280 ± 125 [a]	935 ± 113 [a]	11 400 ± 1650 [a]	8.84 ± 0.183
MFZ 4-87	CO$_2$ CH$_3$, CH$_3$	Cl	12.6 ± 0.399 [a]	528 ± 39.4 [a]	2150 ± 325 [a]	382 ± 36.5 [a]	2.46 ± 0.201 [a]
MFZ 4-86	CO$_2$CH$_2$CH$_3$, CH$_3$	Cl	14.6 ± 0.388 [a]	1560 ± 90.6 [a]	3350 ± 154 [a]	3060 ± 147 [a]	1.52 ± 0.215 [a]
MFZ 2-71	COOCH$_2$CH$_3$, CH$_3$	F	16.8 ± 2.00 [b] 6.87 ± 0.333 [a]	1850 ± 270 [b]	629 ± 31.0 [b]	1890 ± 13 [b]	1.84 ± 0.159 [b]

Table 2 (continued)

Compound	R₁, R₂	X	DAT	SERT	NET	M₁	DAUI
(±)MFZ 1-76R	COOCH₃, CH₃	F	21.3 ± 3.62[b]	1750 ± 243[b]	474 ± 65.9[b]	302 ± 42.8[b]	2.94 ± 0.397[b]
(±)MFZ 4-86R	CO₂CH₂CH₃, CH₃	Cl	22.0 ± 0.844[a]				2.79 ± 0.027
MFZ 2-74	COO-2-propyl, CH₃	F	23.3 ± 3.26[b]	12 000 ± 1280[b]	642 ± 13.4[b]	2680 ± 139[b]	6.24 ± 0.248[b]
(±)MFZ 4-87R	CO₂CH₃, CH₃	Cl	23.4 ± 1.54[a]			382.4 ± 36.5	3.95 ± 0.510
(±)MFZ 2-71R	COOCH₂CH₃, CH₃	F	26.2 ± 3.40[b]	3740 ± 485[b]	1020 ± 117[b]	1860 ± 189[b]	3.61 ± 0.404[b]
MFZ 3-68	COOCH₂CH₂Ph	F	36.7 ± 4.16[c]	3160 ± 461[c]	1030 ± 84.8[c]	354 ± 52.4[c]	2.57 ± 0.180[c]
MFZ 2-82	COOCH₂Ph, CH₃	F	40.2 ± 9.25[b]	2040 ± 303[b]	2230 ± 200[b]	4380 ± 528[b]	2.20 ± 0.340[b]

[1] All binding data are recorded in K_i ± SEM and are published in the cited references [a] [35]; [b] [33]; [c] [34]. If there is no reference cited, the data are previously unpublished but the methods used are identical to those used for the referenced values.

[2] Where there are two values, different radiolabeled ligands or analysis programs or incubation times were used, the details of which can be found in the primary references.

[3] The inhibition of dopamine uptake (DAUI) data is recorded as IC_{50} ± SEM and was obtained in rat brain synaptosomes according to methods detailed in the primary references.

[4] All compounds are optically active and the S-(+)-enantiomer unless otherwise indicated.

binding affinity [36, 37], further underscoring differences in SAR between these two tropane-based classes of dopamine uptake inhibitors.

This complex series of compounds has provided the most DAT-selective agents of all the BZT analogues, with several compounds showing greater than 1000-fold DAT selectivity over muscarinic M_1 receptors (Table 2). Further, both SERT/DAT and NET/DAT selectivity ratios were discovered to be quite high. Nearly all of the compounds in this series prepared so far have demonstrated high affinity for the DAT (K_i values less than 40 nM) and these same compounds have shown remarkably complex pharmacology in vivo. In addition to the S-(+)-stereochemical requirement, at the 2-position, for high DAT binding affinity in this series, the substituents that gave very high DAT affinity (e.g., CH_2OH, MFZ 6-21, in Table 2) contrast to those in the cocaine class, further demonstrating that these DA uptake inhibitors do not share SAR and appear to access non-identical DAT binding domains. This hypothesis is further strengthened by the unique profile of behavioral effects obtained with these compounds [35, 38].

4
In Vivo Findings

As detailed above, BZT and its analogues bind to the DAT with affinities that are related to variations in structure. In addition, most of the BZT analogues are selective for the DAT compared to the other monoamine transporters. These actions should, according to the DAT hypothesis of cocaine's actions, confer upon these drugs cocaine-like behavioral effects. However, when studied in various laboratory models of cocaine action and cocaine abuse, the BZT analogues were typically less effective than cocaine.

4.1
Stimulation of Locomotor Activity

The stimulation of locomotor activity is a benchmark effect of psychomotor stimulant drugs [39]. Most DA uptake inhibitors dose-dependently increase ambulatory activity from low to intermediate doses. At higher doses, activity is often increased to a lesser extent than it is at intermediate doses, or even decreased at the highest doses. In a comparison of various standard DA uptake inhibitors, we found that the maximal effects were generally comparable, if relatively restricted time points for measurement are selected, eliminating duration of action as an influence on the measurement of maximal effects [40]. In contrast, BZT analogues showed variations in effectiveness in stimulating locomotor activity [41, 42]. Figure 3 shows results of initial studies with the first generation of BZT analogues. In those studies cocaine produced a typical bell-shaped dose response for stimulation of locomotor activity. Several of the

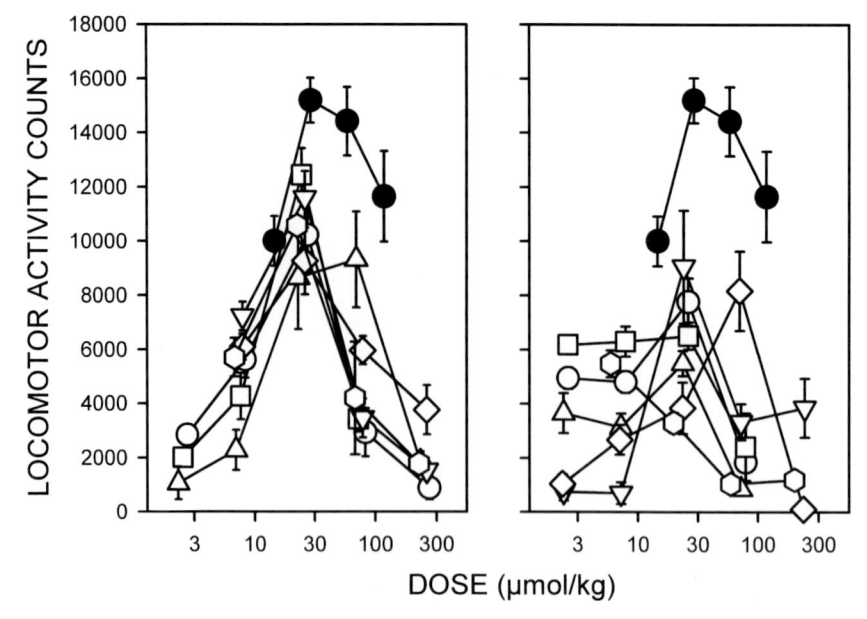

Fig. 3 Dose-dependent effects of 3α-diphenylmethoxytropane analogs on locomotor activity in mice. *Ordinates*: horizontal activity counts after drug administration. *Abscissae*: dose of drug in μmol/kg, log scale. Each *point* represents the average effect determined in eight mice. The data are from the 30-min period during the first 60 min after drug administration, in which the greatest stimulant effects were obtained. Note that the fluoro-substituted compounds (*left panel*) were generally more efficacious than the other compounds. *Left panel symbols: Filled circles* – cocaine; *open circles* – 4′-F-BZT; *squares* – 4′,4″-diF-BZT; *triangles* – 3′,4′-diCl,4″-F-BZT; *downward triangles* – 3′,4′-diF-BZT; *diamonds* – 3′,4′-diF-BZT; *hexagons* – 4′-Br,4″-F-BZT. *Right panel symbols: Filled circles* – cocaine; *open circles* – 4′-Cl-BZT; *squares* – 4′-Cl-BZT (β); *triangles* – 4′,4″-diCl-BZT; *downward triangles* – 3′,4′-diCl-BZT; *diamonds* – 4′-Br-BZT; *hexagons* – 4′,4″-diBr-BZT. Adapted from [42]

BZT analogues had maximal effects that were substantially less than those of cocaine. Compounds with a fluoro-substitution in the *para*-position on either of the phenyl rings had an efficacy that was less than, but approached that of cocaine. Several other structural variants, in particular those with chloro-substitutions, retained relatively high affinity for the DAT in binding assays, but did not stimulate activity to the same extent as did cocaine.

4.2
Cocaine-like Subjective Effects

In rats trained to discriminate cocaine from saline injections, most monoamine uptake inhibitors with affinity for the DAT produce dose-related substitution for cocaine [39, 43]. The potency differences among these DA uptake inhibitors are generally directly related to their affinity for the DAT. In addition monoamine uptake inhibitors with affinity primarily for either the SERT or NET generally do not fully substitute for the cocaine discriminative stimulus [43, 44]. In contrast to those findings with known DA uptake inhibitors, BZT analogues did not fully substitute for cocaine in rats trained to discriminate cocaine from saline [42]. As with the stimulation of locomotor activity, there were differences among the analogues with regard to their effectiveness that were related to their chemical structure. As shown in Fig. 4, *para*-fluoro-substituted analogues tended to be among the most effective, whereas BZT analogues with other *para*-substitutions, despite binding affinities comparable to those of the fluoro-substituted compounds (Table 1), were clearly less effective.

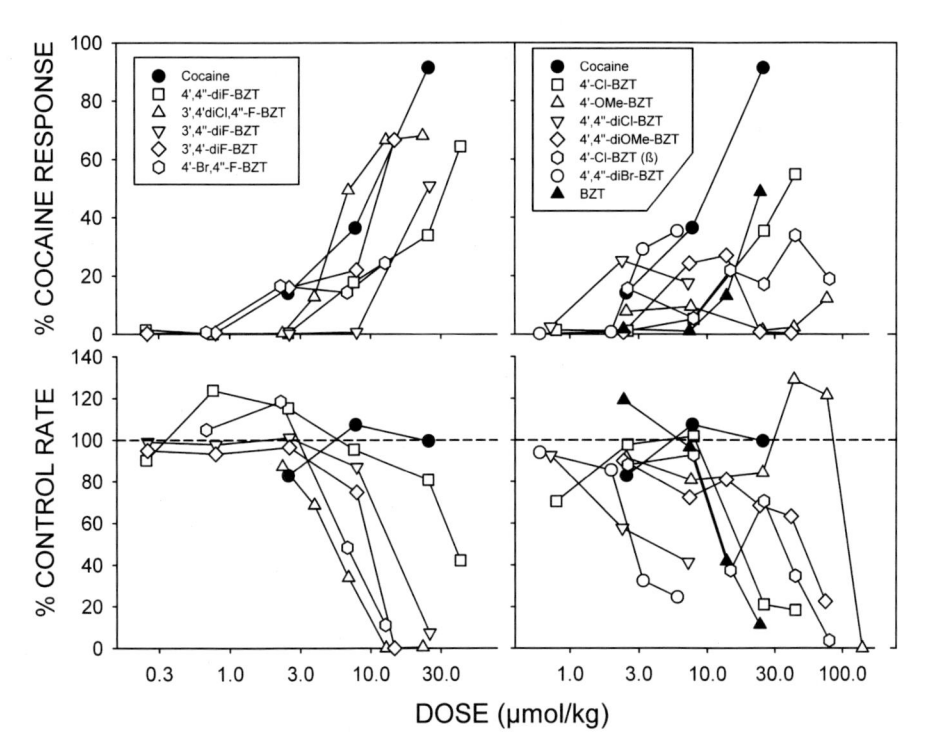

Fig. 4 Effects of 3α-diphenylmethoxytropane analogs in rats trained to discriminate injections of cocaine from saline. Ordinates for top panels: percentage of responses on the cocaine-appropriate key. *Ordinates for bottom panels*: rates at which responses were emitted (as a percentage of response rate after saline administration). *Abscissae*: drug dose in μmol/kg (log scale). Each *point* represents the effect in four or six rats. The percentage of responses emitted on the cocaine-appropriate key was considered unreliable and not plotted if fewer than half of the subjects responded at that dose. Note that the fluoro-substituted compounds (*left panels*) were generally more effective in substituting for cocaine than the other compounds. Adapted from [42]

4.3
Benztropine Self-Administration

Cocaine is well known for its reinforcing effects (e.g., [1]), and indeed there is a wealth of published findings documenting those effects (see [45] for a review). Typically, increases in dose per injection up to intermediate doses increase rates of responding maintained by cocaine. At the highest doses however, rates of responding are typically lower than those maintained at intermediate doses. We examined the reinforcing effects of BZT analogues in two studies, and compared them directly to those of cocaine. In the first study [46] the reinforcing effects of cocaine were compared to those of 3′-Cl- and 4′-Cl-BZT in rhesus monkeys trained to self-administer IV cocaine under a fixed-ratio 10 (FR 10) schedule of reinforcement. Figure 5 shows that self-administration was maintained with cocaine, and that a maximum approaching 60 injections per hour was obtained at a dose of 0.01 mg/kg/injection. Self-administration was also maintained above vehicle levels with both 3′-Cl- and 4′-Cl-BZT, but not to the same extent as by cocaine. In addition, BZT itself did not maintain responding significantly above vehicle levels. In a second study, 3′-Cl-, 4′-Cl-, and 3′,4″-diCl-BZT were further compared to cocaine and

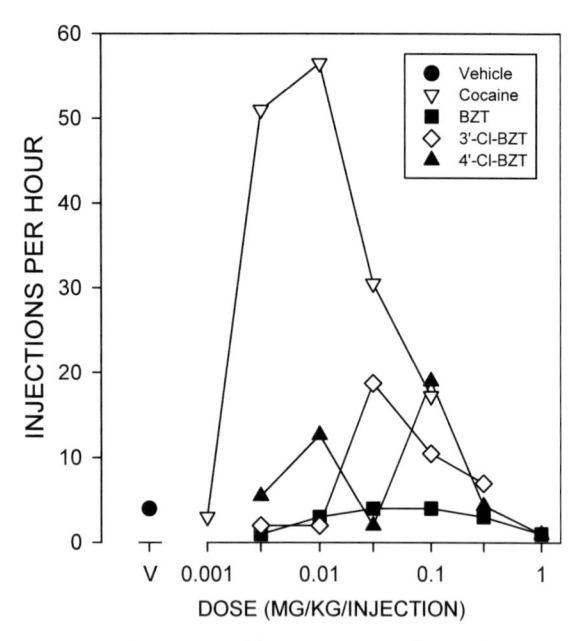

Fig. 5 Self-administration of cocaine and benztropine analogues by rhesus monkeys. *Symbols* represent the mean of the last three sessions of drug availability averaged across subjects. Abscissae: drug dose in mg/kg/injection. *Ordinates*: number of injections per hour. *Points above V* are those obtained with vehicle. Adapted from [46]

GBR 12909 using both FR and progressive-ratio schedules in rhesus monkeys trained to self-administer IV cocaine [47]. As in the previous study, some responding above that maintained by vehicle was obtained under the FR schedule with 3'-Cl- and 4'-Cl-BZT, but not with 3',4''-diCl-BZT. The effects were most pronounced in one of the four subjects, and were generally weaker than those obtained with either cocaine or GBR 12909. Under the progressive-ratio schedule the number of responses required for each successive injection increases in steps, typically until the subject stops responding. The number of steps achieved by the subject is often considered a measure of the effectiveness of the drug as a reinforcing stimulus [48]. Results with the progressive-ratio schedule suggested that the rank order of these compounds for their effectiveness was cocaine > GBR 12909 > 3'-Cl-BZT = 4'-Cl-BZT ≫ 3',4''-diCl-BZT. Though pharmacological or behavioral mechanisms accounting for the variations in effectiveness of these drugs were not examined in this study, the differences in the effectiveness of cocaine and the BZT analogues were clear and consistent with differences displayed with other behavioral end points.

Some BZT analogues have a slower onset of action as compared to cocaine, and it has been suggested that the slow onset may contribute to a decreased reinforcing effectiveness; delays in reinforcement are known to decrease the effectiveness of reinforcers generally, as well as cocaine (e.g., [49–51]).

4.4
Place Conditioning

Delays in onset of action may be readily accommodated in place conditioning procedures by modifying the time between injection and placement of the subject in the conditioning chamber [52], whereas they are less readily accommodated in self administration procedures. The assessment of effects of a drug with delayed onset of effect in place-conditioning procedures may involve selection of an appropriate time between injection and conditioning session.

Li et al. [53] examined place conditioning with BZT analogues administered from immediately to 90 min before conditioning trials. The N-methyl-substituted BZT analogue, AHN 1-055, was without significant effects at doses that ranged from 0.3 to 3.0 mg/kg even when administered up to 90 min before conditioning trials. Effects of AHN 2-005 (0.1–10.0 mg/kg) were significant, and those of JHW 007 (1.0–10.0 mg/kg) approached significance when administered 45 min but not immediately or 90 min before trials. In contrast, the effects of cocaine were robust and reliably dose-dependent from 5.0 to 20 mg/kg when administered immediately before sessions. The results support and extend the previous self-administration results showing reinforcing effects of the BZT analogues that were less than those obtained with cocaine, and further suggest that these differences from cocaine are not entirely accounted for by a slower onset of action.

4.5
Summary

In summary, BZT analogues combine structural features of cocaine and GBR 12909. In addition these compounds selectively bind the DAT among the monoamine transporters and inhibit DA uptake. They generally do not, however, fully reproduce a cocaine-like behavioral effect. The reasons for this disparity between biochemical and behavioral effects have been the subject of various lines of investigation, as described below.

5
Mechanisms for Differences with Cocaine

As mentioned above, the parent compound, BZT, has actions as an antihistaminic and as an antimuscarinic agent [54, 55]. We discuss both antimuscarinic and antihistaminic effects below and provide evidence that these are not likely mechanisms contributing to the differences between BZT analogues and cocaine-like DA uptake inhibitors.

5.1
Histamine Antagonist Effects

If the histamine antagonist effects of the BZT analogues interfere with the expression of their cocaine-like actions, then histamine antagonists should attenuate the effects of cocaine [21]. Rats trained to discriminate injections of 10 mg/kg of cocaine from saline were tested with different doses of cocaine either alone or after pretreatment with histamine antagonists. Increasing doses of promethazine, an H_1 histamine antagonist, did not alter the effects of cocaine. Similarly, other H_1 antagonists, including triprolidine, chlorpheniramine, and mepyramine were not effective in blocking the effects of cocaine. Chlorpheniramine and mepyramine, in contrast to the others, shifted the cocaine dose response curve to the left. This leftward shift is likely due to the actions of these two latter compounds at the DAT [56, 57].

Campbell et al. [21] also examined whether affinity for histamine H_1 receptors relative to affinity for the DAT predicted outcome for the locomotor stimulant effects of BZT analogues. Various assessments of these numbers indicated that the ratios of H_1 to DAT affinities were not significantly related to whether the drugs produced a significant stimulation of activity. Because the affinities of the BZT analogues for H_2 and H_3 receptors were uniformly lower than affinities for the DAT, we did not conduct the same analysis for the other histamine receptors.

In summary, it does not appear that histamine agonist or antagonist actions can substantially interfere with either the discriminative-stimulus or locomotor stimulant effects of cocaine. In addition, studies in the literature suggest that H_1 antagonists have some behavioral effects in common with psychomotor stimulant drugs (e.g., [57]). Thus, it does not appear that actions at histamine receptors interfere in any substantial way with the effects of cocaine. Moreover, if the interactions of H_1 antagonists and stimulants are histamine-receptor mediated, this action by the BZT analogues would not likely interfere with their cocaine-like effects.

5.2
Muscarinic Receptor Antagonist Effects

As with histaminergic effects, if the muscarinic antagonist actions of the BZT analogues interfere with the expression of cocaine-like effects, then a muscarinic receptor antagonist should attenuate the effects of cocaine. Instead, both atropine and scopolamine shifted the dose-effect curve for the discriminative-stimulus effects of cocaine to the left [42]. In addition atropine and other muscarinic antagonists are known to potentiate various effects of stimulant drugs including cocaine [58]. Thus, as with histamine antagonist actions, muscarinic antagonist effects appear to be incapable of interfering in any substantive way with the effects of cocaine, and any muscarinic antagonist actions of BZT analogues would likely enhance their cocaine-like behavioral effects.

However, the BZT analogues have preferential affinity for muscarinic M_1 receptors, whereas atropine and scopolamine are nonselective [59]. Therefore we investigated the effects of the preferential M_1 antagonists, telenzepine and trihexyphenidyl in combination with cocaine. Whereas trihexyphenidyl potentiated the effect of cocaine on locomotor activity, telenzepine attenuated the effect [59]. Both of these M_1 antagonists produced a small leftward shift in the discriminative-stimulus dose-effects of cocaine [60] and potentiated cocaine's effects on DA efflux in the nucleus accumbens [59]. In sum, there was no consistent attenuation of a wide range of effects of cocaine by either telenzepine or trihexyphenidyl, suggesting that the differences between BZT analogues and those of cocaine are not due to preferential M_1 antagonist effects.

In addition to the pharmacological studies, novel BZT analogues were synthesized [22, 23, 25, 35] with reduced affinity at muscarinic M_1 receptors. Substitutions on the tropane nitrogen significantly decreased (> 100-fold) M_1 muscarinic receptor affinity. In addition, these substitutions did not decrease the high affinity of the BZT analogues for the DAT, and selectivity among monoamine transporters was retained. The N-methyl-4′,4″-diF-analogue (AHN 1-055) was the most effective of the BZT analogues at producing a stimulation of locomotor activity that approached that of cocaine. In contrast to that compound, various other N-substituted analogues (e.g., AHN 2-005 and JHW 007) had substantially lower efficacy as either locomotor stimulants or in substituting in rats trained to discriminate cocaine from saline injections [61].

Antagonist activity at muscarinic receptors may also contribute to the differences in the reinforcing effectiveness of BZT analogues and cocaine. In one study [62], rhesus monkeys were trained to self-administer cocaine under fixed-ratio and progressive-ratio schedules. In most cases, combinations of cocaine and scopolamine maintained less self-administration than did cocaine alone. The authors concluded that anticholinergic actions contribute to the diminished self-administration of BZT analogs relative to cocaine and suggested that the mechanism involves either antagonism of the reinforcing effect of cocaine or punishment of the cocaine self-administration by the anticholinergic effect.

Wilson and Schuster [63] previously found atropine to increase rather than decrease rates of responding maintained by cocaine, a result that is not consistent with punishment by the anticholinergic agent. Further the atropine-induced increases in response rates were similar to the effect of lowering the dose of cocaine. Because atropine was administered independently of responding before the experimental session, it was not functioning as a punishing stimulus. All of these considerations imply a pharmacological noncompetitive antagonism of the reinforcing effects of cocaine by the anticholinergic agents.

Li et al. [53] investigated the effects of atropine in the place conditioning procedure. Atropine alone produced a non-significant trend toward conditioned place avoidance, and in combination with cocaine there was a trend towards decreases in effectiveness of cocaine. In contrast to the implications of the Wilson and Schuster results described above, these trends suggest a behavioral rather than pharmacological basis to the interaction of anticholinergics with the reinforcing effects of cocaine, and further suggest a punishing effect of anticholinergic agents. However, because the effects of atropine failed to reach statistical significance and, further, because of the inconsistencies in implications from the Wilson and Schuster findings, firm conclusions regarding the basis for the interaction of anticholinergics with cocaine for reinforcing effects are not possible at this time.

The hypothesis that the anticholinergic effect of the BZT analogues punishes behavior relates specifically to self-administration procedures using the BZT analogues. Noncompetitive pharmacological interactions between cocaine-like and antimuscarinic effects could operate more broadly across the range of behavioral end points. As noted above, there are a wider range of behavioral differences between the BZT analogues and cocaine-like DA uptake inhibitors, as evidenced by the drug discrimination and locomotor activity results. However, the preponderance of data suggests that antimuscarinic effects enhance rather than antagonize many of the effects of cocaine-like stimulants. With the BZT analogues it should be noted that the compounds with reduced M_1 affinity (e.g., AHN 2-005 and JHW 007) have less cocaine-like activity than compounds with greater M_1 affinity (e.g., AHN 1-055, RIK 7). Examining effects across a wide range of compounds and various behavioral end points, suggests that potentiation of the cocaine-like actions of BZT analogues by their antimuscarinic actions operates more broadly. However, whether antimuscarinic actions enhance or attenuate any cocaine-like effects of individual BZT analogues for now will have to be empirically determined, and may depend on the behavioral endpoint and the spectrum of effects of the doses used of the particular analogue.

5.3
Differences Between In Vivo and In Vitro Actions

That the BZT analogues have effects in vitro that predict an effect that is generally not observed in vivo has suggested relatively poor central nervous system penetration of BZT analogues. However, a study of the pharmacokinetics of selected BZT analogues in rats found high concentrations in the brain within minutes after injection, and that the elimination rates of BZT analogues were much slower than that for cocaine [64]. All of the BZT analogues were detected at initial concentrations approximating four to 15 µg per gram of tissue, approximating three to 27 µM concentrations, depending on the drug. These molar concentrations are well above the K_i values for these drugs, that range from 11 to 30 nM [64], see also [65]. In addition the parent compound is known to exhibit reliable CNS activity, and a previous study showed increases in extracellular DA levels assessed by in vivo microdialysis after systemic injection [66]. Thus, several BZT analogues were determined to be available in brain at concentrations that appeared to be well above those necessary for accessing the DAT within a short time after injection.

We compared the effects of 4'-Cl-BZT with those of cocaine on extracellular DA concentrations in the nucleus accumbens [67] and in another study further examined several specific brain regions, including prefrontal cortex and dorsal caudate, and separately examining the nucleus accumbens shell and core [68]. Cocaine and 4'-Cl-BZT produced dose-related increases in the concentration of DA in all regions, and at comparable doses, the effects of 4-Cl-BZT on DA levels in all brain areas except the prefrontal cortex were slightly less than those of cocaine. Possibly, the most notable difference between cocaine and 4'-Cl-BZT was that the latter compound had a long duration of action with DA levels substantially elevated at the higher doses up to five hours after injection. These long-lasting effects are consistent with the slow elimination of BZT analogues noted in the pharmacokinetics studies. Possibly more important, however, was that in addition to long duration of action of 4'-Cl-BZT, the rate of increase in extracellular DA levels was slower than that for cocaine.

The slower onset of effects on extracellular DA concentrations seemingly conflicts with the results of the pharmacokinetic studies which indicated that the BZT analogues were in brain within minutes after injection at concentrations well above their K_i values. We therefore conducted studies of the binding of the analogues in vivo using IV administration of [^{125}I]RTI-121 in mice to label the DAT, and compared those effects to those of cocaine [69, 70]. Figure 6 shows the effects of cocaine and several N-substituted BZT analogues. Cocaine, as has been shown in the literature, displaced [^{125}I]RTI-121 in a dose- and time-dependent manner. Maximal displacement of RTI-121 by cocaine was obtained at 30 min after injection. The BZT analogues also displaced RTI-121 binding in a dose- and time-related manner. Both AHN 1-055 and AHN 2-005 reached their maximum displacement of RTI-121 at 150 min after injection compared to 30 min for cocaine. Consistent with the slow elimination seen in the pharmacokinetic studies, there was little evidence of the dissociation of any of the BZT analogues up to 4 to 4.5 h after injection. There was evidence of a plateau in the in vivo displacement of RTI-121 around 150 min after injection for AHN 1-055 and AHN 2-005, however, for JHW 007 a plateau was not reached up to 4.5 h after injection. When the apparent association rate of JHW 007 was calculated as the displacement of RTI-121 (as a percentage of specific RTI-121 binding) per min over the linear portions of the curves shown in Fig. 6, it was found that the apparent association of JHW 007 was more than 10-fold lower than that for cocaine.

Concurrently with the time course for the in vivo binding to the DAT, we also examined the behavioral effects of the BZT analogues, and compared them to those of cocaine. The onset of effects of cocaine was relatively rapid with maximal effects obtained at 30 min after injection, as in the binding study. The effects of cocaine decreased over the next two hours and were absent at times after that. In contrast, the maximal effects of the BZT analogues, AHN 1-055 and AHN 2-005, were obtained substantially after injection. Because the level of locomotor activity in mice decreases with time in a novel chamber (habituation), it is difficult to determine precisely at what point(s) the BZT analogues produced their maximal effects. However it appears that maximal effects occurred from 90 to 120 min after injection and were essentially sustained throughout the 8-h period of observation. None of the BZT analogues was more effective than cocaine. JHW 007 did not produce significant stimulation of locomotor activity at any time at any of the doses examined.

The relationship between DAT occupancy by cocaine and its stimulation of locomotor activity at the various doses and time points is shown in Fig. 7 [69]. There was a significant correlation of DAT occupancy and stimulant effects, however, the correlation was not nearly as strong as expected, particularly in light of published results (e.g., [1, 71]). One reason that might account for the differences is that previous studies related the in vivo effects of the various compounds to binding constants obtained at equilibrium in vitro.

One interesting aspect of the present cocaine data is that at times soon after injection there was a disproportionately greater effect on locomotor activity than was predicted by DAT occupancy (filled circles in Fig. 7). Because

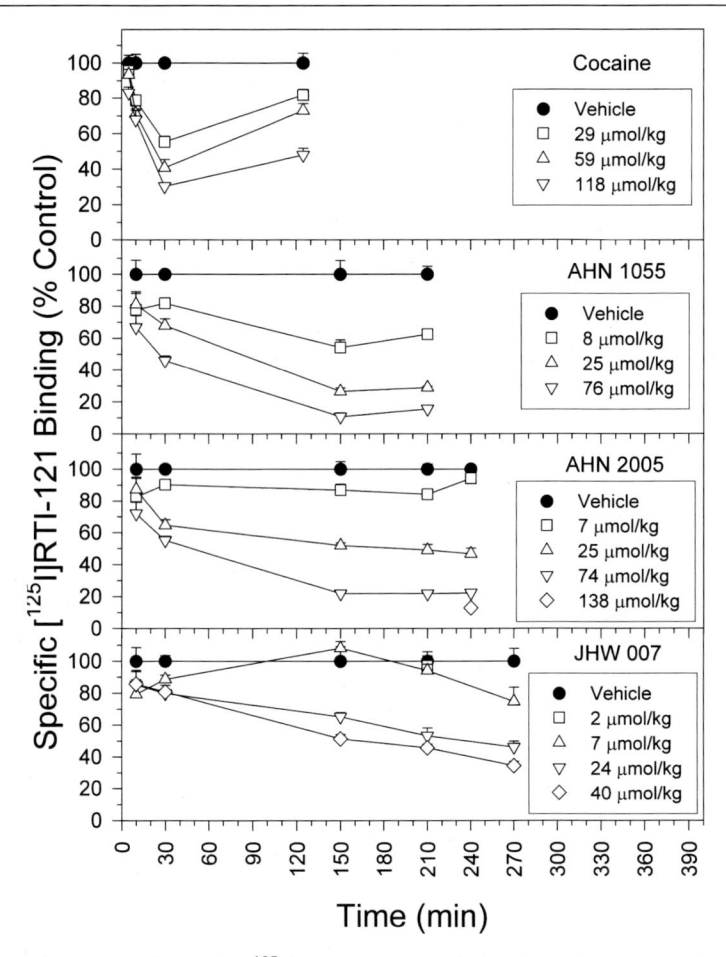

Fig. 6 Displacement of specific [^{125}I]RTI-121 accumulation in striatum at various times following IP injection of cocaine, AHN 1-055, AHN 2-005, and JHW 007. *Ordinates*: specific [^{125}I]RTI-121 binding as a percentage of that obtained after vehicle injection. *Abscissae*: time. For each *point* the number of replicates was from 5 to 10 or 13. Note that maximal displacement of [^{125}I]RTI-121 was obtained with cocaine at 30 min after injection, and at later times with the other compounds. Adapted from [69, 70]

the divergence from the regression was observed with data obtained soon after injection, association rate was suggested to play an important role in the stimulant effects of cocaine (e.g., [72]). As described above, the apparent association rate of JHW 007 was approximately 10-fold lower than that for cocaine, and JHW 007 never reached an apparent equilibrium in the in vivo binding studies (Fig. 6). This drug also had the lowest cocaine-like efficacy among those tested. However at 4.5 h after injection, the displacement of [^{125}I]RTI-121 produced by JHW 007 at the higher doses was comparable to that produced by maximal stimulating doses of cocaine (Fig. 6).

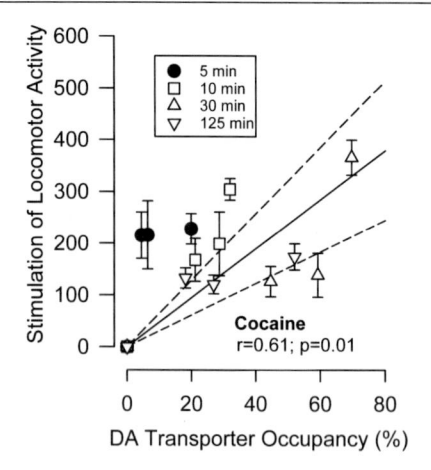

Fig. 7 Relationship between occupancy of the DAT in the striatum by cocaine and its locomotor stimulant effects. *Ordinates*: The amount of stimulation of horizontal locomotor activity expressed as counts after drug administration minus counts after vehicle administration; *Abscissae*: DAT occupancy produced by cocaine (the inverse of specific [^{125}I]RTI-121 binding) expressed as a percentage of vehicle controls. The *solid line* represents the regression of stimulation of locomotor activity on the occupancy of the DAT with the line forced through the origin. The *dashed lines* represent the 95% confidence limits for the regression line. The *error bars on points* represent ±1 SEM. Adapted from [69]

Because JHW 007 lacked robust cocaine-like stimulant effects while producing substantial DAT occupancy, we conducted studies of its potential antagonist effects against cocaine. Saline or JHW 007 was administered four and a half hours before cocaine after which locomotor activity was assessed. Following a saline injection, cocaine produced dose-related increases in locomotor activity that reached a maximum at 40 mg/kg (Fig. 8, filled circles). However, after pretreatment with JHW 007 the effects of cocaine were blocked (Fig. 8, open squares). Other DA uptake inhibitors, such as GBR 12909 (e.g., [13]) or WIN 35,428 (e.g., [61]), when administered in combination with cocaine, produce dose-dependent leftward shifts in the cocaine dose-effect curve. Nonetheless as mentioned above, there are reports of DA uptake inhibitors with unusual effects, some resembling those of JHW 007. Thus, a BZT structure does not appear to be a requirement for high-affinity DAT binding and cocaine antagonist effects.

5.4
Molecular Effects of Benztropine Analogues

There are several reports that suggest that BZT and some of its analogues bind to the DAT in a manner that is different from that of cocaine. Those molecular differences may be related to the actions of BZT analogues that differ from those of cocaine [35]. As described in detail above, there are distinct

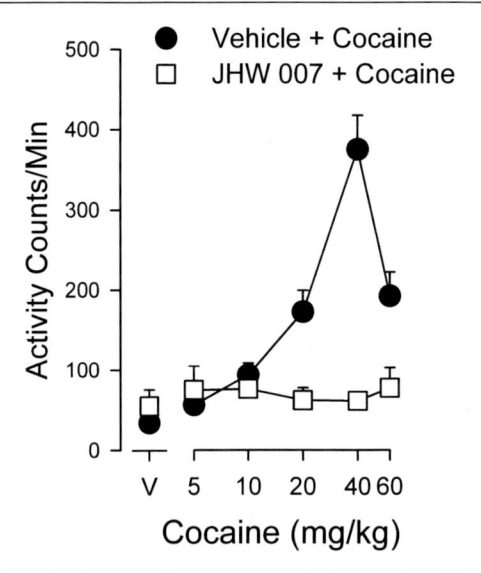

Fig. 8 Blockade of the stimulation of locomotor activity produced by cocaine. *Ordinates*: Locomotor activity counts; *Abscissae*: treatment condition, vehicle (V), or dose of cocaine. Each point represents the average effect determined in eight mice, except $n = 6$ for the combination of 10 mg/kg of JHW 007 with 60 mg/kg cocaine. The *error bars* represent ± 1 SEM. Adapted from [69]

differences between the BZT analogues and the 3-phenyltropane class of DA uptake inhibitors in SAR at the DAT. Specifically, the 3α-diphenylethers with small para- or meta-substituted halogens on one or both phenyl rings give optimal DAT binding affinities, whereas for the cocaine class of molecules a wider variety of 3β-phenyl ring substitutions generally result in high affinity DAT binding. Among the BZT analogues, tropane-N-alkyl- and arylalkyl-substitutions are generally well tolerated at the DAT, and this substitution provides a point of departure from high affinity binding at muscarinic receptors. In contrast, tropane N-substitutions among the 3-phenyltropanes do not typically yield higher DAT affinities, nor has it yielded selectivity across other transporters or receptors to which cocaine binds. A substitution in the 2-position is required for binding at the DAT among the 3β-phenyl tropanes, with R-(−)-2β-substituents being optimal. In contrast, only S-(+)-2β-substitution within the BZT analogues results in high affinity DAT binding [31–35]. Moreover, analogues of BZT without substitution at the 2-position have essentially equal DAT affinities, suggesting that this substituent is not necessary for high affinity binding.

Initial studies using site-directed mutagenesis of the DAT revealed differences in binding modes between cocaine and BZT [73]. Cocaine and BZT differently affected the reaction to the sulfhydryl-reactive cell-impermeant reagent, methanethiosulfonate ethyltrimethyl-ammonium (MTSET), in various DAT mutants expressed in HEK-293 cells. For example, the decrease in the reaction of Cys-135 with MTSET produced by cocaine and other DA uptake inhibitors was not obtained with BZT [73]. In addition, BZT, in contrast to cocaine, had no effect on the reaction of Cys-90 with MTSET. These findings suggested that BZT and standard dopamine uptake inhibitors produce different conformational changes in the DAT.

Several other studies demonstrated differences with BZT or its analogues in their binding to wild-type and mutant DA transporters (e.g., [74]). A recent study with BZT analogues using site-directed mutagenesis of Aspartate 79 in transmembrane (TM) sequence 1 of the DAT, demonstrated that these compounds bind differently than cocaine and its analogues, as well as other structurally diverse standard DAT inhibitors [75]. Studies using photoaffinity labeling techniques have also provided evidence that structurally different DA uptake inhibitors covalently attach to different TM regions of the DAT [26, 27, 76–78]. Together these results support the idea that different DA uptake inhibitors bind to the DA transporter in different ways and suggest that there may be particular conformational changes of the DAT induced by cocaine and cocaine-like drugs that contribute to its substantial abuse liability.

A recent study [38] compared the effects of different DAT inhibitors on the accessibility of MTSET to a cysteine residue inserted into the DAT (I159C) in COS7 cells. This cysteine is thought to be accessible when the extracellular transporter gate is open, but inaccessible when the gate is closed. Preincubation with cocaine and MTSET potentiated the inhibition of DA uptake produced by MTSET alone, consistent with stabilization of the transporter in a conformation open to the extracellular environment. In contrast, preincubation with BZT analogues protected against MTSET inactivation, suggesting that these compounds stabilize the DAT in a closed conformation. Previous studies suggested that Tyr335 in the DAT is critical for regulating the open/closed conformational equilibrium of the DAT, and that mutation of Tyr335 to alanine (Y335A) impairs DA transport capacity and decreases cocaine's potency in the inhibition of DA uptake [38, 79]. Similarly, analogues of cocaine were approximately 100-fold less potent in the inhibition of DA uptake in Y335A mutants compared to WT DAT, whereas BZT analogues were displayed only a 7- to 58-fold loss in potency. Interestingly, one of the BZT analogues, MFZ 2-71, was an exception to the rule, had an 88-fold loss in potency that approached the value obtained with cocaine, and was similar to that of the cocaine analogue, RTI-55. In behavioral studies we also found MFZ 2-71 to be an exception; it not only produced a marked stimulation of locomotor activity, but it also substituted fully in subjects trained to discriminate cocaine from saline injections. Further, for all of the compounds there was a close relationship between the decrease in potencies due to this mutation and their effectiveness in stimulating locomotor activity as well as substituting in cocaine-discriminating rats. Taken together, the data suggest that structurally distinct DAT inhibitors stabilize distinct

DAT conformations, which in turn affects the cocaine-like in vivo effects of the compounds.

6
Preclinical Assessment of Cocaine-Abuse Therapy

Because several of the BZT analogues had DAT affinity and reduced cocaine-like behavioral effects, it was of interest to assess whether the drugs would alter the self-administration of cocaine. Self-administration in laboratory animals is often considered the most appropriate model for preclinical assessment of efficacy for candidate drugs for treatment of drug dependence. Rats that were surgically prepared with indwelling venous catheters were trained to press a lever with IV cocaine infusions functioning as the reinforcer. Experimental sessions in which the subjects self administered cocaine were divided into 30 min intervals in which different doses were available.

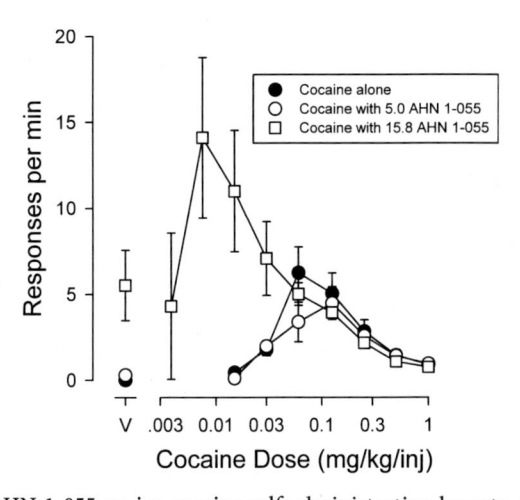

Fig. 9 Effects of AHN 1-055 on i.v. cocaine self administration by rats. *Symbols* represent the mean values from five to six subjects responding under an FR 5 schedule of cocaine injection. During daily experimental sessions subjects had the opportunity to self administer each dose of cocaine during 30 min epochs. The dose of cocaine was changed by altering the injection time or the concentration of the cocaine solution. A 10-min timeout period separated each session epoch and the next available dose, which was one half of the previous dose. Self administration of saline was tested during the last epoch of the session. An injection of the available cocaine dose was delivered independently of responding at the start of each 30-min epoch. All other injections were delivered according to the FR 5 schedule of reinforcement. *Abscissae*: Drug dose in mg/kg/injection. *Ordinates*: number of responses per 30-min epoch. *Points above V* are those obtained in the last epoch during which vehicle could be self administered. *Filled points* show the effects of cocaine dose/injection without pretreatment. *Open points* show cocaine self administration after a pretreatment with one of two doses of AHN 1-055

A separate set of subjects were trained with saccharin rather than cocaine reinforcement. Before selected sessions subjects were given one of several doses of either AHN 1-055 or AHN 2-005.

As is typical of cocaine self-administration, as dose per injection increased, rates of responding first increased and then decreased (Fig. 9, filled symbols), such that cocaine maintained the highest average response rates at an intermediate dose (0.06 mg/kg/injection). Pretreatment with 5.0 mg/kg of AHN 1-055 had no substantive effect on the cocaine dose-effect curve, whereas a higher dose shifted the dose-effect curve leftward, and increased the maximal response rates maintained. In addition, AHN 1-055 produced a dose-related increase in the low response rates obtained when saline rather than cocaine was available for self-administration (compare open points to the filled point above C). This stimulant effect of AHN 1-055 may have contributed to the increase in the maximal rates obtained with cocaine when subjects were pretreated with AHN 1-055.

Effects of AHN 2-005 on cocaine self-administration differed from those of AHN 1-055. This compound produced a dose-related flattening of the cocaine dose-effect curve, without any appreciable shift leftward (compare open symbols to filled circles). In addition, when saline rather than cocaine was available for self-administration (points above C), there was no evidence of a stimulant effect of AHN 2-005 pretreatment. This difference between AHN

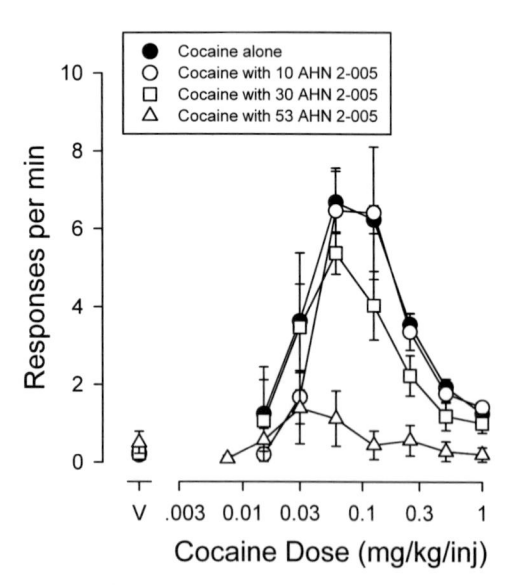

Fig. 10 Effects of AHN 2-005 on i.v. cocaine self administration by rats. *Symbols* represent the mean values from six subjects responding under an FR 5 schedule of cocaine injection. All details are as described for Fig. 9, except that *open points* show cocaine self administration after a pretreatment with one of three doses of AHN 2-005

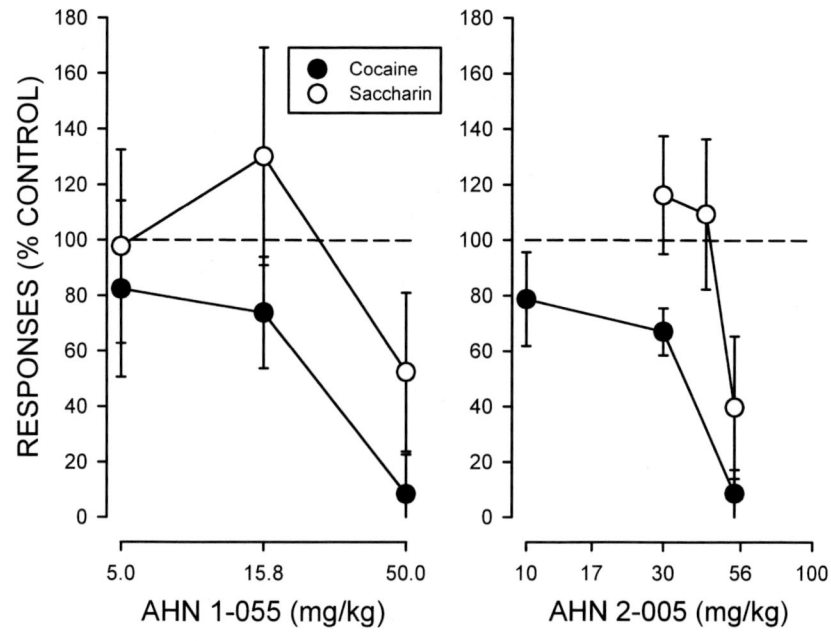

Fig. 11 Comparison of the effects of AHN 1-055 and AHN 2-005 on responding maintained by cocaine and that maintained by saccharin under similar FR 5 schedules of reinforcement that maintained similar rates of responding. The BZT analogues were administered 190 min before the opportunity to respond with saccharin reinforcement so that the time at which their effects were assessed corresponded to the time at which maximal responding was maintained by cocaine. *Abscissae*: Dose of AHN 1-055 or AHN 2-005 in mg/kg. *Ordinates*: Response rates as a percentage of the rates occurring when the BZT analogues were not administered. Note that there was a tendency for both BZT analogues to decrease rates of responding maintained by cocaine at lower doses than those necessary to decrease rates of responding maintained by saccharin reinforcement

1-055 and AHN 2-005 is reminiscent of the initial findings (see above) that AHN 2-005 was less efficacious than AHN 1-055 in various stimulant effects including stimulation of locomotor activity and substitution in rats trained to discriminate cocaine from saline injections.

A drug that severely interferes with appropriate behaviors would not likely be clinically useful, even if it completely eliminated drug-taking and drug-seeking behaviors. Shown in Fig. 11 is a comparison of the effects of both AHN 1-055 and AHN 2-005 on behaviors maintained by cocaine and comparable behavior maintained by saccharin in order to approach an assessment of the specificity by which the drugs might alter cocaine abuse. As can be seen, both compounds decreased rates of responding maintained by cocaine (the cocaine dose/injection was the one that maintained the highest response rates) at lower doses than those that decreased responding maintained by saccharin. Though there was some variability, the data indicate that there is a sensitivity of cocaine-maintained responding to the effects of these BZT analogues that is greater than the sensitivity of responding maintained by saccharin, suggesting that these drugs, or drugs like them may show clinical selectivity as treatments for cocaine abuse.

7
Summary and Conclusions

Additional studies of BZT analogues are ongoing combining molecular and behavioral techniques with the most interesting of the BZT analogues that we have discovered over the years. These studies are directed at the identification and characterization of the DAT binding domains of the BZT analogues in comparison to those for cocaine-like drugs to better understand what contributes to the different behavioral effects of the compounds. Studies addressing association rate and conformational equilibrium of the DAT with BZT analogues, along lines described above, will be important for the testing of various hypotheses regarding mechanisms accounting for the behavioral effects of these compounds.

As has been demonstrated, the BZT analogues are bioavailable, readily penetrate the blood brain barrier and gain high levels of brain to plasma ratios within minutes of injection. Nevertheless, DAT occupancy and subsequent increases in DA, measured by microdialysis, are significantly delayed as compared to cocaine. Further, DAT occupancy and elevated levels of DA are also long lasting. All of these features appear to be important for the lack of abuse liability of the BZT analogues, and their potential as treatments for cocaine abuse.

Future investigations will continue to elucidate the interactions of these drugs in comparison to cocaine at the DAT and to reveal how those differences affect other neural systems and circuitry, so as to provide a better understanding of cocaine's mechanism of action and to feed this information into our drug design. Further, as the BZT analogues do not appear to have significant abuse liability they may also have therapeutic utility in other disorders such as ADHD and obesity, areas in which drug development is an emerging line of investigation.

Acknowledgements We would like to acknowledge the dedicated members of our laboratories, past and present, who have diligently taken on the arduous tasks of synthesizing, characterizing and biologically evaluating all of our novel compounds. We are particularly indebted to Theresa Kopajtic who has performed the vast majority of the in vitro studies on these compounds and has helped in uncountable ways. The authors also acknowledge support from Dr. Jane Acri and Dr. Carol Hubner, NIDA Addiction Treatment Discovery Program, Division of Pharmacotherapies and Medical Consequences of Drug Abuse, NIDA, and data obtained through contract N01DA-6-8062. This work was supported by the National Institute on Drug Abuse – Intramural Research Program.

References

1. Ritz MC, Lamb RJ, Goldberg SR, Kuhar MJ (1987) Science 237:1219
2. Newman AH (1998) Med Chem Res 8:1
3. Newman AH, Kulkarni SS (2002) Med Res Rev 22:1
4. Dutta AK, Zhang S, Kolhatkar R, Reith MEA (2003) Eur J Pharmacol 479:93
5. Runyon SP, Carroll FI (2006) Curr Top Med Chem 6:1825
6. Rothman RB, Baumann MH, Prisinzano TE, Newman AH (2008) Biochem Pharmacol 75:2
7. Rocha BA, Fumagalli F, Gainetdinov RR, Jones SR, Ator R, Giros B, Miller GW, Caron MG (1998) Nat Neurosci 1:132
8. Sora I, Wichems C, Takahashi N, Li XF, Zeng Z, Revay R, Lesch KP, Murphy DL, Uhl GR (1998) Proc Natl Acad Sci USA 95:7699
9. Rothman RB, Greig N, Kim A, De Costa BR, Rice KC, Carroll FI, Pert A (1992) Pharmacol Biochem Behav 43:1135
10. Vaugeois JM, Bonnet JJ, Duterte-Boucher D, Costentin J (1993) Eur J Pharmacol 230:195
11. Colpaert FC, Niemegeers CJ, Janssen PA (1979) Biochem Behav 10:535
12. Husbands SH, Kopajtic T, Izenwasser S, Bowen WD, Vilner BJ, Katz JL, Newman AH (1999) J Med Chem 42:4446
13. Katz JL, Libby TA, Kopajtic T, Husbands SM, Newman AH (2003) Eur J Pharmacol 468:109
14. Cao J, Kulkarni SS, Husbands SM, Bowen WD, Williams W, Kopajtic T, Katz JL, George C, Newman AH (2003) J Med Chem 46:2589
15. Navarro HA, Howard JL, Pollard GT, Carroll FI (2005) Abstract 67th Annual Meeting of the College on Problems of Drug Dependence
16. Carroll FI (2003) J Med Chem 46:1775
17. Newman AH, Allen AC, Izenwasser S, Katz JL (1994) J Med Chem 37:2258
18. Newman AH, Kline RH, Allen AC, Izenwasser S, George C, Katz JL (1995) J Med Chem 38:3933
19. Kline RH, Izenwasser S, Katz JL, Newman AH (1997) J Med Chem 40:851
20. Kulkarni SS, Kopajtic TA, Katz JL, Newman AH (2006) Bioorg Med Chem 14:3625
21. Campbell VC, Kopajtic TA, Newman AH, Katz JL (2005) J Pharmacol Exp Ther 315:631
22. Agoston GE, Wu JH, Izenwasser S, George C, Katz J, Kline RH, Newman AH (1997) J Med Chem 4:4329
23. Robarge MJ, Agoston GE, Izenwasser S, Kopajtic T, George C, Katz JL, Newman AH (2000) J Med Chem 43:1085
24. Newman AH, Robarge MJ, Howard IM, Wittkopp SL, Kopajtic T, Izenwasser S, Katz JL (2001) J Med Chem 44:633
25. Kulkarni SS, Grundt P, Kopajtic T, Katz JL, Newman AH (2004) J Med Chem 47:3388
26. Agoston GE, Vaughan R, Lever JR, Izenwasser S, Terry PD, Newman AH (1997) Bioorg Med Chem Lett 7:3027
27. Vaughan RA, Agoston GE, Lever JR, Newman AH (1999) J Neurosci 19:630
28. Grundt P, Kopajtic TA, Katz JL, Newman AH (2004) Bioorg Med Chem Lett 14:3295
29. Simoni D, Rossi M, Bertolasi V, Roberti M, Pizzirani D, Rondanin R, Baruchello R, Invidiata PFP, Tolomeo M, Grimaudo S, Merighi S, Varani K, Gessi S, Borea PA, Marino S, Savallini S, Bianchi C, Siniscalchi A (2005) J Med Chem 48:337
30. Grundt P, Kopajtic T, Katz JL, Newman AH (2005) Bioorg Med Chem Lett 15:5419
31. Meltzer PC, Liang AY, Madras BK (1994) J Med Chem 37:2001

32. Meltzer PC, Liang AY, Madras BK (1996) J Med Chem 39:371
33. Zou MF, Agoston GE, Belov Y, Kopajtic T, Katz JL, Newman AH (2002) Bioorg Med Chem Lett 12:1249
34. Zou MF, Kopajtic T, Katz JL, Newman AH (2003) J Med Chem 46:2908
35. Zou MF, Cao J, Kopajtic T, Desai RI, Katz JL, Newman AH (2006) J Med Chem 49:6391
36. Meltzer PC, Liang AY, Blundell P, Gonzalez MD, Chen Z, George C, Madras B (1997) J Med Chem 40:2661
37. Meltzer PC, Blundell P, Yong YF, Chen Z, George C, Gonzalez MD, Madras B (2000) J Med Chem 43:2982
38. Loland CJ, Desai RI, Zou MF, Cao J, Grundt P, Gerstbrein K, Sitte HH, Newman AH, Katz JL, Gether U (2008) Mol Pharm 73:813
39. Kelleher RT (1997) Drug Abuse: the Clinical and Basic Aspects. CV Mosby Co, St. Louis
40. Izenwasser S, Terry P, Heller B, Witkin JM, Katz JL (1994) Eur J Pharmacol 263:277
41. Katz JL, Izenwasser S, Newman AH (1997) Pharmacol Biochem Behav 57:505
42. Katz JL, Izenwasser S, Kline RH, Allen AC, Newman AH (1999) J Pharmacol Exp Ther 288:302
43. Baker LE, Riddle EE, Saunders RB, Appel JB (1993) Behav Pharmacol 1:69
44. Kleven MS, Anthony EW, Woolverton WL (1990) J Pharmacol Exp Ther 254:312
45. Johanson CE, Fischman MW (1989) Pharmacol Rev 41:3
46. Woolverton WL, Rowlett JK, Wilcox KM, Paul IA, Kline RH, Newman AH, Katz JL (2000) Psychopharmacology 147:426
47. Woolverton WL, Hecht GS, Agoston GE, Katz JL, Newman AH (2001) Psychopharmacology 154:375
48. Hodos W (1961) Science 134:943
49. Gollub L, Yanagita T (1974) Bull Psychon Soc 4:263
50. Stretch R, Gerber GJ, Lane E (1976) Can J Physiol Pharmacol 54:632
51. Beardsley PM, Balster RL (1993) Drug Alcohol Depend 34:37
52. De Beun R, Jansen E, Geerts NE, Slangen JL, Van De Poll NE (1992) Pharmacol Biochem Behav 42:445
53. Li SM, Newman AH, Katz JL (2005) J Pharmacol Exp Ther 313:1223–1230
54. Richelson E (1981) Antidepressants: Neurochemical, Behavioral and Clinical Perspectives. Raven Press, New York
55. McKearney JW (1982) Psychopharmacology 77:156
56. Tuomisto J, Tuomisto L (1980) Med Biol 58:33
57. Bergman J, Spealman RD (1986) J Pharmacol Exp Ther 239:104
58. Scheckel CL, Boff E (1964) Psychopharmacologia 5:198
59. Tanda G, Ebbs A, Kopajtic T, Campbell B, Newman AH, Katz JL (2007) J Pharmacol Exp Ther 321:332
60. Tanda G, Katz JL (2007) Pharmacol Biochem Behav 87:400
61. Katz JL, Kopajtic T, Agoston GE, Newman AH (2004) J Pharmacol Exp Ther 288:302
62. Ranaldi R, Woolverton WL (2002) Psychopharmacology (Berl) 161:442
63. Wilson MC, Schuster CR (1973) Pharmacol Biochem Behav 1:643
64. Raje S, Cao J, Newman AH, Gao H, Eddington N (2003) J Pharmacol Exp Ther 307:801
65. Othman AA, Syed SA, Newman AH, Eddington ND (2007) J Pharmacol Exp Ther 320:344
66. Church WH, Justice JB, Byrd LD (1987) Eur J Pharmacol 139:345
67. Tolliver BK, Ho LB, Newman AH, Fox LM, Katz JL, Berger SP (1999) J Pharmacol Exp Ther 103:110
68. Tanda G, Ebbs A, Newman AH, Katz JL (2005) J Pharmacol Exp Ther 313:613

69. Desai R, Kopajtic T, Koffarnus M, Newman AH, Katz JL (2005) J Neurosci 25:1889
70. Desai R, Kopajtic T, French D, Newman AH, Katz JL (2005) J Pharmacol Exp Ther 315:397
71. Bergman J, Madras BK, Johnson SE, Spealman RD (1989) J Pharmacol Exp Ther 251:150
72. Volkow ND, Fowler JS, Wang GJ (2002) Behav Pharmacol 13:355
73. Reith MEA, Berfield JL, Wang LC, Ferrer JV, Javitch JA (2001) J Biol Chem 276:29012
74. Chen N, Zhen J, Reith MEA (2004) J Neurochem 89:853
75. Ukairo OT, Bondi CD, Newman AH, Kulkarni SS, Kosikowski AP, Pan S, Surratt CK (2005) J Pharmacol Exp Ther 314:575
76. Vaughan RA, Sakrikar DS, Parnas ML, Foster JD, Duval RA, Lever JR, Kulkarni SS, Newman AH (2007) J Biol Chem 282:8915
77. Zou M, Kopajtic TA, Katz JL, Wirtz S, Justice JJ, Newman AH (2001) J Med Chem 44:4453
78. Parnas ML, Gaffaney JD, Zou MF, Lever JR, Newman AH, Vaughan RA (2008) Mol Pharm e-pub Jan 23rd
79. Loland CJ, Norregaard L, Litman T, Gether U (2002) Proc Natl Acad Sci USA 99:1683

Top Med Chem (2009) 4: 131–154
DOI 10.1007/7355_2007_016
© Springer-Verlag Berlin Heidelberg
Published online: 1 May 2008

The Design, Synthesis and Structure–Activity Relationship of Mixed Serotonin, Norepinephrine and Dopamine Uptake Inhibitors

Zhengming Chen (✉) · Ji Yang · Phil Skolnick

DOV Pharmaceutical, Inc., 150 Pierce Street, Somerset, NJ 08873-4185, USA
zchen@dovpharm.com

Abstract The evolution of antidepressants over the past four decades has involved the replacement of drugs with a multiplicity of effects (e.g., TCAs) by those with selective actions (i.e., SSRIs). This strategy was employed to reduce the adverse effects of TCAs, largely by eliminating interactions with certain neurotransmitters or receptors. Although these more selective compounds may be better tolerated by patients, selective drugs, specifically SSRIs, are not superior to older drugs in treating depressed patients as measured by response and remission rates. It may be an advantage to increase synaptic levels of both serotonin and norepinephrine, as in the case of dual uptake inhibitors like duloxetine and venlafaxine. An important recent development has been the emergence of the triple-uptake inhibitors (TUIs/SNDRIs), which inhibit the uptake of the three neurotransmitters most closely linked to depression: serotonin, norepinephrine, and dopamine. Preclinical studies and clinical trials indicate that a drug inhibiting the reuptake of all three of these neurotransmitters could produce more rapid onset of action and greater efficacy than traditional antidepressants. This review will detail the medicinal chemistry involved in the design, synthesis and discovery of mixed serotonin, norepinephrine and dopamine transporter uptake inhibitors.

Keywords Design · Monoamine · SNDRI · Synthesis and SAR · Triple-uptake inhibitor

Abbreviations

5-HT	Serotonin
DA	Dopamine
DARI/DRI	Selective dopamine reuptake inhibitor

DNRI/NDRI	Balanced dopamine and norepinephrine reuptake inhibitor
MAOI	Monoamine oxidase inhibitor
MDD	Major depressive disorder
NE	Norepinephrine
NERI/NRI	Selective norepinephrine reuptake inhibitor
SAR	Structure activity relationship
SDRI/DSRI	Balance serotonin and dopamine reuptake inhibitor
SNDRI/TRIP/TUI	Balanced serotonin, norepinephrine and dopamine reuptake inhibitor or triple-reuptake inhibitor or triple-uptake inhibitor
SNRI/NSRI	Balanced serotonin and norepinephrine reuptake inhibitor
SSRI	Selective serotonin reuptake inhibitor
TCA	Tricyclic antidepressant

1
Introduction

The first clinically useful tricyclic antidepressant (TCA), imipramine, emerged in the late 1950s (Fig. 1) [1]. Axelrod and co-workers demonstrated that TCAs inhibit the uptake of norepinephrine in brain tissue [2]. In 1965, Schildkraut postulated his catecholamine theory of depression to explain the antidepressant effect of TCAs [3]. Because of subsequent studies demonstrating that imipramine also inhibits the uptake of serotonin [4], Schildkraut's theory eventually evolved into the monoamine hypothesis of depression, now encompassing serotonin as a critical transmitter in the pathophysiology of depression. We now know that TCAs (and other antidepressants) bind to sodium-dependent norepinephrine and/or serotonin transporters [5], as shown in Table 1, inhibiting the uptake of norepinephrine and/or serotonin into presynaptic neurons. Because these transporters are the principal means of removing neurotransmitters from the synaptic cleft, drugs like the TCAs allow NE and 5-HT to remain in the synaptic cleft, activating adrenergic and/or serotonergic receptors. The cellular events produced by elevating synaptic monoamine concentrations that result in a relief of the symptoms of depression have not been elucidated, and indeed, such research may result in new targets for drug discovery [6]. Nonetheless, strategies aimed at enhancing monoaminergic transmission remain an attractive target for developing antidepressants [7].

Because TCAs affect multiple targets (other than NET and SERT), there are many serious side effects associated with their use that can impact patient compliance. These include: anticholinergic effects (dry mouth, dizziness, constipation, difficulty urinating, and blurred vision), adrenergic effects (sweating, and orthostatic hypotension), and antihistaminic effects (sedation and weight gain) [8, 9]. In Table 1, we have listed the in vitro binding affinities for TCAs on the monoamine transporters associated with antidepressive effect, as well as the affinities towards adrenergic, muscarinic, and histamine receports associated with side effects.

Table 1 TCA binding affinities to various receptors, K_i (nM)

Compound	SERT	NET	DAT	$\alpha1$ adrenergic	$\alpha2$ adrenergic	Cholinergic muscarinic	H1 histamine
Imipramine	20	20	>10 000	32	3100	46	37
Desipramine	163	0.63	>10 000	23	1379	66	60
Clomipramine	0.14	53.7	3020	3.2	525	37	31
Amitriptyline	36	19	5300	4.4	114	9.6	0.67
Nortriptyline	279	1.8	11 000	55	2030	37	6.3
Doxepin	220	18	N/A	24	1270	23	0.2
Dothiepin	110	34	N/A	470	2400	25	3.6
Maprotiline	>1000	7	1210	90	>1000	570	2

Data are adapted from PDSP K_i database [10]

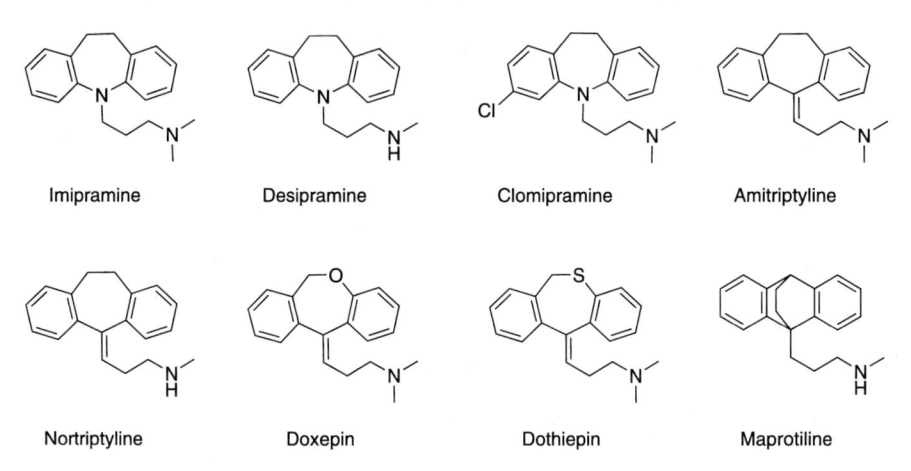

Fig. 1 Structures of tricyclic antidepressants (TCAs)

Several of the early TCAs (e.g., desipramine and nortryptyline) had a high affinity for NET relative to SERT, but also possessed a high affinity at other targets associated with side effects (Table 1). Therapeutic application of moderately selective norepinephrine reuptake inhibitors, such as viloxazine, began as early as 1974, but has not demonstrated an advantage over TCAs (Fig. 2). The first selective norepinephrine reuptake inhibitor (NERI) antidepressant drug, reboxetine, was developed in 1997 with a low side effect profile. This compound has not been marketed in the U.S. A newer member of the group, atomoxetine, has been used primarily in the treatment of attention-deficit hyperactivity disorder (ADHD).

In the late 1960s, Carlsson et al. speculated that selective inhibition of serotonin uptake would be sufficient to elicit an antidepressant action [11]. Thus, selective serotonin reuptake inhibitors (SSRIs) were developed with the

Table 2 SSRI binding affinities to various receptors, K_i (nM)

Compound	SERT	NET	DAT	$\alpha 1$ adrenergic	$\alpha 2$ adrenergic	Cholinergic muscarinic	H1 histamine
Fluvoxamine	11	1119	>10 000	1288	1900	>10 000	>10 000
Fluoxetine	5.7	599	4752	3171	3090	590	5400
Citalopram	9.6	5029	>10 000	1211	>10 000	5600	283
Escitalopram	2.5	6514	>10 000	3870	N/A	1242	1973
Paroxetine	0.4	45	963	2741	3915	108	>10 000
Sertraline	2.8	925	315	188	477	232	>10 000

Data are adapted from PDSP K_i database [10]

aim of removing the undesired interactions with those receptors (muscarinic cholinergic receptor, the histamine receptor, and the $\alpha 1$ adrenoceptor, etc.) from tricyclic antidepressants (TCA). In addition, loss of NET inhibition was thought to minimize or eliminate potential cardiovascular effects. The commercial success of fluoxetine (Prozac®) [12, 13], launched in 1987, validated this platform and resulted in the development of several other SSRI drugs (citalopram, launched in 1989, paroxetine in 1991, sertraline in 1992, and escitalopram, an enantiomer of citalopram, launched in 2002) (Fig. 2). The selectivities of these drugs for SERT relative to NET vary, with escitalopram the most selective (>2000-fold) (Table 2). However, once the selectivity for one transporter exceeds 20-fold, it is unlikely that there will be significant inhibition of the lower affinity target at pharmacologically relevant doses.

This strategy of selectivity at SERT has clearly been beneficial in terms of adverse events, which are lower for SSRIs than for TCAs (Table 2) [14–16]. However, this selectivity for a single monoamine transporter has been shown to afford (albeit not in a consistent fashion) lower efficacy than that of TCAs, especially in severe or endogenous depression. This is suggested, for example, by the demonstration in controlled studies that the TCA clomipramine is significantly superior to citalopram or paroxetine in endogenously depressed patients [17, 18].

New, specific serotonin and norepinephrine reuptake inhibitors (SNRIs) were thus developed with the intention of providing greater antidepressant efficacy than the SSRIs without the side effects of the TCAs [19–23].

2
Overview of Dual-Uptake Inhibitors

Although the difficulties inherent in performing clinical trials makes reliable efficacy comparisons of two specific antidepressants or types of antidepressant problematic, it is believed that dual-acting agents may represent a more

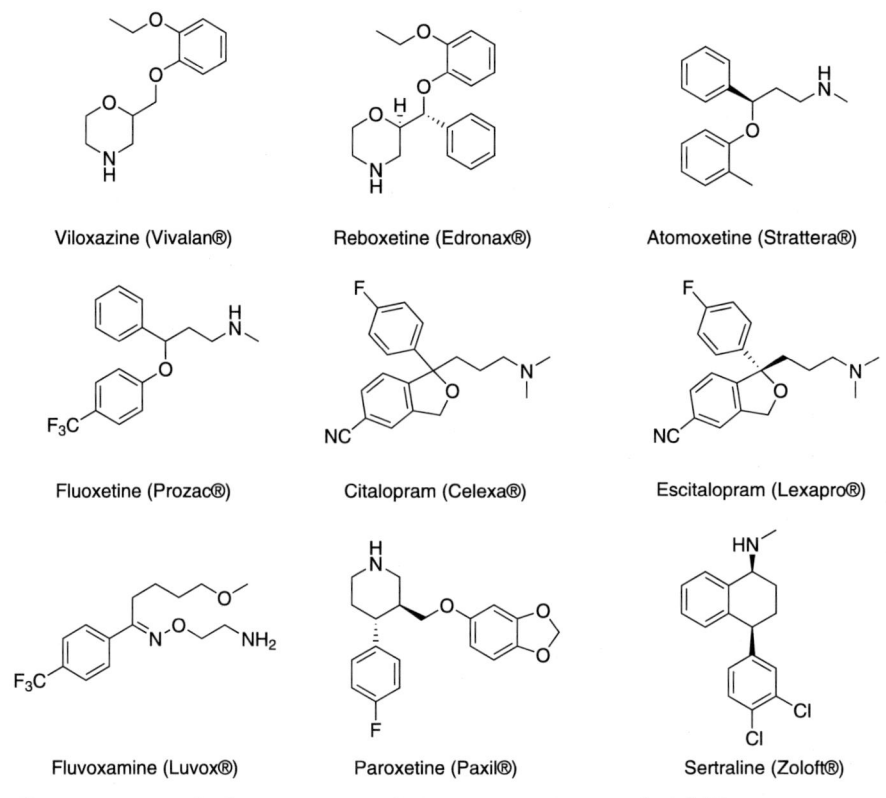

Viloxazine (Vivalan®)   Reboxetine (Edronax®)   Atomoxetine (Strattera®)

Fluoxetine (Prozac®)   Citalopram (Celexa®)   Escitalopram (Lexapro®)

Fluvoxamine (Luvox®)   Paroxetine (Paxil®)   Sertraline (Zoloft®)

Fig. 2 Structures of selective norepinephrine or serotonin reuptake inhibitors

broadly effective treatment option when compared to SSRIs, particularly in treating the physical symptoms and comorbid disorders associated with depression [24].

Venlafaxine, a synthetic derivative of phenethylamine, is the most frequently prescribed SNRI (Fig. 3). Venlafaxine has a single chiral center and exists as a racemic mixture of R-(−)- and S-(+)-enantiomers. The R-enantiomer exhibits dual presynaptic inhibition of serotonin and norepinephrine reuptake, while the S-enantiomer is predominantly a serotonin reuptake inhibitor. In vitro, venlafaxine is >200 fold more potent in binding SERT than NET (Table 3), and approximately 3- to 5-fold more potent at inhibiting serotonin than norepinephrine reuptake. The mechanism of the antidepressant action of venlafaxine in humans is believed to be associated with its potentiation of neurotransmitter activity in the CNS. Venlafaxine and its active metabolite, O-desmethylvenlafaxine (ODV), are potent inhibitors of neuronal serotonin and norepinephrine reuptake. Venlafaxine and ODV have no significant affinity for muscarinic, histaminergic, or α1 adrenergic receptors. The phenylethyl amine-like structure feature of ven-

Table 3 Dual reuptake inhibitor binding affinities to various receptors, K_i (nM)

Compound	SERT	NET	DAT	α1 adrenergic	α2 adrenergic	Cholinergic muscarinic	H1 histamine
Duloxetine	0.8	7.5	240	8300	8600	3000	2300
Venlafaxine	82	2480	7647	>10000	>10000	>10000	>10000
Milnacipran	123	200	>10000	>10000	>10000	>10000	>10000
Bupropion	>10000	1400	570	>10000	>10000	>10000	>10000

Data are adapted from PDSP K_i database [10]

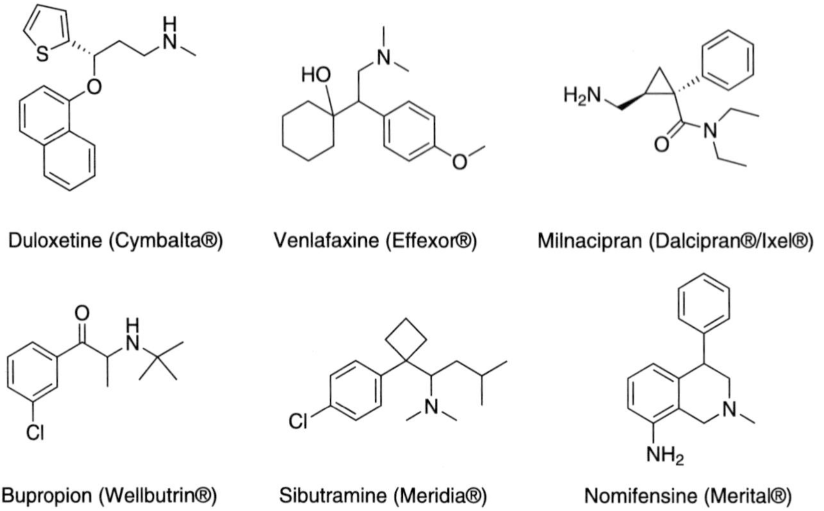

Duloxetine (Cymbalta®)   Venlafaxine (Effexor®)   Milnacipran (Dalcipran®/Ixel®)

Bupropion (Wellbutrin®)   Sibutramine (Meridia®)   Nomifensine (Merital®)

Fig. 3 Structures of dual reuptake inhibitors

lafaxine is also present in other NE or 5-HT uptake inhibitors such as sibutramine [25], and bicifadine. It is interesting to note that sibutramine has been marketed as an antiobesity agent for some time, although it is a dual serotonin/norepinephrine uptake inhibitor.

Duloxetine was developed by Lilly, which also introduced fluoxetine (ProzacR). The structural similarity between these two compounds is apparent. Like venlafaxine, duloxetine is a more potent inhibitor of neuronal serotonin compared to norepinephrine uptake. Duloxetine does not interact with catecholamine, acetylcholine, histamine, opioid, glutamate, or GABA receptors. The antidepressant and analgesic actions of duloxetine are believed to be related to its potentiation of serotonergic and noradrenergic activity in the CNS.

Milnacipran was identified in the early 1980s by scientists from Pierre Fabre studying the pharmacologic activities of bifunctional cyclopropane derivatives. Milnacipran binds to SERT better than NET (Table 3), but inhibits norepinephrine and serotonin reuptake in a 3 : 1 ratio which, practically speaking, is a balanced action upon both transporters. This compound has been marketed in Europe for depression. Recently, Cypress Pharmaceutical announced positive results with milnacipran in reducing the symptoms of fibromyalgia, a chronic condition characterized by pain and tenderness in muscles and joints. Milnacipran exerts no significant actions on postynaptic H1, α1, D1, D2, muscarinic receptors or benzodiazepin/opiate binding sites.

Besides SNRIs, a norepinephrine and dopamine reuptake inhibitor (NDRI, Table 3), bupropion (Fig. 3), is approved for the treatment of depression. Bupropion is an aminoketone, chemically unrelated to TCAs or SSRIs. It is similar in structure to the stimulant cathinone, and to phenethylamines in general. Although the actual mechanism of bupropion's action is not clear, the antidepressant effect is considered to be mediated by its enhancement of dopaminergic and noradrenergic activity. Bupropion exhibits a moderate anticholinergic effect, as it acts as a competitive α3β4 nicotinic antagonist, which may explain its observed efficacy in interrupting nicotine addiction. Mazindol (Fig. 10) is also a NDRI that is available for some time, but not approved for depression. Nomifensine is a NDRI that is effective as an antidepressant, but was not developed because of toxicity.

3
Design, Synthesis and Structure–Activity Relationship of Triple-Uptake Inhibitors

A triple-uptake inhibitor adds the element of dopamine transporter (DAT) blockade to a dual (i.e., serotonin and norepinephrine) uptake inhibitor (an SNRI). A recent comprehensive review compiled a body of pharmacotherapeutic evidence for treating depression by enhancing dopaminergic activity [26]. In addition, a recent randomized controlled clinical trial has demonstrated that augmentation of citalopram (a SSRI) with sustained-release bupropion (a NDRI) has certain advantages in patients who did not respond adequately to a sustained regimen of citalopram, including a greater reduction in the number and severity of symptoms as well as fewer side effects and adverse events [27]. However, treatment with multiple drugs may introduce pharmacokinetic confounds and compliance issues. The ideal drug would be a single molecule that inhibits the uptake of serotonin, norepinephrine, and dopamine.

A number of compounds with the ability to block the reuptake of 5-HT, NE and DA have been identified during the search for selective or mixed uptake inhibitors. While the clinical efficacy of such a "broad spectrum" antidepressant has not yet been fully demonstrated, several compounds have entered

Table 4 Triple uptake inhibitors in preclinical and/or clinical development

Company	Compounds	Clinical trial status
DOV Pharmaceutical, Inc.	DOV 216, 303	Phase II completed
DOV Pharmaceutical, Inc.	DOV 21,947	Phase II in progress 2008
DOV Pharmaceutical, Inc.	DOV 102,677	Phase I
DOV Pharmaceutical, Inc.	Multiple series	Preclinical
GSK/NeuroSearch	NS2359/GSK372475	Phase II
Sepracor	SEP-225289	Phase I
BMS/AMRI	AMR-CNS-1 & 2	Preclinical
Eli Lilly & Co	Multiple series	Preclinical
Acenta Discovery Inc.	One series	Preclinical
Mayo Foundation	PRC-025 and 050	Preclinical

clinical trials for depression and/or ADHD, including DOV 216 303 and 21 947 (DOV), NS 2359 (NeuroSearch/GSK), and SEP-225289 (Sepracor) (Table 4).

3.1
DOV 216 303, DOV 21 947 and DOV 102 677

DOV Pharmaceutical Inc. is developing a series of 3-azabicyclo[3.1.0]hexanes as novel antidepressants. DOV 216 303 [(+/−)-1-(3,4-dichlorophenyl)-3-azabicyclo[3.1.0]hexane hydrochloride, Scheme 1] is the prototype of a class of compounds referred to as "triple"-uptake inhibitors [28–30]. Such compounds inhibit the uptake of 5-HT, NE, and DA, the three neurotransmitters most closely linked to major depressive disorders. The synthesis of DOV 216 303 is illustrated in Scheme 1. The described method is an optimized process for manufacturing DOV 216 303 from the originally published route [31].

The α-bromo-3,4-dichlorophenylacetic acid methyl ester **2** (Scheme 1) is synthesized from 3,4-dichlorophenylacetonitrile by reaction with sodium bromate. The crude α-bromo-3,4-dichlorophenylacetic acid methyl ester then reacts with methyl acrylate to afford the diester **3**, which is converted to 1-(3,4-dichlorophenyl)-1,2-cyclopropanedicarboxylic acid dipotassium salt **4**. The dipotassium salt reacts with urea to provide 1-(3,4-dichlorophenyl)-3-azabicyclo[3.1.0]hexane-2,4-dione **5**, which is reduced by BH_3–THF complex to afford DOV 216 303. Resolution of 216 303 by L-(−)-dibenzoyl tartaric acid gives the (+)-enantiomer DOV 21 947. Resolution of 216 303 by D-(+)-dibenzoyl tartaric acid provides the (−)-enantiomer DOV 102 677.

It is interesting to find out that racemic DOV 216 303 and its two enantiomers all show balanced inhibition of the uptake of 5-HT, NE, and DA (Table 5). DOV 216 303 inhibits [³H]5-HT, [³H]NE, and [³H]DA uptake of the corresponding human recombinant transporters (expressed in HEK 293 cells) with IC_{50} values of approximately 14, 20, and 78 nM, respectively

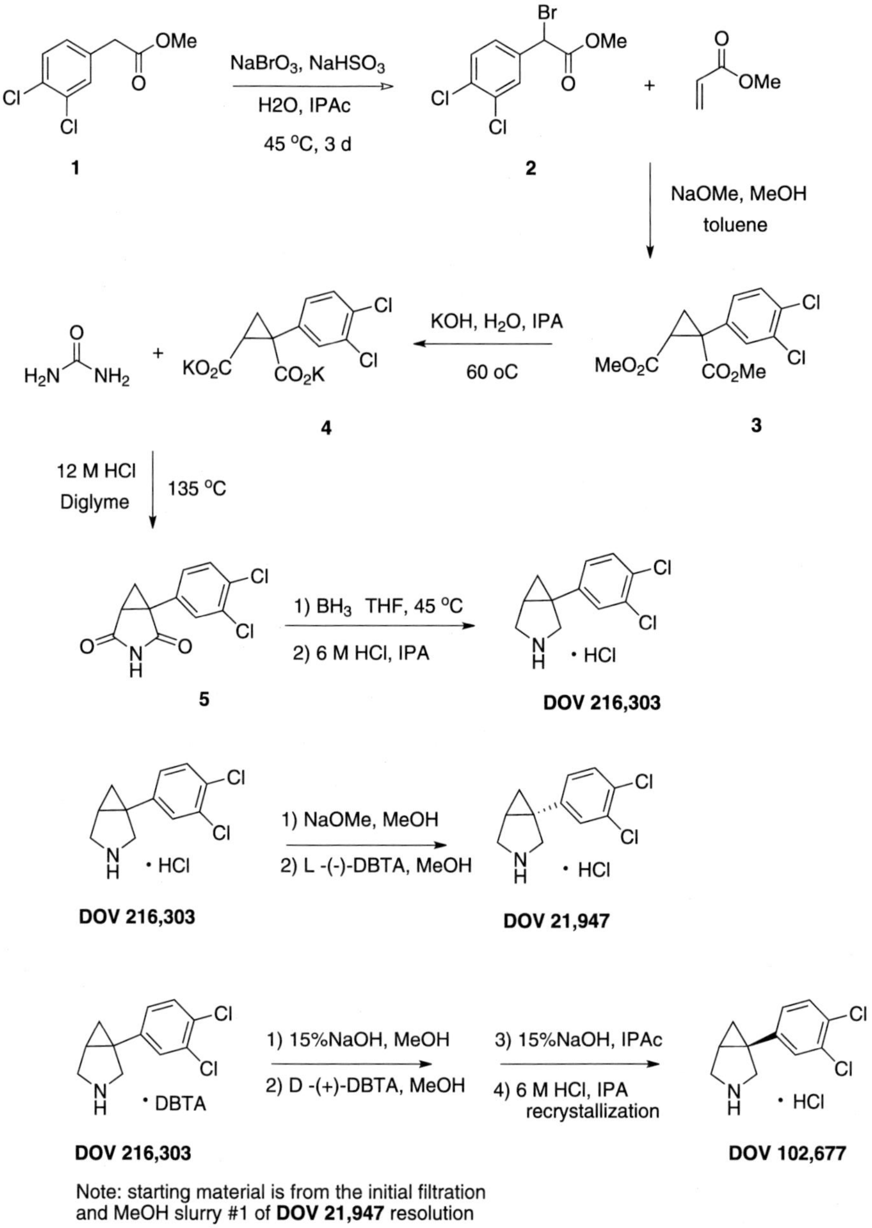

Scheme 1

Note: starting material is from the initial filtration and MeOH slurry #1 of **DOV 21,947** resolution

(Table 5) [32–34]. DOV 216 303 is active in tests predictive of antidepressant activity including the mouse forced swim test and reversal of tetrabenazine-induced ptosis and locomotor depression.

Table 5 In vitro binding and uptake inhibition of DOV compounds

Compound	Inhibition of neurotransmitter uptake, (IC_{50}, nM)			Inhibition of radioligand binding, (K_i, nM)		
	$[^3H]$5-HT	$[^3H]$NE	$[^3H]$DA	SERT	NET	DAT
DOV 216 303	14 ± 1.5	20 ± 6.1	78 ± 15	190 ± 28	380 ± 43	190 ± 40
DOV 21 947	12 ± 2.8	23 ± 3.3	96 ± 20	100 ± 16	260 ± 41	210 ± 56
DOV 102 677	130 ± 26	100 ± 27	130 ± 15	740 ± 140	1000 ± 76	220 ± 43

A phase II trial of DOV 216 303 examined the safety, tolerability and efficacy of DOV 216 303 in severely depressed patients using the SSRI, citalopram as an active control. The trial was of 2 weeks duration and the HAMD (Hamilton Depression Rating Scale) was the primary outcome measure, the Beck Depression Rating Scale and the Zung Self-Rating Depression Scale were the secondary measures. Results from this study indicated that DOV 216 303 is as efficacious as citalopram. Patients treated with either citalopram or DOV 216 303 showed greater than 40% reductions from baseline HAMD scores [29].

DOV 21 947, the (+)-enantiomer, inhibits the uptake of $[^3H]$5-HT, $[^3H]$NE, and $[^3H]$DA in human embryonic kidney (HEK) 293 cells expressing the corresponding human recombinant transporters with IC_{50} values of 12, 23, and 96 nM, respectively. DOV 21 947 also inhibits $[^{125}I]$RTI-55 (3β-(4-iodophenyl)tropane-2β-carboxylic acid methyl ester) binding to the corresponding transporter proteins in membranes prepared from these cells (K_i values of 100, 260, and 210 nM, respectively). DOV 21 947 reduces the duration of immobility in the forced swim test in rats with an oral minimum effective dose (MED) of 5 mg/kg. This antidepressant-like effect manifests in the absence of significant increases in motor activity at doses of up to 20 mg/kg. DOV 21 947 also produces a dose-dependent reduction in immobility in the tail suspension test, with an oral MED of 5 mg/kg. The ability of DOV 21 947 to inhibit the uptake of three biogenic amines closely linked to the etiology of depression may result in a therapeutic profile different from antidepressants that inhibit the uptake of 5-HT and/or NE only [34]. Seven phase I studies of DOV 21 947 have been completed and we intend to initiate a phase II double-blind clinical trial of DOV 21 947 versus placebo in depressed outpatients and inpatients in December of 2007. In addition, at doses similar to those active in models predictive of antidepressant action, DOV 21 947 produced a significant weight loss in two animal models of diet-induced obesity. Rodent models of diet-induced obesity are often used to predict the effectiveness of drugs to produce weight loss in obese individuals.

DOV 102 677 blocks $[^3H]$5-HT, $[^3H]$NE, and $[^3H]$DA uptake in recombinant human transporters with IC_{50} values of 130, 100, and 130 nM, respectively. Radioligand binding to the DA, NE and 5-HT transporters is inhibited with k_i values of 740, 1000, and 220 nM, respectively [35]. DOV 102 677 dose-dependently reduced rat immobility in the forced swim test with an oral MED 20 mg/kg and a maximal efficacy comparable to imipramine. This decrease in immobility time did not appear to result from increased motor activity. Further, without notably altering motor activity, DOV 102 677 was as effective as methylphenidate in reducing the amplitude of the startle response in juvenile mice. DOV 102 677 also potently blocked volitional consumption of alcohol and reduced operant responding for alcohol [36].

DOV scientists and their collaborators have developed efficient, asymmetric synthetic routes for both DOV 102 677 and 21 947 (Scheme 2) [37]. This economical and robust new route uses commercially available 3,4-dichlorophenylacetonitrile and (R)-epichlorohydrin [or (S)-epichlorohydrin] as starting materials to synthesize DOV 102 677 and 21 947, respectively, in 3 steps. Reacting 3,4-dichloro-phenylacetonitrile and (R)-epichlorohydrin [or (S)-epichlorohydrin] in the presence of a base will give the cyclopropyl compounds. The nitrile group is then reduced into the amino alcohol compounds, which are cyclized, with $SOCl_2$ under acidic conditions and subsequently neutralized with base, into DOV 102 677 and 21 947 respectively.

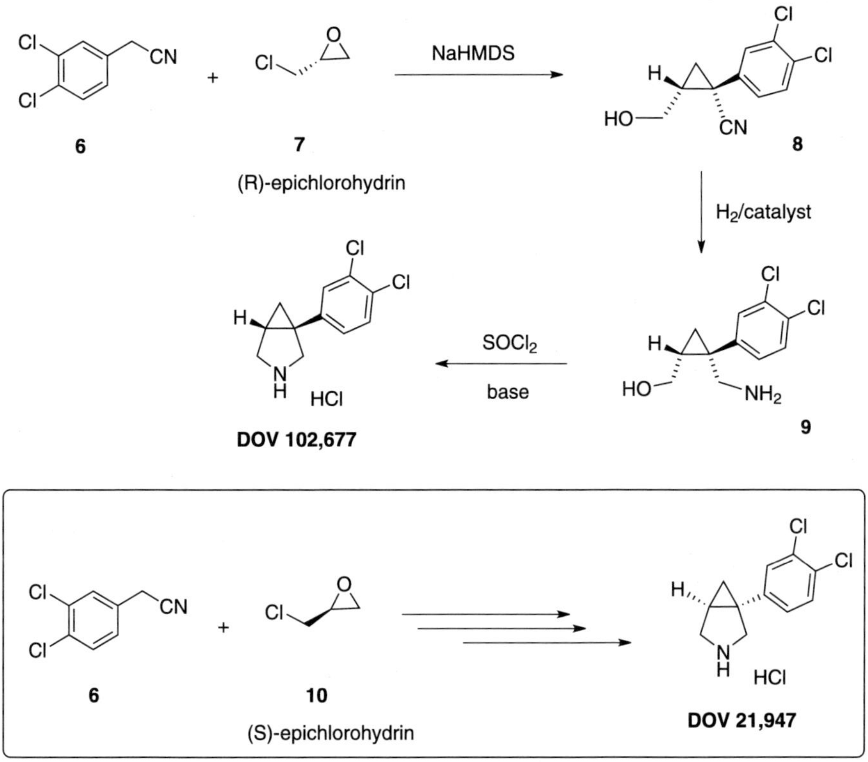

Scheme 2

3.2
PRC-025 and PRC-050

The Mayo Foundation and the Virginia Polytechnic Institute and State University are investigating analogues of venlafaxine as potential antidepressants. The two lead compounds, PRC-025 ((2RS,3RS)-N,N-dimethyl-3-cyclohexyl-3-hydroxy-2-(2'-naphthyl)propylamine) and PRC-050 ((2RS,3RS)-N-methyl-3-cyclohexyl-3-hydroxy-2-(2'-naphthyl)-3-phenylpropylamine), are racemic compounds (Scheme 3) [38, 39]. The similarity among the PRC series, venlafaxine and duloxetine is obvious. Indeed, venlafaxine is the prototype compound for the PRC series as described in the publication [38]. This series of compounds are synthesized from the commercially available arylacetonitriles by a 3-step sequence (Scheme 3). First, Aldol reaction with an appropriate aldehyde provides nitriles 12, followed by reduction of nitrile to afford the primary amines 13, and finally, reductive alkylation of 13 provides the target compounds 14. The following preclinical data on PRC-025 and

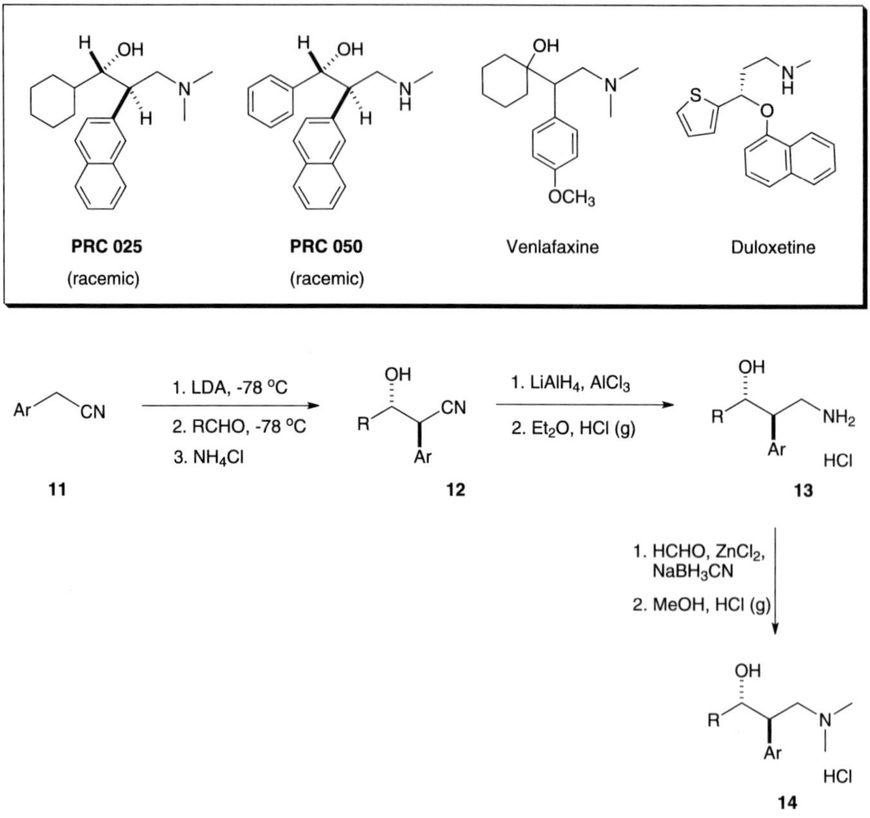

Scheme 3

Table 6 In vitro binding and uptake inhibition of Mayo compounds

Compound	Inhibition of neurotransmitter uptake, (K_i, nM)			(K_d's) for binding to human hSERT, hNET & hDAT (k_d, nM)		
	[^{3}H]5-HT	[^{3}H]NE	[^{3}H]DA	SERT	NET	DAT
PRC-025	6.0 ± 0.8	10 ± 0.5	53 ± 1	6.0 ± 0.8	19 ± 2	100 ± 10
PRC-050	12 ± 2	1.2 ± 0.1	43 ± 7	6.0 ± 0.3	0.40 ± 0.05	120 ± 10
Venlafaxine	39 ± 3	210 ± 20	5300 ± 600	9.0 ± 0.3	1060 ± 40	9300 ± 50

Data are adapted from [39]

PRC-050 were recently reported [39]. The K_i values for inhibition of [^{3}H]5-HT, [^{3}H]NE, and [^{3}H]DA uptake were 6, 10 and 53 nM for PRC-025, and 12, 1.2 and 43 nM for PRC-050, respectively (Table 6). These compounds were tested in animal models predictive of AD activity. In the forced swim test with male Sprague Dawley rats, both PRC-025 and PRC-050 (5 and 10 mg/kg, i.p.) reduced the time spent immobile and increased the time spent swimming, comparable to the effects seen with imipramine (15 mg/kg). In addition, both PRC-025 and PRC-050 were effective in reducing the time spent immobile in the tail suspension test (i.p.), again with effects comparable to imipramine. Studies using the enantiomers of these compounds are reported to be underway.

3.3
Eli Lilly's Triple-Uptake Inhibitors

Martin et al. have reported a series of substituted naphthyl containing chiral [2.2.1] bicycloheptanes as potent triple-uptake inhibitors (Scheme 4) [40]. The dienophiles **17** were prepared by N-acylation of oxazolidinones **16**. Treatment of achiral **17** (R = H, Ar = 3-ClC$_6$H$_4$) with cyclopentadiene at low temperature using Yb(OTf)$_3$ as Lewis acid gave racemic cycloadducts **18** (endo/exo ratio = 4/1). The endo/exo isomers were separated by preparative HPLC or by recrystallization. Hydrogenation of the racemic endo isomer, followed by basic hydrolysis gave the acid derivative **20**, which was then converted to either the dimethyl amino (**21**) or mono methyl amino (**22**) derivatives in two further steps. The racemic endo isomer was separated into its enantiomers by chiral HPLC and then processed through to final materials for test. Typical overall yields for this seven-step sequence were 10–20%.

The N-methyl 2-naphthyl analogue (**26**) and its desmethyl analogue (**28**) (Fig. 4) are active triple-re-uptake inhibitors both in vivo and in vitro [40]. The 2S,3S diastereomer showed the best balance of the desired activities with IC$_{50}$ values (Table 7) of 9, 86 and 28 nM as a reuptake inhibitor of 5-HT, DA

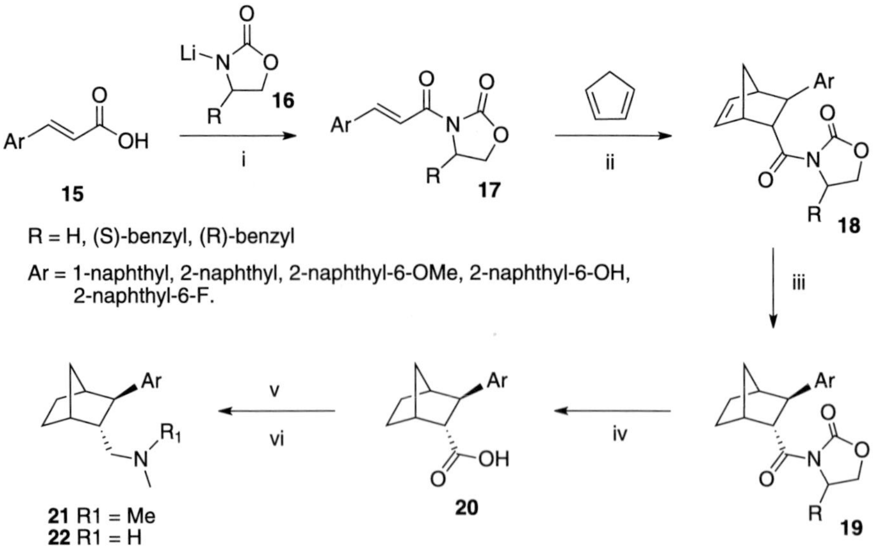

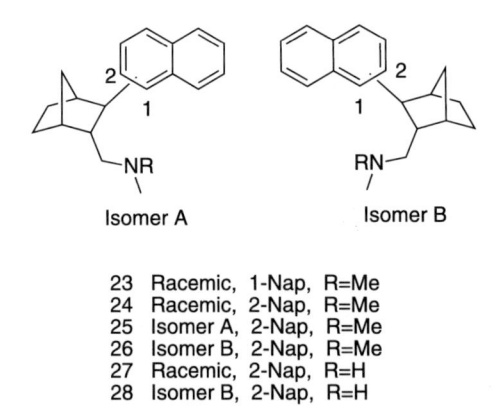

R = H, (S)-benzyl, (R)-benzyl

Ar = 1-naphthyl, 2-naphthyl, 2-naphthyl-6-OMe, 2-naphthyl-6-OH, 2-naphthyl-6-F.

21 R1 = Me
22 R1 = H

Reagents: (i) pivaloyl chloride, Et$_3$N, THF, -78 °C (40–60%); (ii) *[cat], DCM, (80–90%); (iii) H$_2$, Pd/C, EtOAc (80–90%); (iv) LiOH, H$_2$O$_2$ (80–90%); (v) oxalyl chloride, Me$_2$NH or MeNH$_2$ (60–70%); (vi) LiAlH$_4$, Et$_2$O(80–90%).

Scheme 4

Isomer A   Isomer B

23	Racemic, 1-Nap,	R=Me
24	Racemic, 2-Nap,	R=Me
25	Isomer A, 2-Nap,	R=Me
26	Isomer B, 2-Nap,	R=Me
27	Racemic, 2-Nap,	R=H
28	Isomer B, 2-Nap,	R=H

Fig. 4 Structures of compounds in Table 7

and NE, respectively, in rat synaptosomes. It also exhibited oral activity in rat models of depression at oral doses of 20 to 25 mg/kg; the 1-naphthyl analog was much less active, but the 2-naphthyl derivative had only 2 to 3% oral bioavailability due to ring hydroxylation. Exploration of the activity of other bicyclic aromatics led to 5- or 6-substituted benzothiophenes, which provided enhanced potency and improved bioavailability, especially with the

Table 7 Uptake inhibition data reported of Lilly compounds [40]

Compd.	Synaptosomal uptake IC_{50} (nM)		
	5-HT	NE	DA
23	93	297	1514
24	10	45	95
25	20	34	93
26	9	25	76
27	50	45	87
28	9	28	86

use of an N-methylamino substituent. The dimethyl amino derivative of the benzothiophen-5-yl analog has progressed into further development, displaying oral activity of 2.5, 5, and 10 mg/kg in models of 5-HT, DA and NE activity, respectively. However, this compound encountered unspecified problems later in development and was discontinued.

Lilly has also reported the synthesis and discovery of 3-amino piperidine and pyrolidine-based inhibitors of neurotransmitter re-uptake transporters in a 2005 ACS meeting [41]. No further preclinical data on these compounds has been published.

3.4
NS-2359/GSK-372475, SEP-225289, AMRI CNS-1 & 2, and Acenta Series

NeuroSearch has developed potent, tropane-based uptake inhibitors [42–44]. Some of the analogs are shown in Fig. 5 and their inhibition data are shown in Table 8. NS-2359 (GSK-372475) is another triple-uptake inhibitor that went into clinical development. This compound was discovered at NeuroSearch and was subsequently licensed to GlaxoSmithKline (GSK) as part of a 5-year R&D alliance in CNS diseases. Although the structure of NS-2359 (GSK-372475) is not yet published, it is believed that NS-2359 is a tropane analog (Fig. 5). Phase I trial results of NS-2359 are reported in NeuroSearch's press release. In an imaging study involving six healthy volunteers receiving daily doses of NS-2359 (0.25–1.0 mg), SPECT showed very clear and specific binding in relevant areas of the brain and clear dose-dependency [45]. In a single-dose phase I trial in the UK, this compound is well tolerated [46]. In a clinical trial in 54 volunteers, NS-2359 increased attention and improved the ability to recall verbal information. Following these results, NeuroSearch proposed to focus on the development of NS-2359 for treating ADHD [47]. In 2006, NeuroSearch announced that GlaxoSmithKline (GSK) initiated a phase II clinical trial of its drug candidate NS2359 (GSK372475) in patients diagnosed with major depressive disorder. These studies will be conducted in multiple centers worldwide and involve several hundred patients.

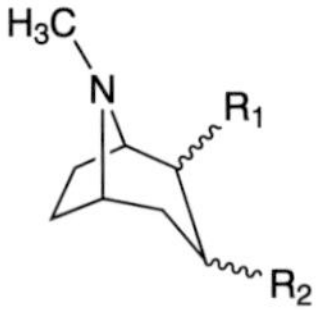

NeuroSearch Series

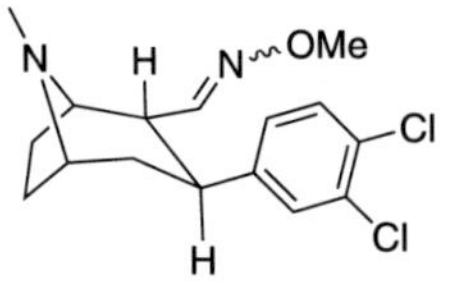

29

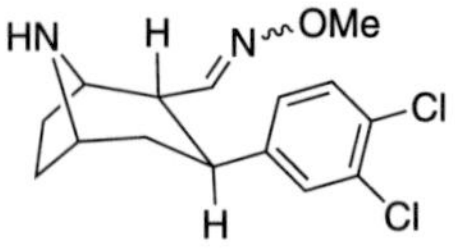

30

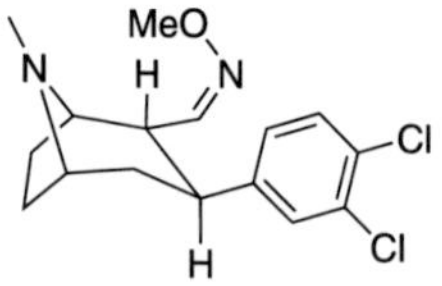

31

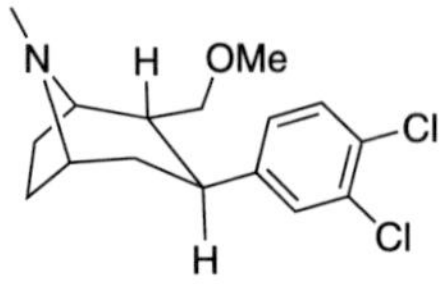

32

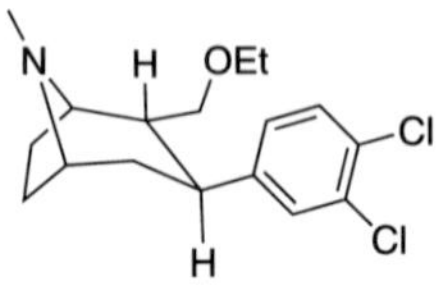

33

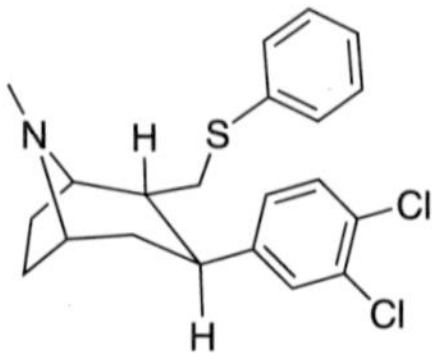

34

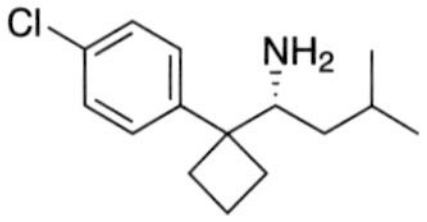

SEP-225289

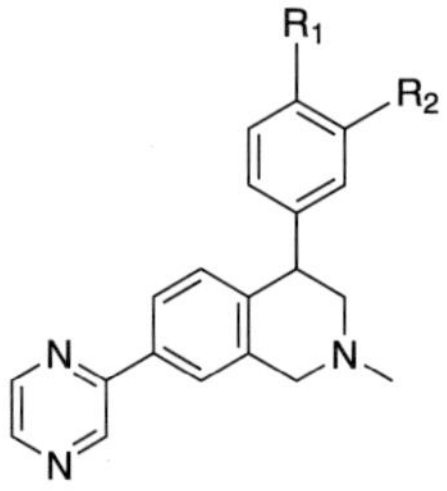

AMRI Series

Fig. 5 Structures of compounds in Tables 8 and 9

Table 8 Uptake inhibition data reported in NeuroSearch patents [45–47]

Compd.	[^{3}H]5-HT uptake IC_{50} (nM)	[^{3}H]NE uptake IC_{50} (nM)	[^{3}H]DA uptake IC_{50} (nM)
29	13	1.3	3
30	1.7	1.3	2
31	nt	1.5	3.4
32	10	2	10
33	11	3.2	8
34	9.2	2.8	4.3

SEP-225289 is another triple-uptake inhibitor that is under development by Sepracor for the treatment of refractory depression and for generalized anxiety disorder [48, 49]. In preclinical studies, SEP-225289 was believed to be a potent and balanced 5-HT, NE, and DA uptake inhibitor, but there is no published uptake inhibition data. SEP-225289 has been put in a randomized, single-blind, placebo-controlled phase I safety, tolerability and pharmacokinetic trial for the treatment of depression [50, 51]. The structure is believed to be as shown in Fig. 5, and is closely related to the active, desmethyl metabolite of sibutramine.

Albany Molecular Research Institute is developing biogenic amine transporter inhibitors for the treatment of a range of CNS disorders. Compounds with various combinations of amine transporter inhibition profiles acting selectively or in combination to increase brain levels of 5-HT, NE, or DA were investigated. Bristol-Myers Squibb has an exclusive worldwide license to develop and commercialize the compounds [52]. Preclinical data for compounds AMR-CNS-1 and 2 were disclosed in a recent presentation [53]. Both AMR-CNS-1 and 2 are believed to be novel 4-phenyl tetrahydroisoquinolines with nomifensine as its prototype compound (Fig. 5 and Table 9).

Acenta Discovery Inc. has synthesized a library of piperidine-based nocaine/modafinil hybrids, and some of which display an improved potency at all three monoamine transporters [54, 55]. Some interesting compounds, which are highly active at blocking multiple transporters, are listed in Fig. 6 and Table 10. Compound 39 is the most potent and balanced triple-uptake inhibitor reported to date. Some of these compounds were reported as more active in tail suspension tests than desipramine. It was reported at a re-

Table 9 In vitro binding and uptake inhibition of AMRI compounds [53]

Compound	Uptake inhibition, (K_i, nM) rat synaptosomes			Binding inhibition, (K_i, nM) rat transporters		
	[³H]5-HT	[³H]NE	[³H]DA	SERT	NET	DAT
AMR-CNS-1	14	12.7	22.4	1.3	8.2	21.6
AMR-CNS-2	24	12.6	24.6	3.5	10.4	31.7
Duloxetine	2.8	3.21	202	0.6	3.9	888

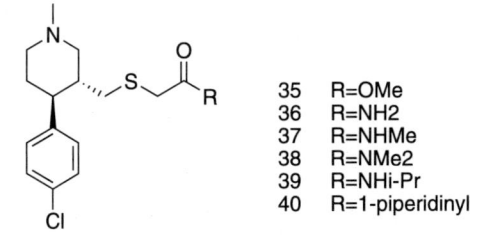

35	R=OMe
36	R=NH2
37	R=NHMe
38	R=NMe2
39	R=NHi-Pr
40	R=1-piperidinyl

Fig. 6 Structures of compounds in Table 10

Table 10 Uptake inhibition data reported in [54, 55]

Compd.	[^{3}H]5-HT uptake K_i (nM)	[^{3}H]NE uptake K_i (nM)	[^{3}H]DA uptake K_i (nM)
35	208 ± 47	25 ± 6	80 ± 23
36	557 ± 150	39 ± 5	159 ± 19
37	110 ± 45	25 ± 2	13 ± 3
38	88 ± 22	27 ± 7	116 ± 46
39	1.1 ± 0.4	0.8 ± 0.1	1.0 ± 0.2
40	4.5 ± 0.8	0.68 ± 0.25	83 ± 1

cent ACS meeting that more systematic in vivo assessments of the selected compounds from this library were on-going to identify potential drug candidates for advancement to pre-development characterization in preparation for a potential IND submission.

3.5
Other Reported Triple-Uptake Inhibitor Series

Dutta et al. have studied the substituted pyran derivatives by carrying out asymmetric synthesis and biological characterization of trisubstituted (2S,4R,5R)-2-benzhydryl-5-benzylaminotetrahydropyran-4-ol and (3S,4R,6S)-6-benzhydryl-4-benzylaminotetrahydropyran-3-ol derivatives and their enantiomers [56]. Biological results indicated that regioselectivity and stereoselectivity played important roles in determining activity for monoamine transporters because only (–)-isomers of 2-benzhydryl-5-benzylaminotetrahydropyran-4-ol derivatives exhibited appreciable potency for the monoamine transporters, in particular for the SERT and NET. Further exploration involved the incorporation of functional groups into the molecular template to promote H bond formation with the transporters. In addition, the des-hydroxyl analogs, disubstituted cis-(6-benzhydryl-tetrahydropyran-3-yl)-benzylamines, were synthesized by a new asymmetric synthesis scheme [57]. The results indicated that the presence of functional groups, such as – OH, – NH$_2$, and the bioisosteric 5-substituted indole moiety in both di- and trisubstituted compounds, significantly increased their potencies for the SERT, and especially for the NET. Among the trisubstituted compounds (Fig. 7 and Table 11), (–)-42 exhibited the highest potency for the NET and the SERT. Compound (–)-41 exhibited the highest selectivity for the NET. Among the disubstituted compounds, a number of compounds, such as (–)-43, (+)-44, (–)-44, and (+)-46, exhibited significant low-nanomolar potencies for the SERT and the NET. Interestingly, compounds 45 and 46 exhibited fairly balanced potencies towards all three transporters, with (–)-46 displaying high potency.

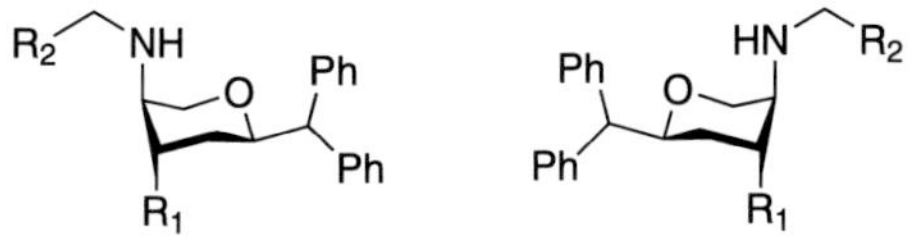

(+) isomer   (-) isomer

41	R1=OH, R2=4-hydroxyphenyl
42	R1=OH, R2=indol-5-yl
43	R1=H, R2=4-hydroxyphenyl
44	R1=H, R2=indol-5-yl
45	R1=H, R2=4-nitrophenyl
46	R1=H, R2=4-aminophenyl

Fig. 7 Structures of compounds in Table 11

Table 11 Uptake inhibition data reported in [56, 57]

Compd.	$[^3H]$5-HT uptake K_i (nM)	$[^3H]$NE uptake K_i (nM)	$[^3H]$DA uptake K_i (nM)
(+)-41	2540 ± 430	94.4 ± 18.5	91.5 ± 7.4
(-)-41	237 ± 12	10.4 ± 1.0	172 ± 20
(+)-42	223 ± 45	13.2 ± 4.0	184 ± 16
(-)-42	15.3 ± 1.3	2.13 ± 0.41	120 ± 8
(+)-43	187 ± 25	40.9 ± 6	67.4 ± 4
(-)-43	37.7 ± 2.6	5.09 ± 0.92	85 ± 5.9
(+)-44	11.8 ± 1.6	15.6 ± 2.3	142 ± 43
(-)-44	14.5 ± 2.7	6.56 ± 1.01	214 ± 11
(+)-45	64.6 ± 11.8	81.6 ± 9.7	59.9 ± 14.2
(-)-45	19.9 ± 1.5	54.3 ± 9.8	$34.6 \pm 6.5\ 8$
(+)-46	45.7 ± 17.7	25.6 ± 3.6	43.9 ± 5.2
(-)-46	16.1 ± 1.6	12.6 ± 3.7	62.4 ± 5.6

Rice and colleagues have reported the synthesis of a series of 3-(3,4-dichlorophenyl)-1-indanamine derivatives as nonselective ligands for biogenic amine transporters [58]. The in vitro data indicate that **47** and **48** displayed high-affinity binding and potent inhibition of uptake at all three biogenic amine transporters (Fig. 8 and Table 12). Particularly, (–)-(**1R,3S**)-**48** was comparable with indatraline in potency. In vivo microdialysis experiments demonstrated that intravenous administration of (–)-(**1R,3S**)-**48** to rats elevated extracellular DA and 5-HT in the nucleus accumbens in a dose-dependent manner [58]. Pretreating rats with 0.5 mg/kg (–)-(**1R,3S**)-**47** elevated extracellular DA and 5-HT by approximately 150% and reduced methamphetamine-induced neurotransmitter release by about 50%. Ex vivo autoradiography, however, demonstrated that iv administration of (–)-(**1R,3S**)-**48** resulted in a dose-dependent, persistent occupation of only 5-HT transporter binding sites.

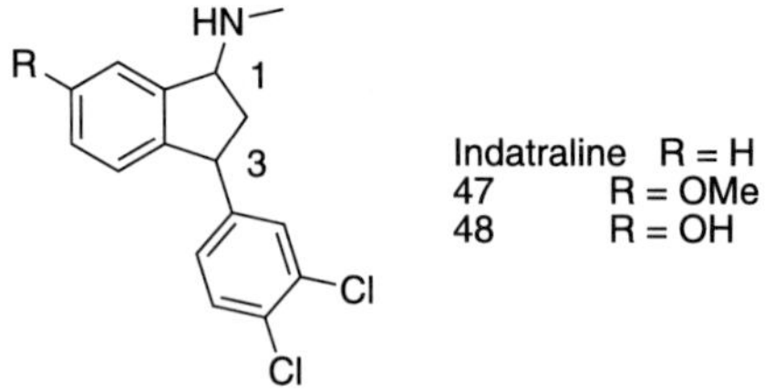

Fig. 8 Structures of compounds in Table 12

Table 12 Uptake inhibition data reported in [58]

Compd.	[³H]5-HT uptake K_i (nM)	[³H]NE uptake K_i (nM)	[³H]DA uptake K_i (nM)
Indatraline	3 ± 0.16	11 ± 1.3	2 ± 0.10
(+)-47-(1R,3S)	19 ± 1.9	34 ± 1.8	12 ± 0.3
(−)-47-(1S,3R)	109 ± 4	650 ± 33	600 ± 23
(−)-48-(1R,3S)	2 ± 0.1	20 ± 2	3 ± 0.1
(+)-48-(1S,3R)	41 ± 1	160 ± 12	32 ± 1

Wang and colleagues have reported a novel class of monoamine transport inhibitors, 3,4-disubstituted pyrrolidines, identified through 3-D database pharmacophore searching using a new pharmacophore model [59]. This class

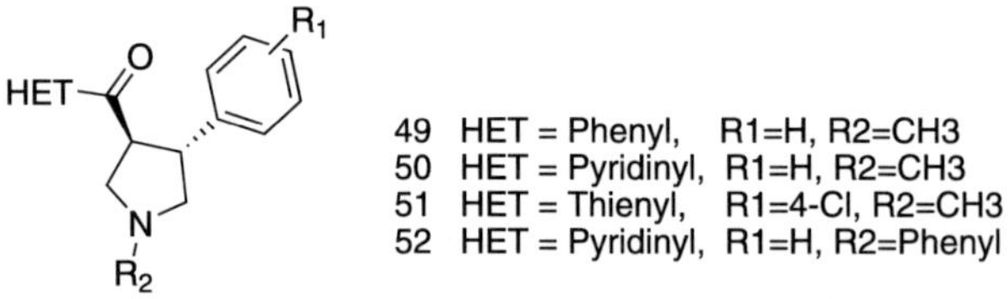

Fig. 9 Structures of compounds in Table 13

Table 13 Uptake inhibition data reported in [59]

Compd.	5-HT uptake K_i (µM)	NET uptake K_i (µM)	DAT uptake K_i (µM)
Cocaine	0.16 ± 0.01	0.19 ± 0.01	0.27 ± 0.02
49	0.75 ± 0.06	0.15 ± 0.02	1.41 ± 0.09
50	0.83 ± 0.05	0.044 ± 0.001	0.63 ± 0.02
51	0.23 ± 0.01	0.031 ± 0.004	0.20 ± 0.01
52	> 10	> 10	> 10

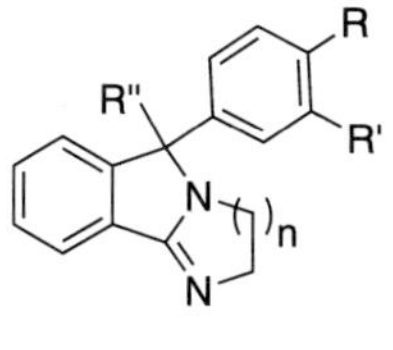

Mazindol	n=1, R=Cl, R'=H, R"=OH
53	n=1, R=Cl, R'=H, R"=H
54	n=1, R=H, R'=H, R"=OH
55	n=1, R,R'=-C4H4-, R"=OH
56	n=2, R=Cl, R'=H, R"=OH
57	n=2, R=OH, R'=H, R"=OH
58	n=2, R=OMe, R'=H, R"=OH
59	n=2, R,R'=-OCH2O-, R"=OH

Fig. 10 Structures of compounds in Table 14

Table 14 Uptake Inhibition data reported in [60, 61]

Compd.	Uptake IC_{50} (nM)		
	hSERT	hNET	hDAT
Mazindol	94 ± 32	4.9 ± 0.5	43 ± 20
53	$15 \pm 5\,6$	6.9 ± 1.5	6.0 ± 0.7
54	2140 ± 450	2.8 ± 0.92	730 ± 180
55	1.8 ± 1.3	4.5 ± 1.5	66 ± 10
56	53 ± 7	4.9 ± 0.5	3.7 ± 0.4
57	60 ± 19	1.9 ± 0.15	59.0 ± 3.6
58	94 ± 34	4.1 ± 1.4	30.4 ± 2.4
59	83 ± 29	0.62 ± 0.25	2.21 ± 0.3

of inhibitors has a selectivity profile different from that of cocaine at the DAT, SERT, and NET (Fig. 9 and Table 13). Among the structures shown, **51** was the most potent analog with K_i values of 0.084 μM against [3H]mazindol binding, and 0.20, 0.23, and 0.031 μM for inhibition of DA, 5-HT, and NE uptake, respectively.

Houlihan et al. have reported on a series of mazindol and homomazindol (**56**) analogues in an effort to develop a selective inhibitor of cocaine binding [60, 61]. This effort identified several fairly balanced triple-uptake inhibitors, e.g., **53**, during the process, shown in Fig. 10 and Table 14.

4
Conclusions

The key approved indications of blockbuster antidepressants often include generalized anxiety disorder (GAD), major depressive disorder (MDD), obsessive-compulsive disorder (OCD), panic disorder (PD), and social anxiety disorder (SAD). In addition, a majority of primary care physicians and psychiatrists also prescribe antidepressants off-label for treatment across a broad range of clinical conditions outside the five key approved indications. It is believed that treatment of neuropathic pain disorders is among the most common off-label uses of antidepressants. Emerging antidepressants in the current pipeline will eventually find success in the antidepressant and related markets. For example, duloxetine, a SNRI, was approved by the FDA in August 2004 for the treatment of major depressive disorder and in September 2004 for the treatment of diabetic peripheral neuropathic pain. On Feb. 27, 2007, duloxetine was also approved for generalized anxiety disorder. Other antidepressants approved for generalized anxiety disorder include Forest Labs' escitalopram, Wyeth's venlafaxine, and GlaxoSmithKline's paroxetine.

Triple-uptake inhibitors or mixed serotonin, norepinephrine, and dopamine uptake inhibitors are expected to be the next generation of drugs for the treatment of major depression. Both preclinical and clinical data indicate this class of compounds will offer clinically significant advantages in efficacy and/or tolerability compared to single and dual uptake inhibitors. Other uses for triple-uptake inhibitors, ranging from treatment of attention deficit hyperactivity disorder (ADHD), smoking cessation, alcohol abuse, to obesity are also under active investigations by the pharmaceutical industry.

References

1. Kuhn R (1958) Am J Psychiatry 115:459
2. Glowinski J, Axelrod J (1964) Nature 204:1318
3. Schildkraut JJ (1965) Am J Psychiatry 122:509
4. Carlsson A, Fuxe K, Ungerstedt U (1968) J Pharm Pharmacol 20:150
5. Blakely RD, DeFelice LJ, Galli A (2005) Physiology 20:225
6. Berten O, Nestler EJ (2006) Nat Rev Neurosci 7:137
7. Schechter LE, Ring RH, Beyer CE, Hughes ZA, Khawaja X, Malberg JE, Rosenzweig-Lipson S (2005) NeuroRx 2:590
8. Schubert-Zsilavecz M, Stark H (2004) Pharm Unserer Zeit 33:282
9. Leonard BE (ed) (2001) Antidepressants. Birkhaeuser Verlag, Basel
10. Roth BL, Kroeze WK, Patel S, Lopez E (2000) Neuroscientist 6:252
11. Carlsson A (1987) Ann Rev Neurosci 10:19
12. Lemberger L, Rowe H, Carmichael R, Oldham S, Horng JS, Bymaster FP, Wong DT (1978) Science 199:436
13. Wong DT, Perry KW, Bymaster FP (2005) Nat Rev Drug Discov 4:764
14. Ban TA (2001) J Neural Transm 108:707
15. Bymaster FP, McNamara RK, Tran PV (2003) Expert Opin Investig Drugs 12:531
16. Holtzheimer PE III, Nemeroff CB (2006) Expert Opin Pharmacother 7:2323
17. Danish University Antidepressant Group (1986) Psychopharmacology 90:131
18. Danish University Antidepressant Group (1990) J Affective Disord 18:289
19. Wong DT, Bymaster FP (2002) Prog Drug Res 58:169
20. Stahl SM, Grady MM, Moret C, Briley M (2005) CNS Spectr 10:732
21. Lopez-Ibor J, Guelfi JD, Pletan Y, Tournoux A, Prost JF (1996) Int Clin Psychopharmacol 11(4):41
22. Vaishnavi SN, Nemeroff CB, Plott SJ, Rao SG, Kranzler J, Owens MJ (2004) Biol Psych 55:320
23. Eckert L, Lancon C (2006) BMC Psychiatry 6:30
24. Richelson E (2003) J Clin Psychiatry 64(13):5

25. Jackson HC, Bearham MC, Hutchins LJ, Mazurkiewicz SE, Needham AM, Heal DJ (1997) Br J Pharmacol 121:1613
26. Papakostas GI (2006) Eur Neuropsychopharmacol 16:391
27. Trivedi MH, Fava M, Wisniewski SR, Thase ME, Quitkin F, Warden D, Ritz L, Nierenberg AA, Lebowitz BD, Biggs MM, Luther JF, Shores-Wilson K, Rush AJ (2006) N Engl J Med 354:1243
28. Skolnick P, Krieter P, Tizzano J, Basile A, Popik P, Czobor P, Lippa A (2006) CNS Drug Rev 12:123
29. Beer B, Stark J, Krieter P, Czobor P, Beer G, Lippa A, Skolnick P (2004) J Clin Pharmacol 44:1360
30. Skolnick P, Popik P, Janowsky A, Beer B, Lippa AS (2003) Eur J Pharmacol 461:99
31. Epstein JW, Brabander HJ, Fanshawe WJ, Hofmann CM, McKenzie TC, Safir SR, Osterberg AC, Cosulich DB, Lovell FM (1981) J Med Chem 24:481
32. Skolnick P, Basile AS (2006) Drug Discov Today 3:489
33. Skolnick P, Basile AS (2007) CNS Neurol Disord Drug Targets 6:141
34. Skolnick P, Popik P, Janowsky A, Beer, Lippa AS (2003) Life Sci 73:3175
35. Popik P, Krawczyk M, Golembiowska K, Nowak G, Janowsky A, Skolnick P, Lippa A, Basile AS (2006) Cell Mol Neurobiol 26:857
36. McMillen BA, Shank JE, Williams HL, Basile AS (2007) Alcohol Clin Exp Res 31:in press
37. Skolnick P, Basile A, Chen Z (2006) Int Patent Appl WO 2006096810
38. Carlier PR, Lo MMC, Lo PCK, Richelson E, Tatsumi M, Reynolds IJ, Sharma TA (1998) Bioorg Med Chem Lett 8:487
39. Shaw AM, Boules M, Zhang Y, Williams K, Robinson J, Carlier PR, Richelson E (2007) Eur J Pharmacol 555:30
40. Axford L, Boot JR, Hotten TM, Keenan M, Martin FM, Milutinovic S, Moore NA, O'Neill MF, Pullar IA, Tupper DE, Van Belle KR, Vivien V (2003) Bioorg Med Chem Lett 13:3277
41. Cases M, Masters JJ, Haughton L, Campbell G, Mann T, Rudyk H, Walter MW, Timms G, Gilmore J, Dobson DR, White C, Boot JR, Findlay JD, Hayhurst L, Kluge AH (2005) 230th ACS Meeting. Washington, DC
42. Moldt P, Wätjen F, Scheel-Krüger J (1998) US Patent 5 736 556
43. Scheel-Krüger J, Moldt P, Wätjen F (2001) US Patent 6 288 079
44. Scheel-Krüger J, Moldt P, Wätjen F (2002) US Patent 6 395 748
45. NeuroSearch (2003) Jul 3 press release
46. NeuroSearch (1999) Ann Rep
47. NeuroSearch (2002) Ann Rep
48. JP Morgan 22nd Ann Healthcare Conference (2004) San Francisco, CA
49. Sepracor (2004) 3rd Quarter Release
50. Sepracor (2005) Apr 25 press releases
51. Sepracor (2005) Oct 19 press releases
52. AMRI (2005) Oct 24 press release
53. AMRI Drug Discovery Symposium (2006) Albany, NY
54. Zhou J, He R, Johnson KM, Ye Y, Kozikowski AP (2004) J Med Chem 47:5821
55. Zhou J, Johnson KM, Giberson KM, McGonigle P, Caldarone BJ, Kozikowski AP (2006) 232nd ACS national meeting. San Francisco, CA
56. Zhang S, Zhen J, Reith MEA, Dutta AK (2005) J Med Chem 48:4962
57. Zhang S, Fernandez F, Hazeldine S, Deschamps J, Zhen J, Reith MEA, Dutta AK (2006) J Med Chem 49:4239

58. Yu H, Kim IJ, Folk JE, Tian X, Rothman RB, Baumann MH, Dersch CM, Flippen-Anderson JL, Parrish D, Jacobson AE, Rice KC (2004) J Med Chem 47:2624
59. Enyedy IJ, Zaman WA, Sakamuri S, Kozikowski AP, Johnsonc KM, Wang S (2001) Bioorg Med Chem Lett 11:1113
60. Houlihan WJ, Kelly L, Pankuch J, Koletar J, Brand L, Janowsky A, Kopajtic TA (2002) J Med Chem 45:4097
61. Houlihan WJ, Ahmad UF, Koletar J, Kelly L, Brand L, Kopajtic TA (2002) J Med Chem 45:4110

Top Med Chem (2009) 4: 155–186
DOI 10.1007/7355_2008_025
© Springer-Verlag Berlin Heidelberg
Published online: 16 October 2008

Molecular Imaging of Transporters with Positron Emission Tomography

Gunnar Antoni[1] (✉) · Jens Sörensen[1] · Håkan Hall[2]

[1]Uppsala Imanet AB, GE Healthcare, P.O. Box 967, 75109 Uppsala, Sweden
Gunnar.antoni@ge.com

[2]Uppsala Applied Science Lab, GE Healthcare, P.O. Box 967, 75109 Uppsala, Sweden

Abstract Positron emission tomography (PET) visualization of brain components in vivo is a rapidly growing field. Molecular imaging with PET is also increasingly used in drug development, especially for the determination of drug receptor interaction for CNS-active drugs. This gives the opportunity to relate clinical efficacy to per cent receptor occupancy of a drug on a certain targeted receptor and to relate drug pharmacokinetics in plasma to interaction with target protein. In the present review we will focus on the study of transporters, such as the monoamine transporters, the P-glycoprotein (Pgp) transporter, the vesicular monoamine transporter type 2, and the glucose transporter using PET radioligands. Neurotransmitter transporters are presynaptically located and in vivo imaging using PET can therefore be used for the determination of the density of afferent neurons. Several promising PET ligands for the noradrenaline transporter (NET) have been labeled and evaluated in vivo including in man, but a really useful PET ligand for NET still remains to be identified. The most promising tracer to date is (S,S)-$[^{18}F]$FMeNER-D$_2$. The in vivo visualization of the dopamine transporter (DAT) may give clues in the evaluation of conditions related to dopamine, such as Parkinson's disease and drug abuse. The first PET radioligands based on cocaine were not selective, but more recently several selective tracers such as $[^{11}C]$PE2I have been characterized and shown to be suitable as PET radioligands. Although there are a large number of serotonin transporter inhibitors used today as SSRIs, it was not until very recently, when $[^{11}C]$McN5652 was synthesized, that this transporter was studied using PET. New candidates as PET radioligands for the SERT have subsequently been developed and $[^{11}C]$DASB and $[^{11}C]$MADAM and their analogues are today the most promising ligands. The existing radioligands for Pgp transporters seem to be suitable tools for the study of both peripheral and central drug–Pgp interactions, although $[^{11}C]$verapamil and $[^{18}F]$fluoropaclitaxel are probably restricted to use in studies of the blood–brain barrier. The vesicular monoamine transporter 2 (VMAT2) is another interesting target for diagnostic imaging and $[^{11}C]$DTBZ is a promising tracer. The noninvasive imaging of transporter density as a function of disease progression or availability following interaction with blocking drugs is highlighted, including the impact on both development of new therapies and the process of developing new drugs. Although CNS-related work focusing on psychiatric disorders is the main focus of this review, other applications of PET ligands, such as diagnosis of cancer, diabetes research, and drug interactions with efflux systems, are also discussed. The use of PET especially in terms of tracer development is briefly described. Finally, it can be concluded that there is an urgent need for new, selective radioligands for the study of the transporter systems in the human brain using PET.

Keywords DAT · NET · PET · Pgp · Positron emission tomography · SERT

Abbreviations

ABC	ATP binding cassette
ADHD	Attention-deficit hyperactivity disorder
BBB	Blood–brain barrier
B_{max}	Total number of binding sites
CNS	Central nervous system
DAT	Dopamine transporter
EAAT	Excitatory amino acid transporter
GAT	GABA transporter
GLUT-1	Glucose transporter type 1
GLYT	Glycine transporter
IC_{50}	Inhibiting concentration, 50% of total
K_D	Dissociation equilibrium constant
KO	Knock-out
MDR	Multidrug resistance
NET	Norepinephrine transporter
PET	Positron emission tomography
Pgp	P-Glycoprotein
SERT	Serotonin transporter
SPECT	Single photon emission tomography

SSRI   Selective serotonin reuptake inhibitor
V_D Volume of distribution
VMAT2   Vesicular monoamine transporter 2

1
Introduction

Neurotransmitter transporters exert an important function in synaptic transmission by removing the neurotransmitters, such as amino acids and biogenic amines, from the synapse and thereby terminating the signal transduction. There are a number of different types of the plasma membrane bound transporters, also called reuptake carriers, of which the sodium/chloride-dependent transporters for serotonin (SERT), dopamine (DAT), norepinephrine (NET), glycine (GLYT), and GABA (GAT), and the sodium-dependent glutamic acid transporters (excitatory amino acid transporters, EAATs) are the most well known. The monoamine transporters are pharmacologically related and have common structural motifs, and consequently neurotransmitters and drugs may interact with more than one type of transporter. Small structural changes in a molecule may thus lead to dramatic changes in selectivity and affinity towards transporters.

Other transporter systems acting both in the central nervous system (CNS) and in the periphery are, for example, the glucose transporters (GLUT-1 and GLUT-2), vesicular monoamine transporter 2 (VMAT2), and the multidrug resistance (MDR) transporters (ATP binding cassette transporters, ABC transporters), of which the P-glycoprotein (Pgp) transporter is of special interest from a drug development perspective. The VMAT2 and GLUT-1 are of interest as tools in a clinical environment for understanding disease at the molecular level.

The transporters are targets for many therapeutic drugs and some are also involved in drug-related abuse. Both the therapeutic and abuse effects are mostly due to the increase in the synaptic levels of the endogenous neurotransmitter, which is a result of the inhibited reuptake.

Visualizing the transporters in vivo using imaging is of great interest in a number of different applications. Neurotransmitter transporters are located presynaptically and imaging can be used for the determination of the density of afferent neurons. The monoamine transporters are, for example, markers for the monoaminergic innervation where the important neurotransmitters dopamine, serotonin, and norepinephrine play important roles in regulation of cognitive functions, movements, and affective states as well as in the pathophysiology of many neuropsychiatric diseases. This is a feature used in imaging of neurodegenerative disorders like Parkinson's disease, where a measure of the DATs in the striatum gives information on the progress of the disease.

Other possible applications include the study of the transporter occupancy during drug abuse or during treatment with antidepressants. Moreover, instead of finding the appropriate individual therapeutic dose of a drug by determining the plasma concentration, a direct visualization of to what extent the transporters are occupied by the given drug gives more accurate information on the correct dosing. It is therefore obvious that imaging of transporters is valuable for research during the development of new pharmaceutical entities, as well as for monitoring during clinical trials or during the disease processes.

Suitable imaging modalities for transporter studies are positron emission tomography (PET) and single photon emission tomography (SPECT). Both techniques are noninvasive methods using compounds (referred to as tracers or radioligands) labeled with short-lived positron emitting radionuclides (PET) or single photon emitting radionuclides (SPECT) for the visualization and quantification of biological and physiological processes in vivo. Although SPECT is also similarly a widely used technique, this review will focus on PET and the most important PET radioligands as tools for the study of neurotransmitter transporters. The main focus is on NET, DAT, SERT, VMAT2, Pgp, and GLUT due to the present research interest worldwide and ongoing drug development in these molecular targets, although other transporters will also be discussed [1, 2].

2
PET Principles

2.1
Equipment and Radionuclides

A PET investigation is based on the administration of a molecule, usually called a "tracer or radioligand", labeled with a positron emitting radionuclide. The main administrative route is via intravenous injection but other routes, such as inhalation, intranasal, and oral, are also used. The PET technique utilizes the unique decay properties of positron emitting radionuclides, where the β^+ decay is followed by an annihilation reaction. In decay by positron emission one proton in the decaying neutron-deficient nucleus is converted to a neutron and an isotope of the corresponding one atom number lower element is formed (see Fig. 1). In most cases a stable nuclide is formed. The positron emitted from one of the most used PET radionuclides, ^{11}C, has a maximum kinetic energy of about 1 MeV which gives a depth of penetration of a few millimeters in tissue. The first interactions between a positron and matter result in the loss of positron energy due to ionizing collisions. When the positron energy has been reduced to a few electron volts it interacts with an electron in the surrounding tissue resulting in an annihilation reac-

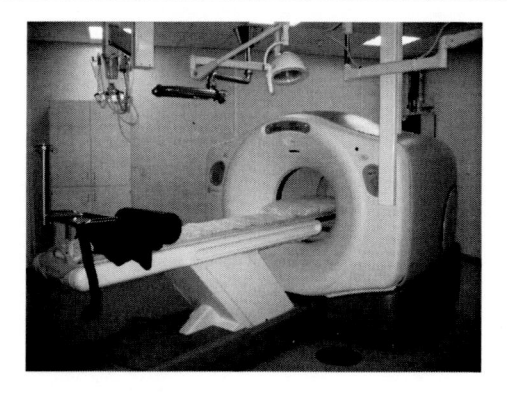

Positron emission

$$^{11}C \longrightarrow {}^{11}B + \beta^+ + \nu + E_{kin}$$

Annihilation reaction

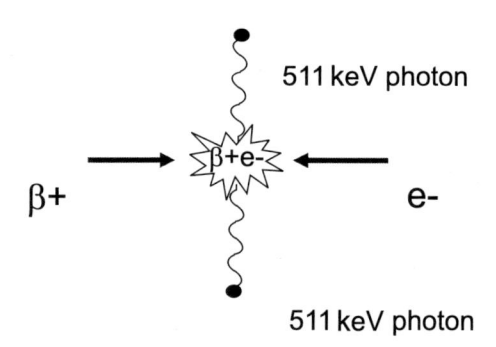

511 keV photon

$\beta+$ $\beta+e-$ e-

511 keV photon

Fig. 1 PET camera—positron emission and annihilation. In the decay of a positron emitting radionuclide a proton is converted to a neutron and a β^+ particle (positron) is emitted together with a neutrino (ν). The kinetic energy (E_{kin}) is distributed between the positron and the neutrino. This is exemplified by ^{11}C decaying to the stable ^{11}B

tion where mass is converted to energy, creating two 511 keV photons from the rest masses of one positron and one electron. The annihilation photons travel in opposite directions, 180° to each other, and contrary to the positron, the photons penetrate tissue and can be detected with external detectors.

The PET camera consists of a ring of detectors surrounding the part of the body to be examined (Fig. 1). The 511 keV annihilation photons that are created in the plane of the detector ring are detected using coincidence requirement. The PET camera will record an annihilation event when one detector is hit by a 511 keV photon and the opposite detector simultaneously (within about 10 ns) detects another 511 keV photon. By recording a large number of events the radioactivity concentration in tissue can be determined and localized with a resolution down to a few millimeters. Reconstruction of the collected data provides a time-resolved three-dimensional image of the distribution of the administered radioligand. For a comprehensive review of PET technology the interested reader could consult suitable textbooks (see, e.g., [3]).

In Table 1 some of the most important PET radionuclides and their main determining physical properties are listed. The PET radionuclides are produced by bombardment of a suitable target material with accelerated charged particles, for example protons, in which an isotope of a new element is created. ^{11}C is today exclusively produced by the $^{14}N(p,\alpha)^{11}C$ reaction in which nitrogen gas is used as target material. The high energy, 17 MeV protons overcome the coulomb barrier in the nitrogen nucleus and the resulting new nucleus is rapidly cleaved in two, an ^{11}C nucleus and an α particle. The ^{11}C nucleus picks up electrons and the resulting naked carbon atom reacts with trace amounts of oxygen or hydrogen present in the target and $[^{11}C]$carbon dioxide or $[^{11}C]$methane are produced, respectively, which are used as starting materials for the synthesis of labeled molecules, tracers.

The radionuclides used in PET can be obtained in very high specific radioactivity and the amount of a radioligand used in a PET study should be sufficiently low to guarantee that the studied system is not, to any measurable extent, disturbed by the radioligand. Typically a PET radioligand is administered in a total amount of a few micrograms or lower, which for most small molecules means 10–100 nmol of the tracer dose. The low amount of substance administered is an advantage in terms of lack of pharmacological or potential toxicological effects.

The short half-lives of the common PET radionuclides, ranging from 2 min to 4.3 days, limit the choice of method used to synthesize a labeled compound to those that give sufficient yield of product within a time frame appropriate for the physical half-life of the radionuclide. As a rule of thumb the labeled compound should be ready for intravenous administration within three half-lives from the end of radionuclide production [4, 5].

Table 1 Some important PET radionuclides

Radionuclide	Production	Decay	$T_{1/2}$
^{11}C	$^{14}N(p,\alpha)^{11}C$	99% β^+	20.4 min
^{13}N	$^{16}O(p,\alpha)^{13}N$	99% β^+	9.9 min
^{15}O	$^{14}N(d,n)^{15}O$	99.9% β^+	2.04 min
^{18}F	$^{18}O(p,n)^{18}F$	97% β^+	109 min
^{68}Ga	$^{68}Ge/^{68}Ga$	90% β^+	68 min
^{124}I	$^{124}Te(p,n)^{124}I$	25% β^+	4.3 days

Despite its short half-life, the synthetic opportunities of ^{11}C ($T_{1/2}$ = 20.4 min) are much greater than those of other more long-lived PET radionuclides and ^{11}C is therefore the main radionuclide used for development of PET tracers [6, 7]. In a drug development study the aim usually is to determine the blocking effect of a drug candidate and at least two PET scans must be performed, before and after treatment with drug. In order to follow the duration of action of the drug on the target protein in favorable cases up to six PET scans can be performed in the same subject with ^{11}C, whereas ^{18}F in most cases is restricted to two or three PET scans due to the higher radiation dose obtained. In addition, to allow the radioactivity to decay before the next PET investigation, the scans with ^{18}F-labeled radioligands have to be performed on separate days. ^{11}C-Labeled radioligands are thus preferably used in drug development PET studies, whereas ^{18}F may be advantageous in a clinical environment where the longer half-life simplifies the logistics.

2.2
PET Tracers and Modeling

Another important aspect of molecular imaging with PET is the opportunity of obtaining quantitative data, since the number of measured photons relates to the amount of substance (density of a specific molecule) in the tissue of interest. However, the PET camera does not provide any information on what chemical entity the photons originate from. In order to relate the dynamic information of the tracer–tissue interaction, a tracer kinetic model is required to interpret the information obtained from the PET scan to quantitatively determine biochemical or physiological processes in vivo [8, 9]. These models, for example those referred to as Logan [8] and Patlak [9] plots, are used to describe the interaction between a molecule and its biological target protein in the brain. The standard three-compartment tracer kinetic model for radioligands is shown in Fig. 2. The equilibrium between F and B, i.e., the time point where the number of tracer molecules dissociating from the receptor protein equals the number of associating molecules, must be established within the time frame for the physical half-life of the radionuclide used and the rate constants k_3 (on rate) and k_4 (off rate) are thus important. For example, a long residence time of the drug on the binding site may be preferred for a drug, whereas a tracer must exhibit a sufficiently fast off rate so that equilibrium may be established during the time frame of the physical half-life of the radionuclide. A special case is of course if there is an irreversible process where k_4 is zero during the time of the investigation. In such a case a simplified modeling method as defined by Patlak et al. [9] might be applied to describe and quantify the kinetics of the tracer.

Any compound that is intended for use in imaging studies in vitro or in vivo should possess certain criteria, some of which are shown in Table 2. A useful PET tracer exhibits suitable pharmacodynamic properties, such as

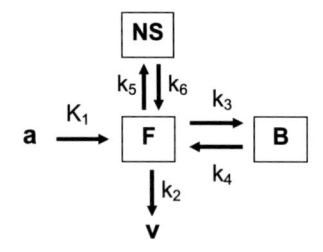

Fig. 2 Three-compartment model. Free radioligand (F), specifically (B) and nonspecifically bound (NS) radioligand, with "a" and "v" denoting arterial and venous blood, respectively, and K_1, k_2, k_3, k_4, k_5, and k_6, representing the unidirectional rate constants to be determined

sufficient affinity and selectivity to give a good image contrast. For example, binding potential, which is related to the ratio of the density of the target protein (B_{max}) and dissociation equilibrium constant (K_D), should preferentially exceed five, and the selectivity should, if possible, be 100-fold over other competing specific protein binding sites. However, these and other data obtained in vitro are not sufficient to guide the selection of a useful tracer molecule for

Table 2 Main receptor and radioligand criteria

Receptor criteria
   Physiological tissue distribution
      Saturability (limited number)
      Linear tissue relationship
   "Expected" pharmacology
      Stereoselectivity
      Treatment affected
      Correlation to other tests
   Sensitive to heat/proteases etc.
   Reversibility

General radioligand criteria
   High affinity (10 pM < K_D < 10 nM)
   Suitable kinetics (steady-state within 1–2 h at RT, dissociation rate low)
   Selectivity (> 100-fold, or possible to selectively block unwanted binding)
   Low nonspecific binding and adsorption in white matter
   Radioligand suitable for labeling with ^{3}H, ^{11}C, ^{18}F, or [125]I
   Metabolically stable or with no labeled metabolites

Additional radioligand criteria for brain PET
   Sufficient passage over blood–brain barrier
   Radioligand suitable for labeling with ^{11}C or [18]F (or others listed in Table 1)

The criteria listed are only a selection of the most important with regard to molecular imaging in vitro and in vivo.

a particular binding site. It must also be recognized that the process of developing diagnostic imaging tools based on labeling of suitable compounds with short-lived radionuclides differs from the normal drug development process aim of finding a therapeutically useful compound. A good pharmaceutical typically lacks some of the biological characteristics needed to be a useful imaging tool and vice versa. Fulfillment of the tracer concept, as originally defined by de Hevesy [10], requires that no biological effect can be measured at the dose used, and methods other than efficacy must, therefore, be used to determine the potential usefulness of the tracer candidates that are used for therapeutic agents. One important parameter is the image contrast, which depends on the ratio of specific to nonspecific binding and is a parameter that can be estimated, although not conclusively, using frozen tissue autoradiography.

One important feature that can be determined using the PET camera is receptor occupancy (RO) of a drug. RO can be estimated for a given dose of drug from calculations performed on serial PET scans, where the first scan is performed before, and the following after, the administration of a drug (Fig. 3). Regions of interest (ROI) are graphically defined in the images, representing the target region (e.g., striatum) and a reference region with no specific binding (e.g., cerebellum). If specific binding cannot be excluded in any region (as is the case when new tracers are validated) the reference tracer activity is defined by arterial blood sampling. The tracer radioactivity concentration is measured from the time of injection onwards for approximately 1 h by acquiring a number of images during the same scan session. The binding potential (BP = B_{max}/K_D) in the target region is estimated by fitting a kinetic model to the time–activity curves from input and reference regions. Receptor occupancy is calculated as RO = $(BP_{baseline} - BP_{drug})/BP_{baseline}100\%$.

The pharmacokinetic properties are more difficult to assess using in vitro methods and usually require experimental verification using in vivo models.

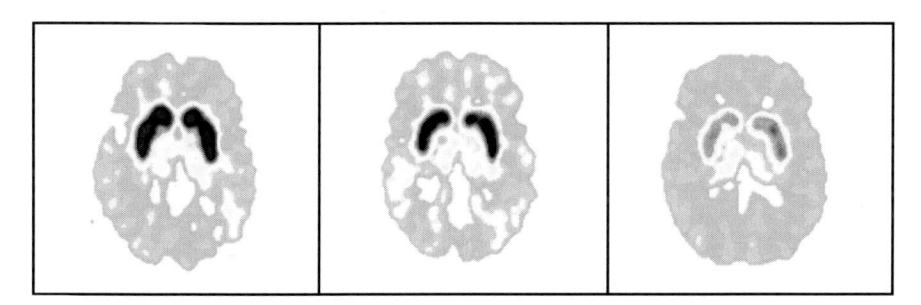

Fig. 3 PET images of the human brain obtained with [^{11}C]β-CIT-FE. Demonstration of receptor occupancy. These images show the relative distribution of [^{11}C]β-CIT-FE, a tracer of the dopamine reuptake transporter, at baseline (*left*), low dose of test drug (*middle*), and higher dose of test drug (*right*). Higher drug dose results in lower binding of tracer

The main factors responsible for limiting the transport of radioligands into the brain are, apart from slow transport over the blood–brain barrier (BBB), too rapid metabolism or clearance from plasma, high plasma protein binding, and exclusion from tissue by efflux pumps. Some attempts to predict pharmacokinetics are found in the literature, for example, the Lipinski rule for absorption and permeability [11] and log$P$ as determining BBB penetration [12].

In contrast to a therapeutic agent, the radioligand is given in very low amounts, tracer dose. It is of crucial importance that the specific radioactivity of the tracer is high enough in order to obtain a sufficient amount of radioactivity to be measured, without a concomitant mass dose that induces a biological effect.

3
PET Studies of Transporters

3.1
Norepinephrine Transporters (NETs)

The catecholamine norepinephrine (noradrenaline) is a neurotransmitter of the sympathetic nervous system, and as such is important in the regulation of the general degree of alertness, and is pivotal in preparing the body for either fight or flight. In the brain, noradrenergic cell bodies are found in the locus coereleus where the highest NET densities are also found. Lower densities are found in the thalamus, hypothalamus, and neocortex. In general, NET density is much lower than that of the other monoamine transporters such as DAT and SERT. Norepinephrine is very much involved in human physiology, such as mood regulation, sleep regulation, and in several behavioral aspects. It also plays an important role in the regulation of glucose metabolism. The CNS effects are mediated by the noradrenergic signaling system found throughout the brain, with several different receptor types (α- and β-adrenoceptors, divided into several subtypes) as well as the presynaptically located NETs [13]. About 70–90% of the norepinephrine released from the neuron is recovered by means of uptake by NETs, and this system is thus to a great extent responsible for the termination of noradrenergic impulse transfer.

The NET is a target for many pharmaceuticals, especially antidepressants, and for a number of drugs of abuse, such as cocaine. A diagnostic imaging tool for the study of NET would provide information on the noradrenergic system in the brain, and a suitable PET ligand would be beneficial in studies of the pathophysiology of a number of neurologic and psychiatric disorders. For example, our knowledge of the central mechanisms underlying disorders such as anxiety, depression, attention-deficit/hyperactivity disorder (ADHD), personality disorders, pain, and Parkinson's and Alzheimer's diseases would be increased by the availability of suitable NET PET tracers [14]. The noradrenergic and dopaminergic systems are strongly interconnected. Dopamine has a higher affinity for NET than norepinephrine itself, and NET may therefore be responsible for the removal of dopamine in brain regions with low DAT densities [15]. NET is also a target for drug abuse as demonstrated in monkey studies where chronic self-administration of cocaine leads to increased NET concentration in the brain [16].

The search for a PET tracer for NET has been intensive during the last decade but has been hampered by the requirement for the high affinity needed [17]. The B_{max} of NET in human insular cortex is below 5 pmol/g, implying that a K_D below 1 nM is needed for a PET ligand. The selectivity criteria are critical and a low affinity for DAT is required, especially in the study of NET in the striatum where the density of DAT is very high (see below).

NET inhibitors can be divided structurally into several classes of compounds where the tricyclic antidepressants were the first examples of clinically useful drugs, followed by the mono- and bicyclic, and the tropane

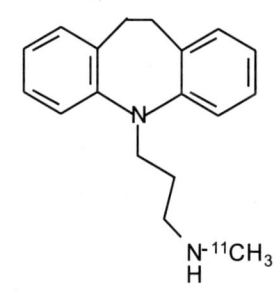

Nisoxetine structures

R=, I, SCH$_3$, CH$_3$, OCH$_3$

Reboxetine structures

Methylreboxetine R= ^{11}CH$_3$
FMeNER-D$_2$ R= CD$_2$^{18}F

Desimipramine

Talupram/Talsupram

X=O,S

Fig. 4 Some NET radioligands

derivatives. Several PET ligands have been developed and evaluated as potential PET tracers for NET and the main chemical structural classes are shown in Fig. 4.

3.1.1
PET Ligands Based on Tricyclic and Bicyclic Structures

The affinity of the tricyclic antidepressant desipramine (desimipramine, DMI) is in principle sufficient to be a good tracer candidate with binding to human NET with a Ki of 0.63 nM [18, 19] and [^{11}C]DMI has been evaluated in cynomolgus monkey [20]. The tracer was readily taken up into monkey brain, but the uptake proved to be insensitive to inhibition by blocking doses of unlabeled DMI. [^{11}C]DMI was also found to be rapidly metabolized with only 20–25% intact tracer remaining after 45 min. The possibility of labeled metabolites entering the brain and thus confounding the interpretation of the PET data could not be ruled out. In spite of its usefulness as an antidepressant drug, labeled DMI is not useful as a PET tracer.

^{11}C-Labeled talsupram and talopram, two other antidepressant drugs acting on the norepinephrine uptake site, were, due to their high affinities (IC$_{50}$ 0.79 and 2.9 nM, respectively), labeled with ^{11}C and investigated as potential NET tracers [20, 21]. Considering the clinical efficacy of these compounds as antidepressants the uptake in NET-rich areas of the brain was, however, surprisingly low for both compounds as measured in cynomolgus and rhesus monkeys [22, 23]. McConathy [21] used the active enantiomers of talopram and talsupram but reported even lower brain uptake than Schou [20] who used the racemates, a result that could be due to species differences. It is not a contradiction that a compound may have clinical efficacy while not being suitable as a tracer. As plausible explanations for the lack of success with these two ligands, the authors suggest that a possible saturable process caused too high a degree of plasma protein binding or efflux by the Pgp transport system. At therapeutic concentrations Pgp may be saturated by a drug which could explain the therapeutic effects of these compounds, in contrast to the tracer concentration used in PET where a very low fraction of the tracers can be distributed to the brain tissue due to efficient efflux by Pgp.

3.1.2
PET Ligands Based on Aryloxy Morpholines

Early attempts to find a NET ligand, such as the ^{11}C-labeled NET inhibitors nisoxetine [24] and an iodido analogue of tamoxetine [25], failed due to high nonspecific binding. However, the methyl analogue of reboxetine (MRB, see Fig. 4) is a promising NET ligand with an IC$_{50}$ value of 2.5 nM, and [^{11}C]MRB has been studied in rat and baboon brain [26–28]. Ding et al. further evaluated [^{11}C]MRB and investigated the racemate as well as the (S,S)- and (R,R)-

enantiomers in baboons [29]. A marked binding and blocking effect of nisoxetine was obtained for the eutomer (S,S)-[^{11}C]MRB in NET-rich areas such as
thalamus and cerebellum. Specific binding was also found in the heart. The
racemate showed specific binding in the brain whereas the (R,R)-enantiomer
of [^{11}C]MRB lacked specific binding, demonstrating an enantioselectivity in
vivo consistent with in vitro data.

A compound within the reboxetine subclass is the desethyl analogue (S,S)-
2-[(2-methoxyphenoxy)phenylmethyl]-morpholine $((S,S)$-MeNER), where
both ^{11}C and ^{18}F labeling have been used (see Fig. 4). The two chiral centers
in MeNER give four stereoisomers, with (S,S)-MeNER being the most potent
with a eudismic ratio of 20 over the (R,R)-enantiomer [30]. (S,S)-[^{11}C]MeNER
readily enters the brain and shows specific binding to NET, as demonstrated
by pretreatment with DMI in monkey PET studies [31]. The kinetics was slow
and the tracer did not reach equilibrium binding within 90 min, i.e., the time
frame for a PET study with ^{11}C. Metabolite analyses in humans showed the
formation of a radiolabeled metabolite more lipophilic than the tracer itself
and it cannot be excluded that this would have a detrimental effect on the
quantification of NET inhibition. The test–retest variability was also found
to be too high for quantification of discrete changes of NET inhibition (An-
drée B, personal communication). The use of ^{18}F, with its longer half-life,
allows a longer time for establishing equilibrium between free tracer and
tracer bound to the target protein as compared to ^{11}C. (S,S)-[^{18}F]MeNER

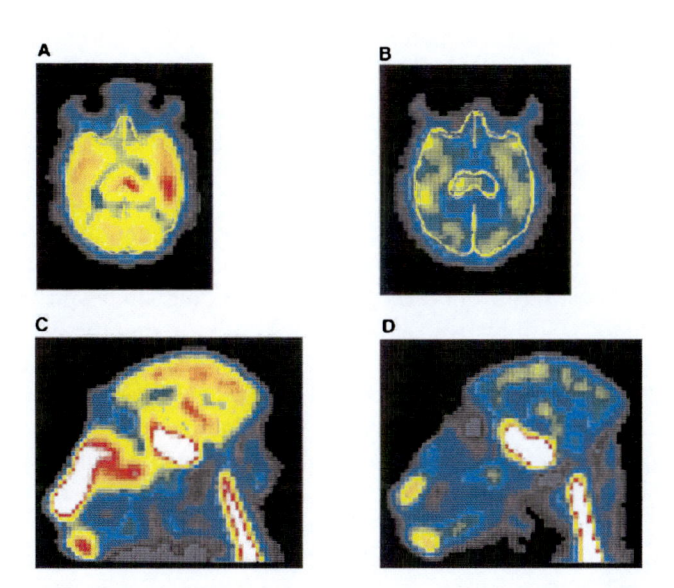

Fig. 5 PET images of the cynomolgus monkey brain obtained using (S,S)[^{18}F]FMeNER-
D_2. The images show binding of (S,S)-[^{18}F]FMeNER-D_2 during baseline (**a,c**) and pretreatment (**b,d**) conditions (DMI 5 mg/kg). **a** and **b** represent horizontal images. **c** and
d represent sagittal images (reprinted from [34] with permission)

was, however, found to be metabolically unstable with loss of the ^{18}F label and formation of free ions of $[^{18}F]$fluoride. To circumvent this problem the ^{18}F-labeled dideuterated analogue (S,S)-$[^{18}F]$FMeNER-D$_2$ was prepared (see Fig. 4). Substitution of hydrogen for deuterium in the fluoromethyl group induces a kinetic isotope effect [32], which may reduce the rate of carbon–fluorine bond breaking and thus increase the metabolic stability of the tracer. (S,S)-$[^{18}F]$FMeNER-D$_2$ has been evaluated in vitro and in monkey [33, 34] showing selective binding to NET (Fig. 5). Moreover, equilibrium was reached later than for the protium analogue and most importantly, it also showed reduced metabolism as manifested by reduced formation of $[^{18}F]$fluoride. In spite of the relatively low affinity of 3.1 nM this tracer is so far the most promising tracer for NET [26, 34].

3.1.3
Conclusions on NET PET Ligands

Although several promising ligands have been labeled and evaluated in vivo, including evaluation in man, currently no really useful PET ligand for NET is available and further research in this field is encouraged. Of the structures investigated, the most promising scaffold is the reboxetine subclass where (S,S)-$[^{18}F]$FMeNER-D$_2$ appears to be a potentially useful tracer. However, further studies in man are needed for the evaluation of (S,S)-$[^{18}F]$FMeNER-D$_2$ as a useful PET ligand for the study of the involvement of NET in health and disease, as well as for determination of the degree of NET inhibition following drug administration [26, 34].

3.2
Dopamine Transporters (DATs)

The neuronal DAT is mainly, but not exclusively, confined to the striatum in spite of the fact that dopamine receptors (especially the D$_1$ subtype) have an extensive localization in the cerebral cortex. Binding studies in vitro and in vivo, as well as in situ hybridization studies, indicate that the distribution of DAT and of mRNA coding for DAT is almost exclusive to nigrostriatal neurons [35, 36]. However, small amounts of DAT mRNA have also been found in extrastriatal regions like the hypothalamus and amygdala.

The importance of DAT can be exemplified by the inherent "brain reward" system where increased dopamine levels mediate the euphoria searched for in drug abuse. One of the most addictive drugs known, cocaine, exerts its effect by inhibiting DAT and thus by increasing synaptic dopamine levels leading to the desired euphoric effect. Several studies have shown that the density of DAT is reduced in patients with degenerative brain disorders, such as in Parkinson's disease [37–40] and in dementia with Lewy bodies [37, 41, 42], whereas DAT density is increased in patients with Tourette's syndrome. However, no marked changes in DAT have been found in Alzheimer's disease [37], and in vivo studies of striatal DAT densities may therefore differentiate between Alzheimer's disease and dementia with Lewy bodies [41, 43]. Less consistent results have been obtained in Huntington's disease and schizophrenia, where an impaired dopamine signaling system has been implicated. In schizophrenia most in vivo PET or SPECT studies fail to show abnormalities in DAT densities in the striatum, although both increases [44] and decreases [45–47] have been shown.

A large number of structurally different compounds have been used for the characterization of regional DAT distribution as well as of its physiology and pharmacology. High binding of DAT radiotracers has been found in the caudate and putamen, regions known to have a high density of the DAT [48]. The striatum is thus a suitable region for determination of changes of DAT binding as a result of drug interactions. Phenyltropane is one of the chemical scaffolds that has been applied for the development of selective tracers for DAT. The cocaine analogues, such as β-CFT (2β-carbomethoxy-3β-(4-fluorophenyl)tropane), β-CIT (2β-carbomethoxy-3β-(4-iodophenyl)tropane), and RTI-121 (3β-(4-iodophenyl)tropane-2β-carboxylic acid isopropyl ester) have been extensively used as markers of DAT in animals or in the human brain (Fig. 6). In vivo studies with both SPECT and PET have been performed in monkeys and humans, particularly using β-[^{123}I]CIT or β-[^{11}C]CIT [49, 50]. However, these compounds were not selective and had substantial affinity for the norepinephrine and serotonin transporters. Compounds with suitable tracer characteristics have been developed by structural modifications in the N-alkyl chain position. This provides an opportunity for labeling with either ^{11}C in the methyl ester or ^{18}F in the alkyl chain, which is a useful strategy in tracer development. The fluoroalkyl analogues β-CIT-FE (N-(fluoroethyl)-2β-carbomethoxy-3β-(4-iodophenyl)tropane) and β-CIT-FP (N-(fluoropropyl)-2β-carbomethoxy-3β-(4-iodophenyl)tropane), which were subsequently developed, showed better in vivo kinetics and higher selectivity for DAT [51]. These radioligands are therefore still used for the study of DAT, and β-CIT-FP (also called ioflupane) is now available as a commercial kit for SPECT (DATScan®) [52, 53]. Both β-[^{11}C]CIT-FE and β-[^{11}C]CIT-FP form labeled metabolites during the course of the PET investigation, one identified as nor-β-[^{11}C]CIT. This compound penetrates the BBB and has a high affinity for SERT, manifested as a radioactive uptake in the thalamus [54, 55]. β-[^{11}C]CIT-FE is superior to β-[^{11}C]CIT-FP due to a more rapidly established plasma to target protein binding equilibrium. β-[^{18}F]CIT-FP is an alternative since the longer half-life gives the option of reaching equilibrium. Another benefit of using β-[^{18}F]CIT-FP is that the metabolite formed, nor-β-CIT which binds to both DAT and SERT, is unlabeled and no interference with measurement of DAT specific binding from labeled metabolites is thus found as shown in monkey PET studies [54].

β-CIT	$^1R=CH_3$	$^2R=I$
β-CIT-FE	$^1R=CH_2CH_2F$	$^2R=I$
β-CIT-FP	$^1R=CH_2CH_2CH_2F$	$^2R=I$
β-CFT	$^1R=CH_3$	$^2R=F$
PE2I	$^1R=CH_2CH=CHCH_2I$ (E)	$^2R=CH_3$

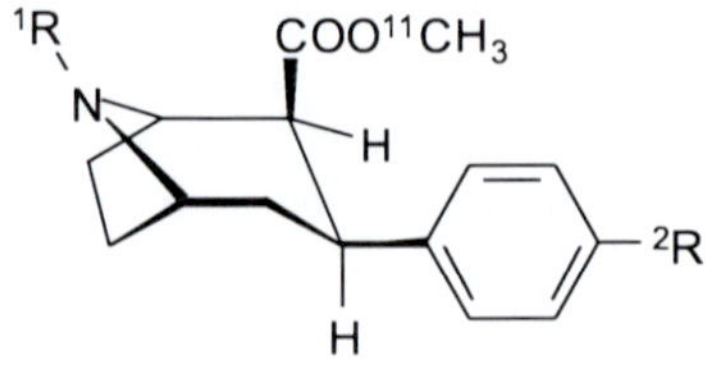

Fig. 6 Some DAT radioligands. In **a** some of the most often used cocaine analogues are shown. In **b** the metabolism of β-CIT-FP is shown, (* = labeled atom), indicating the importance of labeling positions with respect to formation of radiolabeled metabolites and their potential interference with measurement of specific binding

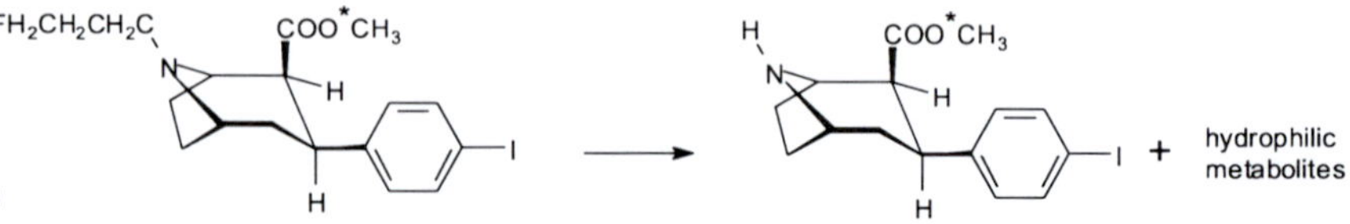

However, as the β-CIT analogues still retain some affinity for the other monoamine transporters, new radioligands have been developed. One promising compound is N-(3-iodoprop-(2E)-enyl)-2β-carboxymethoxy-3β-(4′-methylphenyl)nortropane (PE2I) with a 30-fold lower affinity for the

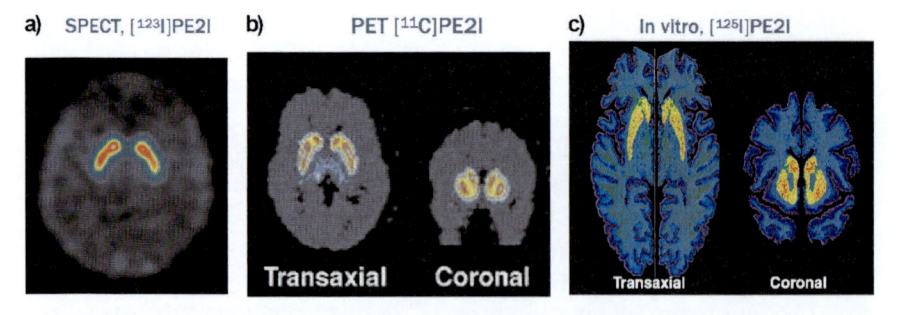

Fig. 7 Comparison of SPECT, PET, and in vitro autoradiography of DATs using PE2I. The images show SPECT (**a**), PET (**b**), and autoradiography (**c**) using PE2I labeled with ^{123}I, ^{11}C, and ^{125}I, respectively, in control human brains. In (**c**) only one of the hemispheres was incubated with $[^{125}I]PE2I$, the mirrored image substitutes the other hemisphere. Images from [59] (SPECT), [62] (PET), and [35] (autoradiography)

other monoamine transporters [35, 56, 57]. PE2I can be labeled either with ^{123}I or with ^{11}C, and is therefore used both in SPECT using [^{123}I]PE2I [38, 58–60] and in PET using [^{11}C]PE2I [61–63] (Fig. 7). Among other compounds within the nortropane scaffold, [^{11}C]LBT999 [64, 65] and [^{18}F]MCL-322 have been suggested from animal PET studies to be suitable ligands for DAT in the human brain.

3.2.1
Conclusions on DAT Ligands

Due to its properties as a marker of presynaptic dopaminergic neurons, the in vivo visualization of DAT is of great value in the study of neurodegenerative diseases of the striatum, and especially of Parkinson's disease. Moreover, as blockade of DAT mediates the euphoric effects of several drugs of abuse, e.g., cocaine and methylphenidate, and is involved in other reward systems, the possibility to study DAT in vivo may help in our understanding of these conditions. Unfortunately, the first PET radioligands based on cocaine were not selective, but more recently several selective tracers such as [^{11}C]PE2I have been characterized and shown to be very useful in animal and human studies.

3.3
Serotonin Transporters (SERTs)

Serotonin (5-hydroxytryptamine, 5-HT) is implicated in a number of homeostasis functions, such as food intake, sleep, and temperature regulation, but also in regulation of mood. The serotonergic neurons project from raphe nuclei and the highest SERT densities are found in thalamus, striatum, and raphe nuclei which thus become suitable targets for measurement of SERT binding potential using PET. Disturbances in the serotonin system may lead to various mental illnesses, such as anxiety and depression. An overstimulation of certain serotonin receptors may result in hallucinations and other psychotic experiences, exemplified by the fact that agonists (e.g., mescaline, d-LSD) have been used in drug abuse.

The most commonly used treatment for depression today is by blocking the SERT using selective serotonin reuptake inhibitors (SSRIs), and SERT is a target for most antidepressant drugs. Although there is a great interest in studying SERT, very few radioligands are available for in vivo or in vitro imaging. Only one of the large number of antidepressant drugs, citalopram (Fig. 8), has been used as a radioligand for SERT [66, 67], however with limited success in vivo. Attempts to use some of the other SSRIs, such as [^{18}F]fluoxetine (Prozac), as radioligands have failed to show any specific binding to SERT [68], in spite of their well-known activity on the transporter. The first useful PET ligand for the visualization of SERT in vivo was [^{11}C]McN5652 (*trans*-1,2,3,5,6,10-β-hexahydro-6-[4-

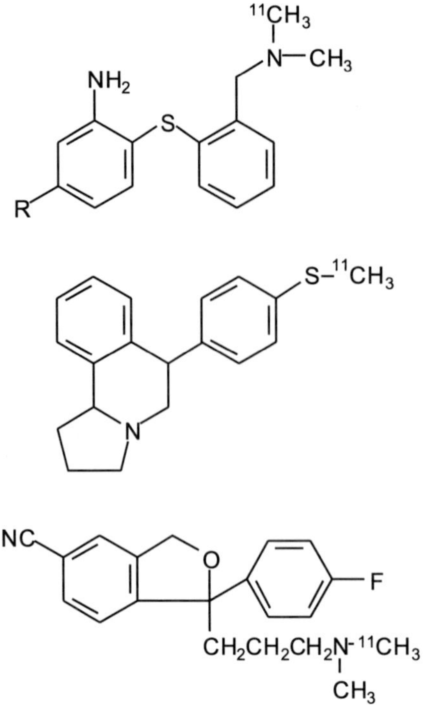

ADAM	R= I
AFM	R= CH_2F
DASB	R= CN
DAPA	R= Br
DAPP	R= OCH_3
MADAM	R= CH_3
HOMADAM	R= OH

McN-5652

Citalopram

Fig. 8 Some SERT radioligands

(methylthio)phenyl]pyrrolo-[2,1-*a*]-isoquinolone) [69–71]. Studies using [^{11}C]McN5652 have shown, for example, increases in SERT in several cortical and subcortical regions of patients with depressive disorders [72, 73]. More recently, an [^{18}F]methyl analogue of McN5652, i.e., [^{18}F]FMe-McN5652, has been developed and was found to be superior to the original [^{11}C]McN5652 in the in vivo visualization of SERT in the human brain [74, 75].

The above mentioned cocaine derivatives, i.e., β-CIT and analogues, are mainly used as radioligands for DAT, but due to their low selectivity some of these compounds may also be used as tracers for SERT in most brain regions outside the striatum [76]. The most promising compound in this series appears to be nor-β-CIT, which has so far, however, only been used in SPECT [54, 77].

Another promising radioligand for SERT is [^{11}C]-3-amino-4-(2-dimethylaminomethylphenylsulfanyl)benzonitrile, or [^{11}C]DASB (Fig. 8) [78–80]. [^{11}C]DASB is a very selective ligand, with a 1000-fold selectivity versus NET and DAT [80] and with favorable properties for use in the clinical setting [81]. This compound has also been used to determine the amount of serotonin in the synapse [81–83] in depression and, for example, following treatment with SSRIs. It has been suggested that the extracellular concentration of 5-

HT is low in depression, which should be normalized by treatment with drugs blocking SERT. One study has shown increased SERT binding potential in patients with major depression as compared to healthy subjects, indicating an overactive transport from the synapse. However, other studies have failed to verify this (reviewed in [81]). Studies with PET using [^{11}C]DASB, [^{11}C]McN5652, and [^{11}C]MADAM have shown the occupancy of SERT in depressed patients on treatment with various SERT inhibitors to be very high, normally in the order of 80% [81], values that are similar to those obtained in previous SPECT studies.

Several other compounds evaluated as potential radioligands for SERT (see Fig. 8) are the benzylamines within the MADAM/ADAM group (e.g., MADAM [84, 85], ADAM [86, 87], HOMADAM [88], and EADAM [89]), of which only [^{11}C]MADAM has been used in human PET so far ([^{123}I]ADAM has been used in SPECT in humans). The first studies with [^{11}C]MADAM indicate that this radioligand may be as suitable as [^{11}C]DASB for the quantification of SERT in the human brain.

3.3.1
Conclusions on SERT

Although there are a large number of SERT inhibitors used today in the clinic as SSRIs, it was not until very recently, when [^{11}C]McN5652 was synthesized, that it was possible to study this transporter using PET. None of the selective SSRIs is suitable as a PET radioligand. Today, [^{11}C]DASB appears to be the preferred PET radioligand for SERT, although the newer compounds in the [^{11}C]MADAM series may also be well suited for imaging of SERT in the human brain.

3.4
Vesicular Monoamine Transporter 2

VMAT2 is expressed in the CNS as well as in the β-cells of the islets of Langerhans in the pancreas. The primary function of the VMAT2 protein is translocation of monoamine neurotransmitters from the cytosol into synaptic vesicles. This function is regulated by exchange between actively cycling synaptic terminals and a reserve pool leading to a relatively stable number of vesicles in the neurons. This offers potential for general molecular imaging tools of serotonergic, dopaminergic, and noradrenergic innervation in the CNS, using an angle other than the earlier described use of selective tracers for the different monoamine transporters [90, 91].

It is clear that an objective marker for the severity of Parkinson's disease would be of significant value for the understanding of disease progression and the study of potential new neuroprotective therapies. DAT is apparently regulated by dopamine receptor activation and the distribution of DAT molecules

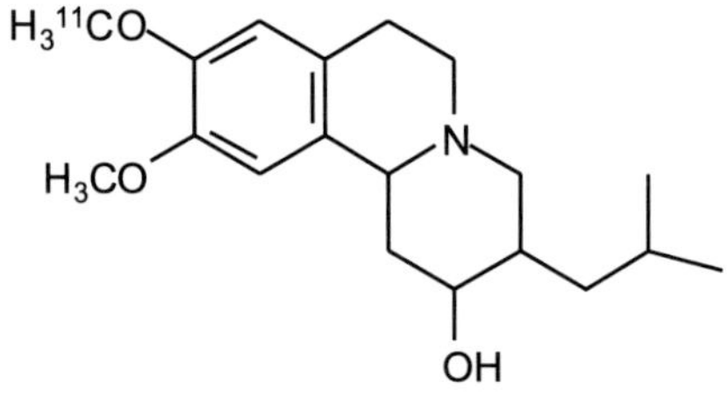

Fig. 9 [^{11}C]Dihydrotetrabenazine, a tracer for VMAT2

between the cell membrane and endosome may thus be affected by interactions with drugs [92]. Since VMAT2 binding in dopaminergic neurons is supposed to be unaffected by interaction with drugs, changes in dopamine receptor activation, or DAT expression, it may be more predictive in determining dopaminergic neuronal density.

(+)[^{11}C]Dihydrotetrabenazine (Fig. 9) ((+)[^{11}C]DTBZ) [93] has been investigated as a tracer aimed at estimating the striatal VMAT2 binding site density [94] in both healthy subjects and Parkinson patients. An age-related (0.5%/year) decline in VMAT2 binding in healthy subjects has been described [94, 95]. In the Parkinson patients a clear reduction in striatal (+)[^{11}C]DTBZ binding was found which correlated with the severity of the disease.

Interestingly, VMAT2 is also expressed in pancreatic β-cells and (+)[^{11}C]-DTBZ has also been used to probe the β-cell mass. The total β-cell mass in the pancreas influences, and is a measure of, the amount of insulin secretable. A noninvasive imaging tool that can give important information about the progression of diabetes types I and II is highly relevant both in drug development and in clinical routine. The use of (+)[^{11}C]DTBZ and micro-PET for determination of β-cell mass in rats has been reported and the tracer found to discriminate between euglycemic rats and rats with chemically induced diabetes [96]. Further research in the field of monoaminergic vesicular transporter 2 will hopefully validate the usefulness of (+)[^{11}C]DTBZ as a tracer for the determination of β-cell mass.

3.4.1
Conclusions on VMAT2

It appears that (+)[^{11}C]DTBZ is a useful tracer for the study of striatal dopaminergic neuronal density. However, it still has to be investigated if it can be applied as a general tool for studying CNS monoamine innervations, as it can only be used in tissues which selectively express a particular transporter system having a limited distribution. Use of (+)[^{11}C]DTBZ in rodents indicates that it may be a useful tool for quantification of pancreatic β-cell mass.

3.5
ABC Transporters

The BBB is comprised of brain endothelial and ependymal cells, with the luminal layer facing the capillary and the abluminal facing the interstitial fluid which creates an efficient barrier that prevents diffusion of molecules between the cells. Pgp has been identified as one of the most important of several energy-dependent transporters belonging to the ATP binding cassette (ABC) transporters [97]. Pgp is a cell-surface transporter localized in the luminal membrane mediating baso-lateral to apical transport of hydrophobic molecules. This efflux of substances together with metabolism accounts for the major detoxification pathways in our body, and is thus an important defense mechanism and protects the body from xenobiotics. This might also be a disadvantage, as Pgp, with its increased expression in cancer cells, also induces MDR in cancer chemotherapy [98]. Moreover, Pgp-mediated efflux may also be responsible for the lack of efficacy of many potentially CNS-active drugs, in spite of the fact that the in vitro binding properties of the drug suggest specific interactions with molecular targets in the brain. It cannot be ruled out that differences in Pgp expression may also be responsible for interindividual susceptibility to progressive neurodegenerative disorders such Parkinson's and Alzheimer's diseases [99].

Furthermore, drug–drug interactions with Pgp inhibition may lead to increased BBB penetration and unwanted CNS side effects or overdosing, and classifying a drug as either a Pgp substrate or an inhibitor is of importance. The opportunity to study Pgp transport mechanisms and identify drugs that are Pgp substrates using simple assays is becoming increasingly recognized as pivotal in drug development. Radioligands for the study and quantification of Pgp expression and drug-related interactions are thus of value in drug development as well as in clinical investigations. Several PET and SPECT tracers have been investigated as potential tools for visualization of Pgp in vivo [100].

^{99m}Tc-labeled SPECT tracers labeling Pgp, such as sestamibi [101], tetrofosmin [102, 103], and furifosmine [104], have been applied in human cancer-related studies. It was found that uptake of the labeled Pgp substrate correlated with tumor response to treatment: the higher the uptake the more drug sensitive the tumor. Although interesting results have been obtained using SPECT imaging, the difficulty in obtaining quantitative data with SPECT limits its usefulness. PET may offer an advantage due to the quantitative measurement of the radioactivity that can be translated to concentration of tracer in tissue.

3.5.1
^{67}Ga/^{68}Ga Compounds for SPECT and PET Studies of Pgp Expression in Tumors

Gallium radionuclides, such as ^{67}Ga and ^{68}Ga, are available for diagnostic use in SPECT and PET, respectively. A ^{67}Ga/^{68}Ga complex, gallium (bis(3-ethoxy-2-hydroxybenzylidene)-N,N'-bis(2,2-dimethyl-3-aminopropyl)ethylenediamine) (Ga-[3-ethoxy-ENBDMPI]$^{+}$) [105] has been explored as a potential tracer for Pgp using SPECT and PET (Fig. 10) [106, 107]. It was found that ^{67}Ga-[3-ethoxy-ENBDMPI]$^{+}$ readily accumulated in drug-sensitive cells, whereas multidrug-resistant cells showed a 100-fold lower accumulation, which demonstrates the interaction of ^{67}Ga-[3-ethoxy-ENBDMPI]$^{+}$ with Pgp. Micro-PET biodistribution studies in KO and wild-type mice were also performed using ^{68}Ga-[3-ethoxy-ENBDMPI]$^{+}$ and showed increased penetration and retention of the tracer in KO mice as compared with wild-type mice, which were consistent with the ex vivo biodistribution data. The results suggest that the nonmetabolized $^{67/68}$Ga-[3-ethoxy-ENBDMPI]$^{+}$ is a promising probe of Pgp-mediated drug transport in vivo and by using ^{68}Ga quantitative biodistribution data could be obtained. To continue the validation of ^{68}Ga-[3-ethoxy-ENBDMPI]$^{+}$, human PET studies are needed.

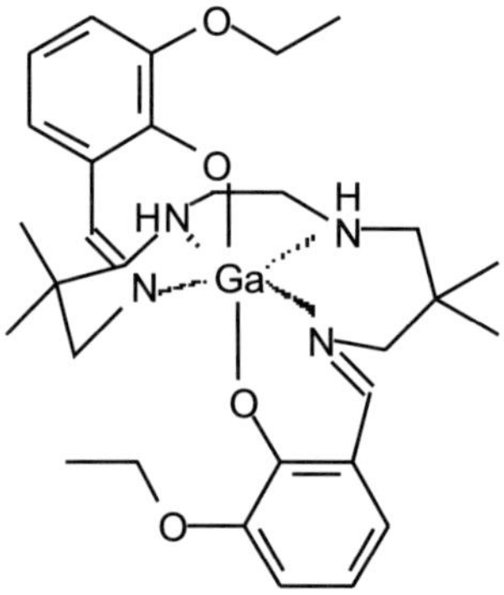

Fig. 10 Ga-[3-ethoxy-ENBDMPI]$^{+}$, a tracer for Pgp

3.5.2
^{11}C Tracers for the Study of Pgp

Several ^{11}C-labeled compounds have been suggested and investigated as potential tracers for the study of Pgp, for example, [^{11}C]carvedilol [109] and [^{11}C]verapamil [110, 111]. Most work has been done to evaluate the interaction of the calcium channel blocker verapamil with Pgp in the BBB. Verapamil is a chiral compound and both the racemate and the pure (R)- and (S)-enantiomers have been investigated as potential Pgp tracers

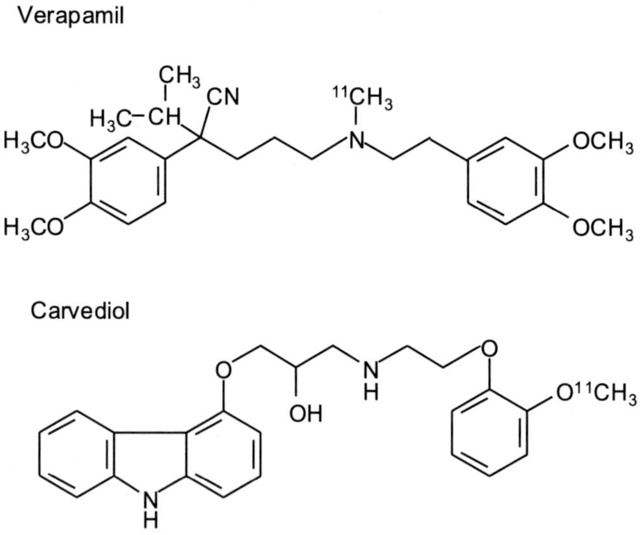

Verapamil

Carvediol

Fig. 11 Pgp tracers, [^{11}]verapamil and [^{11}C]carvedilol

(Fig. 11). The interaction of [^{11}C]verapamil with Pgp is not stereoselective, but as the (S)-enantiomer is more sensitive to hepatic metabolism and has higher plasma clearance, the more metabolically stable enantiomer (R)-verapamil may be the preferred enantiomer [112, 113]. A comparison between [^{11}C]verapamil interaction with Pgp in rats and humans has shown that there are remarkable similarities with respect to blocking with the selective Pgp inhibitor cyclosporine A, which suggests that rat is a suitable animal model to predict Pgp interactions in humans [114, 115]. There are several different models available to evaluate [^{11}C]verapamil PET data. The timing of drug and tracer administration may be critical and kinetic information related to this is needed for a correct interpretation of data. This was examined in rat micro-PET studies where a bolus-infusion administration route was used to extrapolate the degree and duration of cyclosporine A induced Pgp blockade [116–118]. By this approach information such as onset of action as well as kinetics for the blocking effect could be determined, information that is important when designing PET studies to investigate potential drug–Pgp interactions of a new drug. The infusion model avoids misinterpretation of data by choosing the incorrect timing of drug and tracer administration. The potential of using [^{11}C]verapamil for the assessment of drug–Pgp interactions in humans has been verified by determining brain [^{11}C]verapamil concentrations before and after cyclosporine A treatment [114, 118].

Another interesting aspect is the effect of aging on Pgp expression. [^{11}C]Verapamil was used to determine a correlation between age and Pgp function and a decrease with age was found which may support the hypothesis that development and progression of neurodegenerative disorders to some extent may be explained by loss of Pgp function leading to higher exposure to toxic substances [119, 120].

An alternative tracer for Pgp is [11C]carvedilol which in animals showed favorable tracer kinetics and might be a more sensitive Pgp probe than [11C]verapamil [109].

3.5.3
[18F]Fluoropaclitaxel as a Tracer for the Study of Pgp

The well-known anticancer drug paclitaxel, belonging to the taxane family derived from the yew tree (*Taxus brevifolia*), is a Pgp substrate and has been investigated using PET [121]. Paclitaxel and the analogue [18F]fluoropaclitaxel (FPAC) (Fig. 12) can be assumed to exhibit similar affinity to Pgp considering the small structural difference introduced by the fluorine [122]. The volume of distribution (V_D) in different organs was calculated before and after administration of the noncompetitive Pgp blocker XR9576 (tariquidar) as a measure of Pgp-mediated efflux. It was possible to detect a significant increase in V_D following pretreatment with the Pgp modulator XR9576. A consistent finding was increased retention of FPAC in the liver and lung, and a small increase in the efflux rate in the kidney. The latter could not be explained by decreased delivery to the kidneys and possibly other mechanisms affect the V_D in the kidneys. It has to be pointed out that the numbers of animals and PET scans were small and it is likely that a repeated larger study

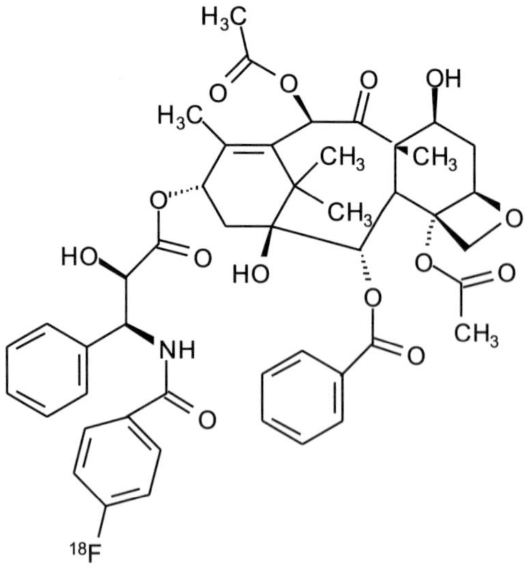

Fig. 12 [18F]Fluoropaclitaxel, a tracer for Pgp

would provide further proof for the use of FPAC as a tracer for the study of functional aspects of Pgp expression.

3.5.4
Conclusions on Pgp Tracers

The existing tracers for Pgp seem to be suitable tools for the study of both peripheral, in tumors for example, and CNS drug–Pgp interactions.

3.6
Glucose Transporters (GLUTs)

During recent years there has been a tremendous growth in the use of PET in the clinical setting. The rationale for this large-scale introduction into healthcare is the evidence formed over two decades that the tracer [^{18}F]fluorodeoxyglucose (FDG) has a highly elevated retention in some common cancers, and today routine PET imaging in cancer patients using FDG changes the treatment decisions in approximately 30% of all cases [123].

Glucose is also a fundamental metabolic substrate for cellular energy production in normal tissues. Subsequently FDG-PET has also been found valuable for imaging of the brain and the heart. Normal neurons consume glucose in huge amounts and neurodegenerative processes, such as Alzheimer's disease, can be diagnosed and followed by evaluating the distribution and degree of regional hypometabolism with PET [124]. Normal cardiac myocytes utilize both fatty acids and carbohydrates for energy production and preferences are mainly dictated by plasma levels. The detection of elevated myocardial glucose transport using PET is interesting from a diagnostic perspective, because elevated FDG uptake in a part of the cardiac muscle known to be damaged (reduced blood flow or reduced motility) is indicative of viable tissue that could be salvaged if blood flow is restored [125].

While the reason why aggressive tumors prefer glucose has not been fully elucidated, the actual uptake mechanisms have been studied in more detail. FDG enters the cell via dedicated GLUTs, a family of transmembrane macromolecules, of which four are currently known (GLUT-1–4). Inside the cell FDG is phosphorylated by hexokinase II to FDG-6-phosphate (FDG-6P), which is not a substrate of the glycolytic pathway. Dephosphorylation and backward diffusion of FDG-6P is slow and the molecule is effectively trapped in the cytosol. With this information at hand it is clear that FDG-PET measures glucose uptake rather than glucose consumption, although both processes are highly interrelated. Glucose uptake depends on the amount of GLUT and hexokinase proteins in tissue, their rates, and the amount of substrate available. The PET image integrates these factors.

The predominantly expressed GLUT in malignant cells is GLUT-1, which is insulin-independent. Massive overexpression, both on the mRNA and the protein level, as well as increased rates have been found and correspond to survival [126–129]. GLUT-3 is the predominant transporter in cerebral neurons. [130, 131]. Normal cardiac myocytes rely mainly on insulin-dependent GLUT-4 for transport, but GLUT-1 is increased in proportion to myocyte derangement and disease progression in heart failure [132].

Since PET enables noninvasive evaluation of tissue metabolism, it is understandable that quantitative FDG-PET has been advocated as a biomarker for drug development [133–136]. It is likely that FDG-PET as an indicator of overall glucose flux will continue to be used for both diagnosis of disease and evaluation of drug effects in the future, but the more precise imaging of GLUTs requires new specific tracers [137].

4
Conclusions

PET is today used as a tool for the study of several transporter proteins, e.g., DAT, SERT, GLUT, VMAT2, and Pgp and to some extent NET. The "golden" tracers for these systems have so far not yet been discovered and research is ongoing at many PET centers worldwide, to evaluate existing and search for new tracers as well as to develop new models for the interpretation of PET data.

The task of developing a new tracer to measure a specific molecular process in vivo is maybe as difficult as designing a new drug. Although the key aspects to some extent coincide, they also differ at some crucial points. Affinity and specificity as well as on and off rate constants for substance–protein interaction are some important parameters. The processes in the development of a suitable PET tracer are similar to drug development in several aspects. For example, it is not possible from preclinical in vitro data and structure–activity relationships to determine tracer or drug properties to a higher level than to select candidates for further experimental evaluation. In tracer development further preclinical in vivo studies and finally clinical PET experiments are needed to find the few compounds that have the desired kinetic characteristics necessary for a radiotracer. The main differences between drug and tracer are the time, concentration, and specificity perspectives. The time constraint infers that the tracer must reach the target organ and bind to the target protein in sufficient amount while, at the same time, the amount of free tracer in blood and tissue should be low to allow a high signal to noise level. This must be achieved within the short time frame for the half-life of the radionuclide. Moreover, since the tracer concept requires that the tracer does not occupy more than 5% of the target protein, it is not possible to use too high a concentration of the tracer to increase tissue uptake. As the influx parameter K_1 is a function of plasma concentration, the unidirectional influx is determined by the inherent rate constant and cannot be influenced without compromising the tracer concept. Specificity is another important parameter and tracer requirements are, apart from high ratio of binding to the target versus other receptor or enzyme proteins, also dependent on nonspecific tissue binding. A too high nonspecific binding precludes a correct quantitative determination of the tracer–target protein interaction. As can be understood, tracer development is both similar to but also different from drug development. Some aspects are common, such as affinity and selectivity studies, whereas other aspects only apply for either of the two processes.

In this chapter we have discussed some of the transporters where useful or at least potentially useful tracers exist. There are several transporters of physiological importance not covered here, and we foresee developments in the future where PET tracers also become available for the study of other transporters that may give insight into different brain disorders in terms of etiology or disease progress. Molecular targets of interest lacking a suitable PET imaging tool are, for example, the glycine, glutamate, and GABA transporters. The joint collaboration between scientists in the field of drug development, with access to structure templates with affinity for specific transporter proteins, and PET experts, with competence within labeling synthesis and PET methodology, is the key for further successful innovations in this area of research.

References

1. Laakso A, Hietala J (2000) Curr Pharm Des 6:1611
2. Guilloteau D, Chalon S (2005) Curr Pharm Des 11:3237
3. Valk PE, Bailey DL, Townsend DW, Maisey MN (2003) Positron emission tomography—basic science and clinical applications. Springer, London
4. Långström B, Bergson G (1980) Radiochem Radioanal Lett 43:47
5. Långström B, Kihlberg T, Bergström M, Antoni G, Björkman M, Forngren BH, Forngren T, Hartvig P, Markides K, Yngve U, Ögren M (1999) Acta Chem Scand 53:651
6. Antoni G, Kihlberg T, Långström B (2003) In: Vértes A, Nagy S, Klencsár Z (eds) Handbook of nuclear chemistry, vol 4. Radiochemistry and radiopharmaceutical chemistry in life sciences. Kluwer, The Netherlands, p 119
7. Antoni G, Långström B (2003) In: Welch MJ, Redvanly CS (eds) Handbook of radiopharmaceuticals: radiochemistry and applications. Wiley, UK, p 141
8. Logan J (2000) Nucl Med Biol 27:661
9. Patlak CS, Blasberg RG (1985) J Cereb Blood Flow Metab 5:584
10. De Hevesy G (1923) Biochem J 17:439
11. Lipinski CA, Lombardo F, Dominy BW, Feeney PJ (2001) Adv Drug Deliv Rev 46:3
12. Waterhouse RN (2003) Mol Imaging Biol 5:376
13. Blakely RD, De Felice LJ, Hartzell HC (1994) J Exp Biol 196:263
14. Zhou J (2004) Drugs Future 29:1235
15. Paczkowski FA, Bryan-Lluka LJ, Porzgen P, Bruss M, Bonisch H (1999) J Pharmacol Exp Ther 290:761

16. Macey DJ, Smith HR, Nader MA, Porrino LJ (2003) J Neurosci 23:12
17. Ding YS, Lin KS, Logan J (2006) Curr Pharm Des 12:3831
18. Owens MJ, Morgan WN, Plott SJ, Nemeroff CB (1997) J Pharmacol Exp Ther 283:1305
19. Van Dort ME, Kim JH, Tluczek L, Wieland DM (1997) Nucl Med Biol 24:707
20. Schou M, Sovago J, Pike VW, Gulyas B, Bogeso KP, Farde L, Halldin C (2006) Mol Imaging Biol 8:1
21. McConathy J, Owens MJ, Kilts CD, Malveaux EJ, Camp VM, Votaw JR, Nemeroff CB, Goodman MM (2004) Nucl Med Biol 31:705
22. Zavidovskaia GI (1969) Zh Nevropatol Psikhiatr Im S S Korsakova 69:594
23. Strömgren LS, Friderichsen T (1971) Nord Psykiatr Tidsskr 25:119
24. Haka MS, Kilbourn MR (1989) Int J Radiat Appl Instrum B 16:771
25. Chumpradit S, Kung MP, Panyachotipun C, Prapansiri V, Foulon C, Brooks BP, Szabo SA, Tejani-Butt S, Frazer A, Kung HF (1992) J Med Chem 35:4492
26. Wilson AA, Johnson DP, Mozley D, Hussey D, Ginovart N, Nobrega J, Garcia A, Meyer J, Houle S (2003) Nucl Med Biol 30:85
27. Logan J, Ding YS, Lin KS, Pareto D, Fowler J, Biegon A (2005) Nucl Med Biol 32:531
28. Severance AJ, Milak MS, Kumar JS, Prabhakaran J, Majo VJ, Simpson NR, Van Heertum RL, Arango V, Mann JJ, Parsey RV (2007) Eur J Nucl Med Mol Imaging 34:688
29. Ding YS, Lin KS, Garza V, Carter P, Alexoff D, Logan J, Shea C, Xu Y, King P (2003) Synapse 50:345
30. Öhman D, Norlander B, Peterson C, Bengtsson F (2001) Ther Drug Monit 23:27
31. Schou M, Halldin C, Sovago J, Pike VW, Gulyas B, Mozley PD, Johnson DP, Hall H, Innis RB, Farde L (2003) Nucl Med Biol 30:707
32. Fowler JS, Wang GJ, Logan J, Xie S, Volkow ND, MacGregor RR, Schlyer DJ, Pappas N, Alexoff DL, Patlak C et al (1995) J Nucl Med 36:1255
33. Schou M, Halldin C, Pike VW, Mozley PD, Dobson D, Innis RB, Farde L, Hall H (2005) Eur Neuropsychopharmacol 15:517
34. Schou M, Halldin C, Sovago J, Pike VW, Hall H, Gulyas B, Mozley PD, Dobson D, Shchukin E, Innis RB, Farde L (2004) Synapse 53:57
35. Hall H, Halldin C, Guilloteau D, Chalon S, Emond P, Besnard J, Farde L, Sedvall G (1999) NeuroImage 9:108
36. Hurd YL, Pristupa ZB, Herman MM, Niznik HB, Kleinman JE (1994) Neuroscience 63:357
37. Piggott MA, Marshall EF, Thomas N, Lloyd S, Court JA, Jaros E, Burn D, Johnson M, Perry RH, McKeith IG, Ballard C, Perry EK (1999) Brain 122:1449
38. Prunier C, Payoux P, Guilloteau D, Chalon S, Giraudeau B, Majorel C, Tafani M, Bezard E, Esquerre JP, Baulieu JL (2003) J Nucl Med 44:663
39. Tissingh G, Bergmans P, Booij J, Winogrodzka A, Van Royen EA, Stoof JC, Wolters EC (1998) J Neurol 245:14
40. Shih MC, Hoexter MQ, Andrade LA, Bressan RA (2006) Sao Paulo Med J 124:168
41. Costa DC, Walker Z, Walker RW, Fontes FR (2003) Mov Disord 18(Suppl 7):S34
42. O'Brien JT, Colloby S, Fenwick J, Williams ED, Firbank M, Burn D, Aarsland D, McKeith IG (2004) Arch Neurol 61:919
43. Small GW (2004) Dement Geriatr Cogn Disord 17(Suppl 1):25
44. Sjöholm H, Bratlid T, Sundsfjord J (2004) Psychopharmacology (Berl) 173:27
45. Dean B, Hussain T (2001) Schizophr Res 52:107
46. Hsiao MC, Lin KJ, Liu CY, Tzen KY, Yen TC (2003) Schizophr Res 65:39
47. Laakso A, Bergman J, Haaparanta M, Vilkman H, Solin O, Syvälahti E, Hietala J (2001) Schizophr Res 52:115

48. Marcusson J, Eriksson K (1988) Brain Res 457:122
49. Innis R, Baldwin R, Sybirska E, Zea Y, Laruelle M, al-Tikriti M, Charney D, Zoghbi S, Smith E, Wisniewski G, Hoffer PB, Wang S, Milius RA, Neumeyer JL (1991) Eur J Pharmacol 200:369
50. Farde L, Halldin C, Müller L, Suhara T, Karlsson P, Hall H (1994) Synapse 16:93
51. Neumeyer JL, Wang SY, Gao YG, Milius RA, Kula NS, Campbell A, Baldessarini RJ, Zea-Ponce Y, Baldwin RM, Innis RB (1994) J Med Chem 37:1558
52. Benamer TS, Patterson J, Grosset DG, Booij J, De Bruin K, Van Royen E, Speelman JD, Horstink MH, Sips HJ, Dierckx RA, Versijpt J, Decoo D, Van Der Linden C, Hadley DM, Doder M, Lees AJ, Costa DC, Gacinovic S, Oertel WH, Pogarell O, Hoeffken H, Joseph K, Tatsch K, Schwarz J, Ries V (2000) Mov Disord 15:503
53. Catafau AM, Tolosa E (2004) Mov Disord 19:1175
54. Bergström KA, Halldin C, Hall H, Lundkvist C, Ginovart N, Swahn CG, Farde L (1997) Eur J Nucl Med 24:596
55. Lundkvist C, Halldin C, Ginovart N, Swahn CG, Farde L (1997) Nucl Med Biol 24:621
56. Chalon S, Garreau L, Emond P, Zimmer L, Vilar MP, Besnard JC, Guilloteau D (1999) J Pharmacol Exp Ther 291:648
57. Guilloteau D, Emond P, Baulieu JL, Garreau L, Frangin Y, Pourcelot L, Mauclaire L, Besnard JC, Chalon S (1998) Nucl Med Biol 25:331
58. Ziebell M, Thomsen G, Knudsen GM, De Nijs R, Svarer C, Wagner A, Pinborg LH (2007) Eur J Nucl Med Mol Imaging 34:101
59. Pinborg LH, Ziebell M, Frokjaer VG, De Nijs R, Svarer C, Haugbol S, Yndgaard S, Knudsen GM (2005) J Nucl Med 46:1119
60. Prunier C, Bezard E, Montharu J, Mantzarides M, Besnard JC, Baulieu JL, Gross C, Guilloteau D, Chalon S (2003) NeuroImage 19:810
61. Jucaite A, Odano I, Olsson H, Pauli S, Halldin C, Farde L (2006) Eur J Nucl Med Mol Imaging 33:657
62. Halldin C, Erixon-Lindroth N, Pauli S, Chou YH, Okubo Y, Karlsson P, Lundkvist C, Olsson H, Guilloteau D, Emond P, Farde L (2003) Eur J Nucl Med Mol Imaging 30:1220
63. Leroy C, Comtat C, Trebossen R, Syrota A, Martinot JL, Ribeiro MJ (2007) J Nucl Med 48:538
64. Dolle F, Emond P, Mavel S, Demphel S, Hinnen F, Mincheva Z, Saba W, Valette H, Chalon S, Halldin C, Helfenbein J, Legaillard J, Madelmont JC, Deloye JB, Bottlaender M, Guilloteau D (2006) Bioorg Med Chem 14:1115
65. Saba W, Valette H, Schollhorn-Peyronneau MA, Coulon C, Ottaviani M, Chalon S, Dolle F, Emond P, Halldin C, Helfenbein J, Madelmont JC, Deloye JB, Guilloteau D, Bottlaender M (2007) Synapse 61:17
66. Mantere T, Tupala E, Hall H, Särkioja T, Räsänen P, Bergström K, Callaway J, Tiihonen J (2002) Am J Psychiatry 159:599
67. Smith HR, Daunais JB, Nader MA, Porrino LJ (1999) Ann NY Acad Sci 877:700
68. Mukherjee J, Das MK, Yang ZY, Lew R (1998) Nucl Med Biol 25:605
69. Szabo Z, Scheffel U, Mathews WB, Ravert HT, Szabo K, Kraut M, Palmon S, Ricaurte GA, Dannals RF (1999) J Cereb Blood Flow Metab 19:967
70. Szabo Z, Kao PF, Scheffel U, Suehiro M, Mathews WB, Ravert HT, Musachio JL, Marenco S, Kim SE, Ricaurte GA et al (1995) Synapse 20:37
71. Suehiro M, Scheffel U, Ravert HT, Dannals RF, Wagner HN Jr (1993) Life Sci 53:883
72. Ichimiya T, Suhara T, Sudo Y, Okubo Y, Nakayama K, Nankai M, Inoue M, Yasuno F, Takano A, Maeda J, Shibuya H (2002) Biol Psychiatry 51:715
73. Reivich M, Amsterdam JD, Brunswick DJ, Shiue CY (2004) J Affect Disord 82:321

74. Brust P, Hinz R, Kuwabara H, Hesse S, Zessin J, Pawelke B, Stephan H, Bergmann R, Steinbach J, Sabri O (2003) Neuropsychopharmacology 28:2010
75. Kretzschmar M, Brust P, Zessin J, Cumming P, Bergmann R, Johannsen B (2003) Eur Neuropsychopharmacol 13:387
76. Reneman L, Booij J, Lavalaye J, De Bruin K, De Wolff FA, Koopmans RP, Stoof JC, Den Heeten GJ (1999) Synapse 34:77
77. Hiltunen J, Åkerman KK, Kuikka JT, Bergström KA, Halldin C, Nikula T, Räsänen P, Tiihonen J, Vauhkonen M, Karhu J, Kupila J, Lansimies E, Farde L (1998) Eur J Nucl Med 25:19
78. Ginovart N, Wilson AA, Meyer JH, Hussey D, Houle S (2001) J Cereb Blood Flow Metab 21:1342
79. Houle S, Ginovart N, Hussey D, Meyer JH, Wilson AA (2000) Eur J Nucl Med 27:1719
80. Wilson AA, Ginovart N, Hussey D, Meyer J, Houle S (2002) Nucl Med Biol 29:509
81. Meyer JH (2007) J Psychiatry Neurosci 32:86
82. Lundquist P, Wilking H, Höglund AU, Sandell J, Bergström M, Hartvig P, Långström B (2005) Nucl Med Biol 32:129
83. Ginovart N, Wilson AA, Meyer JH, Hussey D, Houle S (2003) Synapse 47:123
84. Lundberg J, Halldin C, Farde L (2006) Synapse 60:256
85. Lundberg J, Odano I, Olsson H, Halldin C, Farde L (2005) J Nucl Med 46:1505
86. Fang P, Shiue GG, Shimazu T, Greenberg JH, Shiue CY (2004) Appl Radiat Isot 61:1247
87. Huang Y, Hwang DR, Narendran R, Sudo Y, Chatterjee R, Bae SA, Mawlawi O, Kegeles LS, Wilson AA, Kung HF, Laruelle M (2002) J Cereb Blood Flow Metab 22:1377
88. Jarkas N, Votaw JR, Voll RJ, Williams L, Camp VM, Owens MJ, Purselle DC, Bremner JD, Kilts CD, Nemeroff CB, Goodman MM (2005) Nucl Med Biol 32:211
89. Jarkas N, McConathy J, Votaw JR, Voll RJ, Malveaux E, Camp VM, Williams L, Goodman RR, Kilts CD, Goodman MM (2005) Nucl Med Biol 32:75
90. Scherman D, Raisman R, Ploska A, Agid Y (1988) J Neurochem 50:1131
91. Kilbourn MR (1997) Nucl Med Biol 24:615
92. Batchelor M, Schenk JO (1998) J Neurosci 18:10304
93. DaSilva JN, Carey JE, Sherman PS, Pisani TJ, Kilbourn MR (1994) Nucl Med Biol 21:151
94. Frey KA, Koeppe RA, Kilbourn MR, Vander Borght TM, Albin RL, Gilman S, Kuhl DE (1996) Ann Neurol 40:873
95. Bohnen NI, Albin RL, Koeppe RA, Wernette KA, Kilbourn MR, Minoshima S, Frey KA (2006) J Cereb Blood Flow Metab 26:1198
96. Simpson NR, Souza F, Witkowski P, Maffei A, Raffo A, Herron A, Kilbourn M, Jurewicz A, Herold K, Liu E, Hardy MA, Van Heertum R, Harris PE (2006) Nucl Med Biol 33:855
97. Schinkel AH, Jonker JW (2003) Adv Drug Deliv Rev 55:3
98. Tan B, Piwnica-Worms D, Ratner L (2000) Curr Opin Oncol 12:450
99. Kortekaas R, Leenders KL, Van Oostrom JC, Vaalburg W, Bart J, Willemsen AT, Hendrikse NH (2005) Ann Neurol 57:176
100. Hendrikse NH, Franssen EJ, Van der Graaf WT, Vaalburg W, De Vries EG (1999) Eur J Nucl Med 26:283
101. Del Vecchio S, Ciarmiello A, Pace L, Potena MI, Carriero MV, Mainolfi C, Thomas R, D'Aiuto G, Tsuruo T, Salvatore M (1997) J Nucl Med 38:1348
102. Sikic BI, Fisher GA, Lum BL, Halsey J, Beketic-Oreskovic L, Chen G (1997) Cancer Chemother Pharmacol 40(Suppl):S13

103. Marian T, Szabo G, Goda K, Nagy H, Szincsak N, Juhasz I, Galuska L, Balkay L, Mikecz P, Tron L, Krasznai Z (2003) Eur J Nucl Med Mol Imaging 30:1147
104. Fukumoto M, Yoshida D, Hayase N, Kurohara A, Akagi N, Yoshida S (1999) Cancer 86:1470
105. Sharma V, Beatty A, Wey SP, Dahlheimer J, Pica CM, Crankshaw CL, Bass L, Green MA, Welch MJ, Piwnica-Worms D (2000) Chem Biol 7:335
106. Sharma V (2004) Bioconjug Chem 15:1464
107. Sharma V, Prior JL, Belinsky MG, Kruh GD, Piwnica-Worms D (2005) J Nucl Med 46:354
108. Levchenko A, Mehta BM, Lee JB, Humm JL, Augensen F, Squire O, Kothari PJ, Finn RD, Leonard EF, Larson SM (2000) J Nucl Med 41:493
109. Bart J, Dijkers EC, Wegman TD, De Vries EG, Van der Graaf WT, Groen HJ, Vaalburg W, Willemsen AT, Hendrikse NH (2005) Br J Pharmacol 145:1045
110. Hendrikse NH, Vaalburg W (2002) Novartis Found Symp 243:137
111. Hendrikse NH, De Vries EG, Franssen EJ, Vaalburg W, Van der Graaf WT (2001) Eur J Clin Pharmacol 56:827
112. Luurtsema G, Molthoff CF, Schuit RC, Windhorst AD, Lammertsma AA, Franssen EJ (2005) Nucl Med Biol 32:87
113. Nelson WL, Olsen LD, Beitner DB, Pallow RJ Jr (1988) Drug Metab Dispos 16:184
114. Hsiao P, Sasongko L, Link JM, Mankoff DA, Muzi M, Collier AC, Unadkat JD (2006) J Pharmacol Exp Ther 317:704
115. Bart J, Willemsen AT, Groen HJ, Van der Graaf WT, Wegman TD, Vaalburg W, De Vries EG, Hendrikse NH (2003) NeuroImage 20:1775
116. Hendrikse NH, Schinkel AH, De Vries EG, Fluks E, Van der Graaf WT, Willemsen AT, Vaalburg W, Franssen EJ (1998) Br J Pharmacol 124:1413
117. Syvänen S, Blomquist G, Sprycha M, Höglund UA, Roman M, Eriksson O, Hammarlund-Udenaes M, Långström B, Bergström M (2006) NeuroImage 32:1134
118. Sasongko L, Link JM, Muzi M, Mankoff DA, Yang X, Collier AC, Shoner SC, Unadkat JD (2005) Clin Pharmacol Ther 77:503
119. Toornvliet R, Van Berckel BN, Luurtsema G, Lubberink M, Geldof AA, Bosch TM, Oerlemans R, Lammertsma AA, Franssen EJ (2006) Clin Pharmacol Ther 79:540
120. Vogelgesang S, Cascorbi I, Schroeder E, Pahnke J, Kroemer HK, Siegmund W, Kunert-Keil C, Walker LC, Warzok RW (2002) Pharmacogenetics 12:535
121. Kurdziel KA, Kiesewetter DO, Carson RE, Eckelman WC, Herscovitch P (2003) J Nucl Med 44:1330
122. Kiesewetter DO, Jagoda EM, Kao CH, Ma Y, Ravasi L, Shimoji K, Szajek LP, Eckelman WC (2003) Nucl Med Biol 30:11
123. Gambhir SS, Czernin J, Schwimmer J, Silverman DH, Coleman RE, Phelps ME (2001) J Nucl Med 42:1S
124. Mosconi L (2005) Eur J Nucl Med Mol Imaging 32:486
125. Di Carli MF (1998) Cardiol Rev 6:290
126. Kunkel M, Reichert TE, Benz P, Lehr HA, Jeong JH, Wieand S, Bartenstein P, Wagner W, Whiteside TL (2003) Cancer 97:1015
127. Kato H, Takita J, Miyazaki T, Nakajima M, Fukai Y, Masuda N, Fukuchi M, Manda R, Ojima H, Tsukada K, Kuwano H, Oriuchi N, Endo K (2003) Anticancer Res 23:3263
128. Mamede M, Higashi T, Kitaichi M, Ishizu K, Ishimori T, Nakamoto Y, Yanagihara K, Li M, Tanaka F, Wada H, Manabe T, Saga T (2005) Neoplasia 7:369
129. Reske SN, Grillenberger KG, Glatting G, Port M, Hildebrandt M, Gansauge F, Beger HG (1997) J Nucl Med 38:1344
130. Simpson IA, Carruthers A, Vannucci SJ (2007) J Cereb Blood Flow Metab

131. Vannucci SJ, Clark RR, Koehler-Stec E, Li K, Smith CB, Davies P, Maher F, Simpson IA (1998) Dev Neurosci 20:369
132. Paternostro G, Pagano D, Gnecchi-Ruscone T, Bonser RS, Camici PG (1999) Cardiovasc Res 42:246
133. Oquendo MA, Krunic A, Parsey RV, Milak M, Malone KM, Anderson A, Van Heertum RL, Mann JJ (2005) Neuropsychopharmacology 30:1163
134. Nordberg A, Lilja A, Lundqvist H, Hartvig P, Amberla K, Viitanen M, Warpman U, Johansson M, Hellstrom-Lindahl E, Bjurling P et al (1992) Neurobiol Aging 13:747
135. Kelloff GJ, Hoffman JM, Johnson B, Scher HI, Siegel BA, Cheng EY, Cheson BD, O'Shaughnessy J, Guyton KZ, Mankoff DA, Shankar L, Larson SM, Sigman CC, Schilsky RL, Sullivan DC (2005) Clin Cancer Res 11:2785
136. Dutka DP, Camici PG (2003) Heart Fail Rev 8:167
137. Landau BR, Spring-Robinson CL, Muzic RF Jr, Rachdaoui N, Rubin D, Berridge MS, Schumann WC, Chandramouli V, Kern TS, Ismail-Beigi F (2007) Am J Physiol Endocrinol Metab 293:E237

Top Med Chem (2009) 4: 187–222
DOI 10.1007/7355_2008_026
© Springer-Verlag Berlin Heidelberg
Published online: 16 October 2008

Pharmacology of Glutamate Transport in the CNS: Substrates and Inhibitors of Excitatory Amino Acid Transporters (EAATs) and the Glutamate/Cystine Exchanger System x_c^-

Richard J. Bridges (✉) · Sarjubhai A. Patel

Center for Structural and Functional Neuroscience, Department of Biomedical and Pharmaceutical Sciences, University of Montana, Missoula, MT 59812, USA
richard.bridges@umontana.edu

Abstract As the primary excitatory neurotransmitter in the mammalian CNS, L-glutamate participates not only in standard fast synaptic communication, but also contributes to higher order signal processing, as well as neuropathology. Given this variety of functional roles, interest has been growing as to how the extracellular concentrations of L-glutamate surrounding neurons are regulated by cellular transporter proteins. This review focuses on two prominent systems, each of which appears capable of influencing both the signaling and pathological actions of L-glutamate within the CNS: the sodium-dependent excitatory amino acid transporters (EAATs) and the glutamate/cystine exchanger, system x_c^- (Sx_c^-). While the family of EAAT subtypes limit access to glutamate receptors by rapidly and efficiently sequestering L-glutamate in neurons and glia, Sx_c^- provides a route for the export of glutamate from cells into the extracellular environment. The primary intent of this work is to provide an overview of the inhibitors and substrates that have been developed to delineate the pharmacological specificity of these transport systems, as well as be exploited as probes with which to selectively investigate function. Particular attention is paid to the development of small molecule templates that mimic the structural properties of the endogenous substrates, L-glutamate, L-aspartate and L-cystine and how strategic control of functional group position and/or the introduction of lipophilic R-groups can impact multiple aspects of the transport process, including: subtype selectivity, inhibitory potency, and substrate activity.

Keywords Aspartate · Aspartylamide · Benzyloxyaspartate · Pyrrolidine-dicarboxylate · Uptake

Abbreviations

#24	3-{[3′-Trifluoromethyl-2-methyl-1,1′-biphenyl-4-yl)carbonyl]amino}-L-alanine
4-MG	(2S,4R)-4-Methyl glutamate
2,4-MPDC	2,4-Methanopyrrolidine-2,4-dicarboxylate
(2S,3R)-3-Me-L-*trans*-2,3-PDC	(2S,3R)-3-Methyl-L-*trans*-2,3-pyrrolidine dicarboxylate
(2S,4R)-4-Me-L-*trans*-2,4-PDC	(2S,4R)-4-Methyl-L-*trans*-2,4-pyrrolidine dicarboxylate
(4S,2′S)-APOC	(4S,2′S)-2-[2′-Aminopropionate]-4,5-dihydro-oxazole-4-carboxylic acid
(4S,5S)-POAD	(4S,5S)-2-Phenyl-4,5-dihydro-oxazole-4,5-dicarboxylate
AHTP	(R,S)-2-Amino-3-(1-hydroxy-1,2,3-triazol-5-yl)propionate
AMPA	α-Amino-3-hydroxy-5-methyl-4-isoxazolepropionic acid
CNB-TBOA	(2S,3S)-3-[3-(4-Cyanobenzoylamino)benzyloxy]aspartate
DHK	Dihydrokainate
EPSC	Excitatory postsynaptic current
ETB-TBOA	(2S,3S)-3-{3-[4-Ethylbenzoylamino]benzyloxy}aspartate
KA	Kainic acid
L-*anti-endo*-3,4-MPDC	L-*anti-endo*-3,4-Methanopyrrolidine dicarboxylate
L-AP4	L(+)-2-Amino-4-phosphonobutyric acid
L-β-THA	L-β-*threo*-Hydroxy-aspartate
L-SOS	L-Serine-O-sulphate
L-TBOA	L-*threo*-Benzyloxy aspartate
L-*trans*-2,3-PDC	L-*trans*-2,3-Pyrrolidine dicarboxylate
L-*trans*-2,4-PDC	L-*trans*-2,4-Pyrrolidine dicarboxylate
NBI 59159	N-4-(9H-Fluoren-2-yl)-L-asparagine
NMDA	N-Methyl-D-aspartate
PMB-TBOA	(2S,3S)-3-[3-(4-Methoxybenzoylamino)benzyloxy]aspartate
(S)-4-CPG	(S)-4-Carboxy-phenylglycine
T3MG	*threo*-3-Methylglutamate
TDPA	S-2-Amino-3-(3-hydroxy-1,2,5-thiadiazol-5-yl)propionic acid
TFB-TBOA	(2S,3S)-3-(3-[4-(Trifluoromethyl)-benzoylamino]-benzyloxy)aspartate
WAY-855	3-Amino-tricyclo[2.2.1.0[2.6]]heptane-1,3-dicarboxylate
WAY 212922	N[4]-[7-(Trifluoromethyl)-9H-fluoren-2-yl]-L-asparagine

| WAY 213394 | N[4]-(2'-Methyl-1,1'-biphenyl-4-yl)-L-asparagine |
| WAY 213613 | N4-[4-(2-Bromo-4,5-difluorophenoxy)phenyl]-L-asparagine |

1
Introduction

As our understanding of the physiological and pathological significance of glutamate-mediated neurotransmission has increased over the past 40 years, so too has our interest in the development of the excitatory amino acid (EAA) analogues that can be used to study the structure, function, and therapeutic relevance of the participating proteins. Now recognized as the primary excitatory neurotransmitter in the mammalian CNS, this dicarboxylic amino acid participates not only in standard fast synaptic communication, but also contributes to the cellular signaling that underlies higher order processes such as development, synaptic plasticity, learning and memory [1]. To a large extent, the ability of L-glutamate (L-Glu) to participate in such a wide variety of functions is a result of the large diversity of EAA receptor subtypes and splice variants available for signal mediation. These receptors include those directly associated with ion channels (ionotropic receptors, iGluRs), such as NMDA, KA and AMPA receptors, as well as those coupled to second messenger systems (metabotropic receptors, mGluRs). The iGluRs have garnered significant attention from both mechanistic and therapeutic perspectives, because the over-activation of these receptors and subsequent influx of excessive levels of ions (e.g., Ca^{2+}, Na^{+}) is now accepted as the fundamental pathological pathway (i.e. excitotoxicity) underlying neuronal injury in both acute insults (e.g., stroke, trauma, hypoglycemia) and chronic neurological and neuropsychiatric disorders (e.g., amyotrophic lateral sclerosis, epilepsy, schizophrenia, Alzheimer's disease, Huntington's disease) [2–8]. The mGluR's, on the other hand, are more typically associated with the modulation of excitatory transmission through a variety of G-protein coupled signal transduction pathways [9, 10]. In addition to neurodegenerative pathways, interest in these receptors has been linked to a number of disorders where a therapeutic benefit may be found in establishing new balance points in excitatory signaling through regulatory mechanisms, including epilepsy, anxiety and pain management [9, 11, 12]. Not surprisingly, much of the progress made in understanding the pharmacology and physiological roles of EAA receptors have been dependent upon strong medicinal chemistry efforts that have yielded growing libraries of receptor specific agonists and antagonists.

More recently, attention has focused on another family of proteins that are also becoming accepted as integral to excitatory signaling i.e., the transport systems that regulate the extracellular and intracellular concentrations of glutamate in the CNS. Given the ability of the transporters to efficiently control the movement of glutamate into and out of cellular and subcellular compartments, these proteins have the potential to influence the amount and time-course of the excitatory transmitter that reaches synaptic and extrasynaptic receptors, thereby influencing both physiological signaling and pathological injury. Not surprisingly, this has also created a need for additional EAA analogues with which to selectively characterize these transport proteins. The present review focuses on two cellular transport systems that appear to have a significant impact on excitatory transmission and/or excitotoxicity: the sodium-dependent excitatory amino acid transporters (EAATs) and the glutamate/cystine exchanger, system x_c^- (Sx_c^-). The EAATs have long been recognized for the ability to efficiently sequester L-Glu in neurons and glia, thereby providing a route to clear this transmitter from the extracellular space and regulate its access to EAA receptors. Sx_c^-, on the other hand, functions as an obligate exchanger that typically couples the import of L-cystine (L-Cys$_2$) with the export of L-Glu. While the uptake of L-Cys$_2$ has been the more common focus of studies on Sx_c^-, recent evidence suggests that the resulting export of L-Glu may represent a non-synaptic source of this excitatory transmitter that may also contribute to both physiological and pathological signaling. This review is intended to complement other works examining the structure, regulation, and physiology of these transporters [13–19], by placing an emphasis on the small molecules that have been developed to delineate the pharmacological specificity of these two transport systems, as well as be exploited as probes with which to assess their potential contributions to CNS signaling and pathology.

2
Excitatory Amino Acid Transporters (EAATs)

Among the uptake systems addressed in this review the excitatory amino acid transporters (EAATs) have been studied for the longest period of time and, not surprisingly, are the most thoroughly characterized. Interest in glutamate transport initially arose within the context of establishing that this dicarboxylic acid was indeed an excitatory transmitter, as it provided a mechanism to meet the requisite criteria of transmitter inactivation in the absence of a synaptically localized degradative enzyme akin to acetylcholinesterase. The identification of the high affinity glutamate uptake systems was also key to unraveling the complexities of its metabolic compartmentalization in the brain and the recognition of the glutamine cycle, in which synaptically released glutamate is rapidly transported into astrocytes and converted to glutamine by glutamine synthetase [20]. In turn, the glutamine is shuttled back to the presynaptic terminal, where it is reconverted into glutamate by glutaminase for repackaging into synaptic vesicles. The significance of this pathway, and particularly the maintenance of low extracellular levels of glutamate, became increasingly evident as the pathological properties of L-Glu were delineated. Referred to as excitotoxicity, the over activation of ionotropic EAA receptors by excitatory agonists produces an excessive influx of ions (particularly Na^+ and Ca^{2+}) that triggers a number of pathological pathways [6, 21–23]. Glutamate-mediated neuronal injury has become increasingly recognized as a fundamental mechanism of CNS pathology that is now linked to acute insults (e.g., stroke, head trauma, spinal cord injury) as well as chronic neurodegenerative diseases (e.g., amyotrophic lateral sclerosis, Alzheimer's, Parkinson's and Huntington's disease [5, 24–27].

2.1
Properties

Initially characterized on a basis of activity, the transport system was distinguished by its sodium dependency and high affinity ($K_m \approx 1$–$5\,\mu M$) for L-Glu [28]. To date, five isoforms of the EAATs have been isolated and characterized. The first three were identified almost simultaneously: GLAST (EAAT 1) [29] and GLT1 (EAAT 2) [30], which are principally considered glial transporters, from rat brain and EAAC1 (EAAT 3) [31], a neuronal transporter, from rabbit intestine. In addition to the subsequent identification of the three homologous transporters from human brain [32], EAAT 4 and EAAT 5 were isolated from cDNA libraries of human cerebellum and human retina, respectively [33, 34]. On the basis of substrate specificity, EAATs 1–5 are considered as members of System X_{AG}^-, while molecular analysis reveals that the EAATs are members of a novel gene family [Human Genome Organization solute carrier family (SLC1)] that also includes the sodium-dependent neutral amino acid transporters ASCT1 and ASCT2 [35, 36]. EAAT 1–3 have been reported to be present outside the nervous system, while EAAT 4 is believed to be restricted to cerebellar and EAAT 5 to the retina [13, 37].

The EAATs are thought to function using an "alternate access" gating mechanism in which the substrate initially interacts with an outwardly facing binding domain [38, 39]. The binding of the substrate and requisite ions then produces a conformational change that orients the protein such that the substrate now has access to the intracellular compartment. In the instance of the EAATs, the transport of one molecule of L-Glu is coupled to the inward movement of 3 Na^+ ions and a H^+ [40]. Interestingly, recent crystallographic data (see below) suggest that a minimum of two sodium ions are needed for binding, while the third may be more closely linked to the transition/translocation step [39]. The reorientation that allows the binding domain to again be accessed extracellularly is coupled to the export of one K^+ ion. The Na^+ and K^+ gradients generated in CNS cells allows L-Glu to be concentrated intracellularly more than 5 orders of magnitude and provides a way to maintain L-Glu extracellular concentrations at a level that do not induce excitotoxic injury. Further, the stoichiometric ratio of substrate and its co-transported ions also makes the uptake of L-Glu electrogenic [41]. Advantageously, this property enables the transporter's activity, pharmacology and biophysics to be investigated using electrophysiological recording techniques. Using such approaches, the transporters have also been found to exhibit ion conductances that are stoichiometrically independent of uptake and more consistent with channel-like properties [33, 34, 42].

The publication of a crystal structure of an archaeal glutamate transporter homologue from *Pyrococcus horikoshii* (Glt$_{Ph}$) by Gouaux and colleagues in 2004 was a major turning point in understanding the molecular structure of the EAATs [43]. While a number of previous studies supported the conclusion that the EAATs exist as homomultimers, the number of proposed noncovalently associated subunits ranged from dimers to pentamers [44–46]. In the instance of Glt$_{Ph}$, the transporter is composed of three subunits that assemble into a bowl-like configuration with its solvent-filled basin facing the extracellular surface. Each subunit is composed of eight α-helical transmembrane segments and two helical hairpin loops. Subsequent studies on Glt$_{Ph}$, have yielded more clearly resolved binding sites for the substrate, as well as two of three requisite sodium ions needed for uptake [39]. Access to these sites appears to be gated by one of the two hairpin loops, HP 2. In the mechanistic model emerging from these studies, the binding of substrate and sodium prompts the movement and stabilization of the HP 2 loop in a closed configuration. Although not yet demonstrated, the HP 1 loop may play an analogous role in the subsequent gating of the substrate and access of ions to the intracellular compartment. Interestingly, when Glt$_{Ph}$ was crystallized with L-TBOA, rather than L-aspartate, the aryl group of this competitive inhibitor appeared to interact with the HP 2 and prevent this presumed gate from closing. Such an effect would explain the action of TBOA, as well as a number of other analogues discussed in the pharmacology section, as a non-substrate inhibitor; i.e., a ligand that binds to the substrate site but cannot be translocated across the membrane. Having such structural models in hand provides an important new contextual framework with which to evaluate our understanding of the pharmacological specificity of the EAATs that emerged from traditional SAR studies.

2.2
Pharmacology

Much of the pharmacological characterization of the EAATs have revolved around traditional competition assays, where EAA analogues are assayed for the ability to reduce the uptake of a radiolabeled substrate, typically ^{3}H-L-Glu, ^{3}H-L-aspartate or ^{3}H-D-aspartate. The latter of these three has been extensively employed because it affords the advantage of being metabolically inert. Early studies relied on a variety of CNS preparations, such as synaptosomes tissue slices, or primary cell cultures, while current experiments more often employ cells transfected to selectively express one of the five isolated subtypes. *Xenopus* oocytes have played an especially important role in this regard, not only as an expression system for cloning the subtypes, but also because of the ease with which electrophysiological techniques can be used to exploit the electrogenic character of EAAT-mediated uptake and quantify substrate-induced currents in these cells [47]. This approach has allowed inhibitors previously identified in radiolabeled flux studies to be further differentiated into alternative substrates or non-substrate inhibitors [48]. Indeed, it is now common to find substrates characterized on a basis of maximum current produced (I_{max}), rather than the traditional V_{max} value generated in radiolabel flux experiments.

Initial studies delineating the pharmacology of EAAT-mediated uptake were carried out prior to the isolation of the individual subtypes and therefore relied upon CNS preparations containing a mixture of isoforms. Nonetheless, a number of fundamental structure–activity relationships emerged from these experiments that may generalize to ligands capable of binding to most or all of the EAATs [28, 49, 50]. These include being an α-amino acid with a second acidic/charged group separated from the α-carboxylate by 2–4 carbon atoms. The uptake systems exhibit an enantioselectivity that is observed with the longer length ligands, such as L-Glu, but diminishes with the shorter length analogues; e.g., D-aspartate and D-cysteate are active not only as inhibitors, but also as substrates. While little if any variability is tolerated with respect to the α-carboxyl group, the distal carboxylate can be replaced with sulfinic and sulfonic, but not phosphate-containing, charged groups. Some limited modification can also be made to the distal carboxylate without loss of inhibitory activity, such as derivatization to a hydroxamate. Additional steric bulk appended directly on the carbon backbone also appears to be tolerated in the transporter binding site, as illustrated by the inhibitory activity of *threo*-OH-aspartate, 3-methyl-, and 4-methyl-glutamate [51]. In particular the hydroxy-aspartate provided a template for the development of the very potent TBOA library of inhibitors [52].

Further progress in the development of more potent inhibitors, as well as the first major inroads into subtype-selective ligands, emerged through the development and application of conformationally constrained analogues [53]. In these compounds, ring systems and/or steric bulk are introduced so as to limit the conformational flexibility of the compounds. As a result of this increased rigidity, the compounds mimic fewer of the conformations attainable by glutamate and, consequently, interact with a more limited range of glutamate binding sites. In addition to conformational bias, appending alkyl or aryl groups to the carbon backbone of substrates such as L-Glu or L-aspartate also serve to facilitate interactions between the ligands and lipophilic protein domains adjacent to the substrate binding site. As will be discussed below, such side group interactions may be key to enhancing the potency and/or subtype specificity of the inhibitors, as well as a determining factor in the differentiation of alternative substrates from non-substrate inhibitors. Another benefit of the conformational restrictions, particularly evident in compounds in which a glutamate or aspartate mimic is embedded in a ring system, is that the positions of the functional groups can be accurately mapped in 3D space and incorporated into SAR-based pharmacophore models [54–56]. Examples of analogues designed using both acyclic and cyclic templates can be found in the groups of inhibitors discussed below.

2.2.1
Substituted Aspartate Analogues

Long recognized as an alternative substrate, L-aspartate (Fig. 1a) has proven to be of particular value as a "backbone" template in the design of EAAT inhibitors. Indeed, two of the most potent families of blockers identified to date can be traced back to two aspartate analogues characterized in some of the earliest SAR work focused on glutamate uptake: β-*threo*-hydroxyaspartate (β-THA) and L-aspartate-β-hydroxamate (Fig. 1b,c) [50]. β-THA was shown to be an effective uptake blocker in tissue slices, synaptosomes and cultured astrocytes, while more recent studies with the cloned EAATs demonstrate that it acts as a competitive inhibitor at all five EAAT subtypes [47, 57]. Kinetic studies yielded K_i values that range from a low of about 1 μM (EAAT 4 and 5) to a high of about 40 μM at EAAT 3. Interestingly, β-THA also exhibits a wide range of activities with respect to its ability to serve as a substrate and be translocated across the plasma membrane. Thus, based on substrate-induced currents (i.e., I_{max}), β-THA is similar to L-Glu at EAATs 3 and 4,

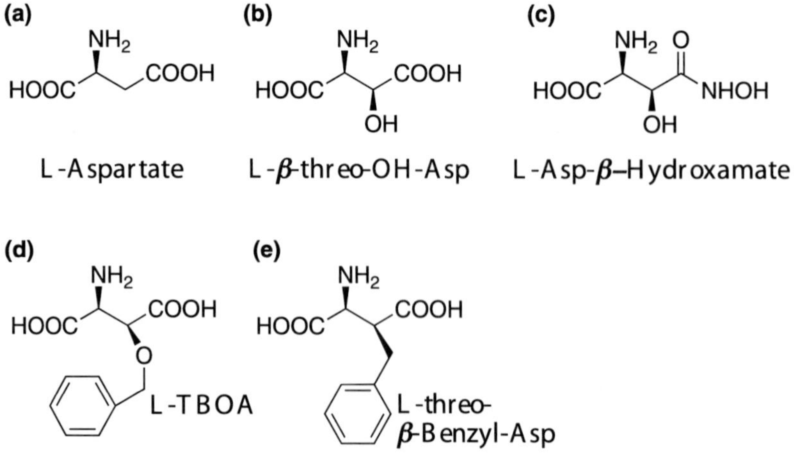

Fig. 1 β-Substituted aspartate analogues

about 30–70% as active as L-Glu at EAATs 1 and 2, and a non-substrate inhibitor of EAAT 5.

A significant advance in the development of much more potent inhibitors came from Shimamoto and colleagues who discovered that the hydroxy group of β-THA could serve as a linkage point for attachment of a variety of side chains. These compounds were initially synthesized with ester linkages and included the acetoxy, propionyloxy, benzoyloxy, and (1-, 2-napthoyl) oxy derivatives [52]. While these compounds proved to be effective competitive inhibitors when tested at the EAATs (most often EAAT 1 or 2), their use was limited because of chemical instability attributable to ester cleavage or acyl migration. Importantly, however, these results led to the development of comparable analogues that incorporated an ether linkage. Especially noteworthy was D,L-TBOA (Fig. 1d), which was identified at the time as one of the most potent EAAT 2 inhibitors yet recognized, exhibiting K_i values in the low to sub-μM range, depending upon the assay employed [58]. Further characterization revealed that: (i) more inhibitory activity resided with the L-enantiomer, (ii) TBOA competitively blocked all five EAATs, and (iii) in each instance TBOA was acting as a non-substrate inhibitor. Continued work by the Shimamoto group has produced a growing library of TBOA derivatives that included modifications to the benzyl moiety, as well as its replacement with a variety of aryl groups [59]. In addition to demonstrating that the presumed lipophilic pocket with which these side groups were binding could accommodate substantial steric bulk, some of the analogues were found to be significantly more potent than the parent compound. Thus, (2S,3S)-3-[3-(4-methoxybenzoylamino)benzyloxy]aspartate (PMB-TBOA) (Fig. 2a), (2S,3S)-3-{3-[4-(trifluoromethoxy)benzoylamino]benzyloxy}aspartate CF$_3$O-Bza-TBOA and

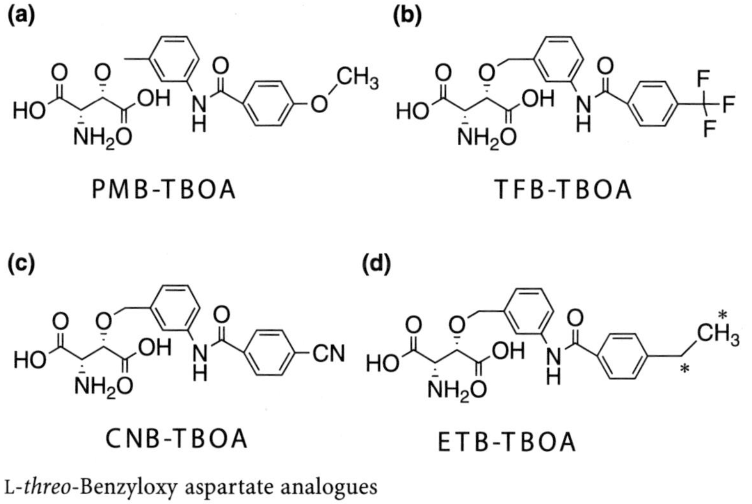

(a) PMB-TBOA

(b) TFB-TBOA

(c) CNB-TBOA

(d) ETB-TBOA

Fig. 2 L-*threo*-Benzyloxy aspartate analogues

(2S,3S)-3-{3-[4-(trifluoromethyl)benzoylamino]benzyloxy}aspartate (TFB-TBOA) (Fig. 2b) each exhibited IC_{50} values in the 1–10 nM range. Equally significant, these modifications also increased the selectivity of some of the analogues. For example, PMB-TBOA, (2S,3S)-3-[3-(4-cyanobenzoyl-amino)benzyloxy]aspartate (CNB-TBOA) (Fig. 2c) and (2S,3S)-3-[3-(4-fluro-benzoylamino)benzyloxy}aspartate (F-BzA-TBOA) are each 20 to 40-fold more potent at EAAT 2 relative to EAAT 3. Most recently a tritiated deriva-tive of (2S,3S)-3-{3-[4-ethylbenzoylamino]benzyloxy}aspartate (ETB-TBOA) (Fig. 2d) was synthesized and used as a radioligand to selectively investi-gate the binding properties of the EAATs [60]. This compound exhibited K_d values of 10 to 30 nM for EAATs 1, 2, 4 and 5, and 320 nM for EAAT 3. Beyond their value in distribution studies, such radioligands can serve as standards to cross-correlate SAR data that is strictly based upon ligand binding, much in the same way as has been done in studies of monoamine transporters [61, 62].

Esslinger and coworkers recently prepared the β-substituted aspartate analogue in which the benzyl group is attached to the carbon backbone via a methylene bond, rather than the ether linkage used in TBOA [54]. Significantly, L-β-benzyl-aspartate (Fig. 1e) proved to be one of the first com-petitive inhibitors identified that preferentially blocked EAAT 3. The *threo* diastereoisomer, which was more potent than the *erythro*, yielded a K_i value in C17 cells expressing EAAT 3 (0.8 µM) that was about 10-fold lower than found with either EAAT 1 or 2. Electrophysiological recording in oocytes ex-pressing EAAT 3 confirmed that L-β-*threo*-benzyl-aspartate is a non-substrate inhibitor. Computational modeling suggested that while the position of the carboxylate and amino groups are very similar, the differential activity at EAAT 3 may reside in the orientation of the benzyl group and/or the pres-ence of the ether oxygen and how these specifically interact with the binding domains of the individual EAATs.

2.2.2
Substituted Glutamate Analogues

While it has not proved quite as fruitful in terms of producing highly po-tent or specific inhibitors, similar modifications to the carbon backbone of L-Glu (Fig. 3a) has yielded a number of compounds that exhibit intriguing activities at the EAATs. Initial studies, many of which were carried out by Vandenberg and colleagues, focused on the addition of methyl or hydroxyl groups to the 3 and 4 positions of glutamate. In this manner, the ($\pm$)-*erythro* diastereoisomer of 3-methyl-glutamate was shown to exhibit little or no ac-tivity at either EAAT 1 or 2, while ($\pm$)-*threo*-3-methyl-glutamate (Fig. 3b) was found to be inactive at EAAT 1, yet act as a non-substrate inhibitor of EAAT 2 ($K_i \approx 20$ µM; [51]). In the instance of substitutions at the 4-position, both hydroxy and methyl groups are tolerated, but exhibit stereo-specific differ-ences [51, 63]. Thus, (2S,4S)-4-hydroxy-glutamate is an alternative substrate

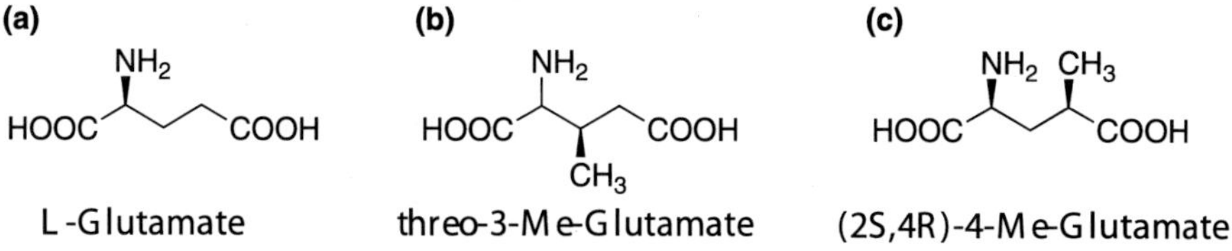

(a)

NH$_2$

HOOC⁀⁀COOH

L-Glutamate

(b)

NH$_2$

HOOC⁀⁀COOH

CH$_3$

threo-3-Me-Glutamate

(c)

NH$_2$ CH$_3$

HOOC⁀⁀COOH

(2S,4R)-4-Me-Glutamate

Fig. 3 3-, 4-Substituted glutamate analogues

of both EAAT 1 ($K_m \approx 61\,\mu M$) and EAAT 2 ($K_m \approx 48\,\mu M$), while the $2S,4R$ isomer appears inactive. In contrast, ($2S,4S$)-4-methylglutamate is inactive at EAATs 1 and 2, while the $2S,4R$ isomer (Fig. 3c) is an alternative substrate of EAAT 1 ($K_m \approx 54\,\mu M$) and a non-substrate inhibitor of EAAT 2 ($K_i \approx 3\,\mu m$). When substitutions at the 4-position of L-Glu were markedly expanded to include a variety of alkyl groups, the resulting SAR data was consistent with TBOA-based analogues with respect to the conclusion that the binding sites of EAATs 1–3 could accommodate considerable steric bulk, although the modifications did not yield a comparable increase in inhibitory potency. With the exception of the ability of EAAT 1 to use ($2S,4R$)-methylglutamate as a substrate, all of the alkyl-substituted analogues proved to be non-substrate inhibitors at all three transporters. ($2S,4R$)-4-Benzyl-glutamate was also of particular note because it preferentially inhibited EAAT 1, compared to either EAAT 2 or 3 [63].

2.2.3
Amide Derivatives of Aspartate and Diaminopropionate

Another valuable series of EAAT inhibitors has emerged from collaborative work of Dunlop, Foster, Butera and colleagues, using L-aspartate and diaminopropionate as structural templates [64, 65]. Rather than TBOA-like molecules, in which additions were made to the C3 position, this new series of analogues employed amide linkages to mimic the distal carboxylate group and as a point of attachment for a wide variety of aryl groups. This modification is particularly interesting as it suggests that the distal carboxylate, unlike its α counterpart, can be partially "masked". While the ability of such a modification to be tolerated in the binding site was suggested by early studies examining the activity of aspartate and glutamate hydroxamates, this library of analogues is particularly interesting because it contains inhibitors that both exhibit IC$_{50}$ values in the sub-μM range and display increased subtype selectivity. For example, WAY-213394 (IC$_{50} \approx$ 0.1 μM), WAY-2129222 and WAY-0213613 (Fig. 4a–c) are reported to be between 20 to 100-fold more potent as inhibitors of EAAT 2 (IC$_{50} \approx 0.1\,\mu M$), then either EAAT 1 or 3. Equally interesting are analogues such as NBI-59159 (IC$_{50} \approx 0.1\,\mu M$) and 3-{[3′-trifluoromethyl-2-methyl-1,1′-biphenyl-4-yl)carbonyl]amino}-L-alanine cited by Greenfield and co-workers as #24 [64]

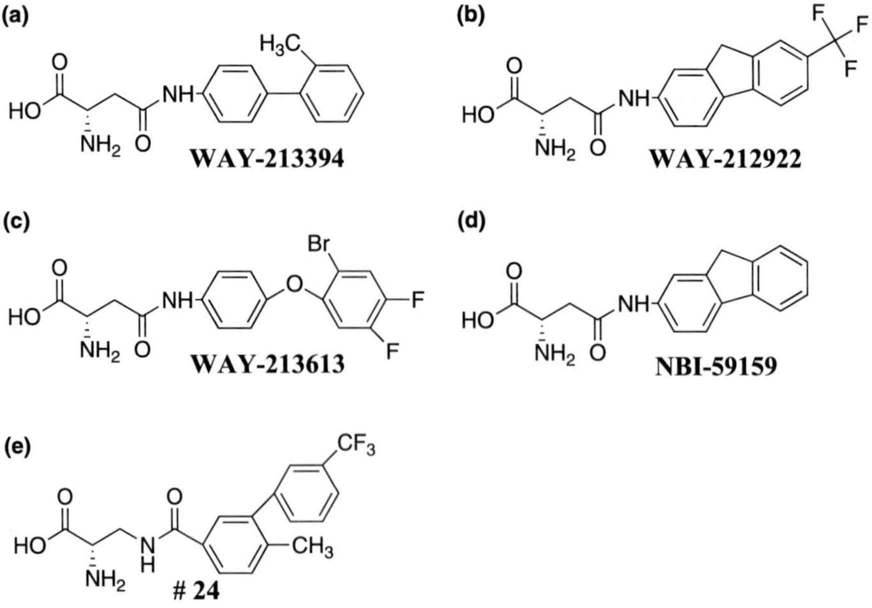

Fig. 4 Amide derivatives of aspartate and diaminopropionate

($IC_{50} \approx 9\,\mu M$) (Fig. 4d,e) that are among the first to show preferential activity at EAAT 3.

2.2.4
Carboxy-cyclopropyl and Cyclobutyl Analogues

As an alternative to appending functional groups directly to a glutamate or aspartate template, a number of inhibitor collections have exploited ring systems to bias, and often at times limit, the spatial configuration of the required carboxylic and amino function of the excitatory amino acids. The carboxy-cyclopropylglycines have proven quite advantageous in this respect, as this parent structure has led to pharmacological probes of not only transporters, but also ionotropic and metabotropic receptors [66, 67]. Three of the isomers, CCG-II, III and IV, effectively inhibited the transport of L-Glu, although these initial studies were primarily carried out in heterogeneous CNS preparations (e.g., synaptosomes, glial plasmalemmal vesicles), rather than with isolated EAAT subtypes [68]. In later studies employing expressed EAATs, both CCG-III and CCG-IV (Fig. 5a,b) were shown to inhibit EAAT 2 with IC_{50} values of 0.3 and $1\,\mu M$ respectively, while CCG-II produced no inhibitory activity when tested at $10\,\mu M$ [69]. Subsequent studies reported that CCG-III blocked EAAT 1 and EAAT 3, as well as EAAT 2 (although the K_i values were somewhat higher, e.g., $2.5-10\,\mu M$), but that CCG-IV was considerably less potent at EAATs 1-3 (K_i or IC_{50} values ranging from 170 to $900\,\mu M$) [58, 70].

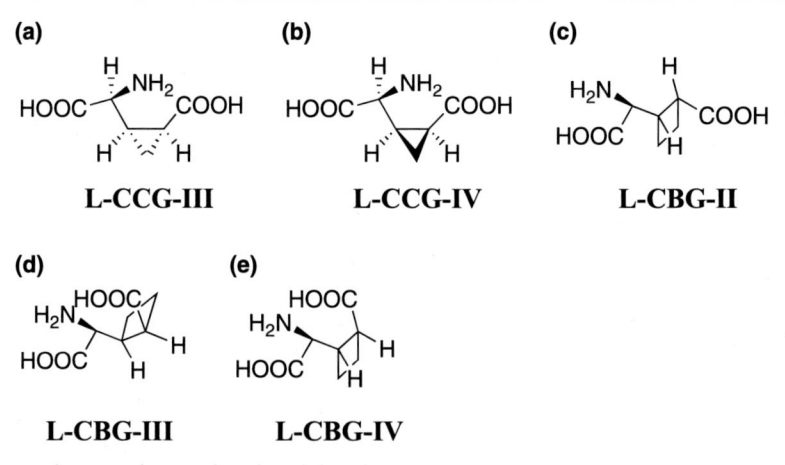

Fig. 5 Carboxy-cyclopropyl and cyclobutyl analogues

A related series of L-2-(2-carboxycyclobutyl)glycines (CBG's) was recently prepared using an aspartate aminotransferase in a chemo-enzymatic synthesis and assessed for EAAT activity with a FLIPR-based membrane potential assay [71, 72]. While none of the CBG isomers tested proved to be exceptionally potent inhibitors, L-CBG-I was found to be a substrate for EAAT 1–3, albeit with K_m values 10 to 20-fold greater than for L-Glu. In contrast, L-CBG-II (Fig. 5c) inhibits all three EAATs, acting as an alternative substrate of EAAT 1 (K_m = 96 μM) and a non-substrate inhibitor of EAATs 2 and 3 (K_m = 22 and 49 μM, respectively). A marked subtype selectivity (up to about 30-fold) was observed with L-CBG-III and -IV (Fig. 5d,e), which are non-substrate inhibitors of EAAT 2 and 3, but exhibit markedly less activity at EAAT 1. Of all the CBG's characterized, CBG-IV was reported to be the most potent, yielding K_i values of 7 and 10 μM at EAATs 2 and 3, respectively [71].

2.2.5
Pyrrolidine Dicarboxylate (PDC) Analogues

One of the earliest cyclic analogues of glutamate to garner attention as an uptake inhibitor was dihyrokainate (DHK) (Fig. 6a), the reduced form of the classic iGluR agonist kainate [73]. While first characterized as a transport blocker in heterogeneous CNS preparations, interest in DHK has remained, as studies with the isolated EAAT subtypes identified it as a selective inhibitor of EAAT 2 [32]. The same five-membered pyrrolidine ring present in DHK is also found in a series of analogues in which the distal carboxylate group is directly attached to the pyrrolidine ring rather than via a methylene linkage. In the first of this series, carboxylate groups were placed at the 2 and 4 positions to mimic glutamate. Of the four diastereoisomers, L-*trans*-2,4-pyrrolidine dicarboxylate (L-*trans*-2,4-PDC) (Fig. 6b) was found to be the most potent

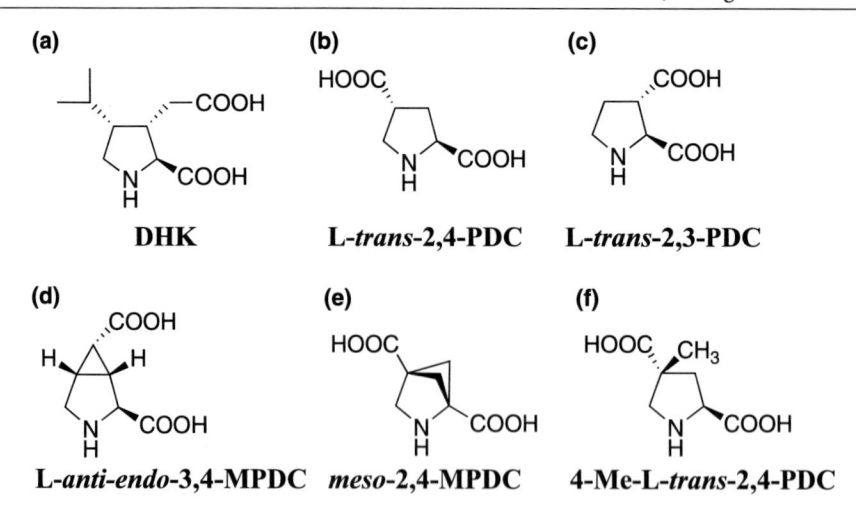

Fig. 6 Pyrrolidine dicarboxylate (PDC) analogues

inhibitor of synaptosomal L-Glu transport ($K_i \approx 2-5\,\mu M$) [74]. Significantly, L-*trans*-2,4-PDC also exhibited little if any activity at EAA ionotropic receptors. When later examined against the individual EAAT subtypes, L-*trans*-2,4-PDC was shown to: (i) inhibit all five subtypes, (ii) act as an alternative substrate of EAATs 1–4 ($K_m \approx 28$, 7, 27 and 3 μM, respectively) and (iii) be a non-substrate inhibitor of EAAT 5 ($EC_{50} \approx 6\,\mu M$) [32–34]. Repositioning the distal carboxylate mimic from the 4 to the 3 position on the pyrrolidine ring yielded a more "aspartate-like" compound that was void of substrate activity at EAAT 2, yet acted at this subtype as a non-substrate inhibitor ($K_i \approx 10\,\mu M$) with an enhanced selectivity (≈ 10 to 20 fold), compared to EAATs 1–4 ([48] and M.P. Kavanaugh, personal communication). With a functional group configuration more similar to aspartate, it was not surprising that L-*trans*-2,3-PDC (Fig. 6c), unlike L-*trans*-2,4-PDC, is a potent NMDA agonist and excitotoxin [75].

Further restricting the PDC template through the addition of methylene bridges has also demonstrated that subtle structural changes can produce compounds with very divergent properties at the EAATs, particularly as related to the ability of analogues to serve as alternative substrates. For example, L-*anti-endo*-3,4-MPDC (Fig. 6d) is a non-substrate inhibitor of EAAT 2, exhibiting K_i values of $\approx 5\,\mu M$ synaptosomes (in which the EAAT 2 subtype is predominant) and $\approx 9\,\mu M$ in oocytes expressing EAAT 2. The closely related 2,4-methanopyrrolidine-2,4-dicarboxylate (2,4-MPDC) (Fig. 6e) also binds and competitively inhibits synaptosomal uptake of ^3H-D-aspartate ($K_i \approx 7\,\mu M$), yet appears to be translocated quite effectively by EAAT 2, as it yielded an I_{max} value that is actually greater (115%, $K_m \approx 45\,\mu M$) than that produced by the endogenous substrate L-Glu [48]. Interestingly, these results suggest that conformational restriction does not necessarily attenuate substrate suitability. Analogous changes in activity can also be produced by introducing methyl groups into the PDC template. Thus, both L-*trans*-2,4-PDC and (2S,4R)-4-methyl-2,4-PDC (Fig. 6f) act as competitive inhibitors of EAAT 2-mediated uptake, (K_i = 2–3 µM in synaptosomes), although the presence of the methyl group essentially converts L-*trans*-2,4-PDC from a substrate to a non-substrate inhibitor ([55], also confirmed in oocytes expressing EAAT 2 by M.P. Kavanaugh). Taken together, these SAR studies highlight that the value of the PDC's rests not as much with their potency as inhibitors, but with their constrained structures that allow requisite functional group positions to be identified and exploited in the generation of pharmacophore models for both binding and translocation.

2.2.6
Heterocyclic and Carbocyclic Analogues

A number of other alternative ring systems have also been employed in the development of EAAT inhibitors, several of which were initially used as probes of EAA receptor pharmacology. For example, the library of azole-containing heterocyclic analogues used in the development of AMPA receptor agonists and antagonists by Krogsgaard-Larsen, Stensbol, Madsen, Brauner-Osborne and colleagues have also yielded compounds that exhibit activity at the EAATs [76, 77]. In particular, (R,S)-2-amino-3-(1-hydroxy-1,2,3-triazol-5-yl)propionate (AHTP) (Fig. 7a) inhibited ^{3}H-D-aspartate uptake into Cos-7 cells expressing EAAT 1 and EAAT 2 with IC_{50} values of about 100 and 300 µM, respectively. The compound appeared inactive at EAAT 3. This activity is notable not so much in terms of potency, but because this triazole derivative is one of the few compounds that preferentially acts at EAAT 1, rather than EAAT 2. In contrast, S-2-amino-3-(1-hydroxy-1,2,5-thiadiazol-4-yl)propionic acid (TDPA) (Fig. 7b), a related analogue in which the central nitrogen of the triazole group was replaced with sulfur, proved to be a selective inhibitor of EAAT 2 ($K_i \approx 100$ µM) [77]. Comparisons between these compounds and related AMPA analogues also led to the supposition that AMPA itself does not bind to the EAATs because its 5-methyl substituent is not tolerated within the binding site.

A similar series of analogues incorporating oxazole and oxazoline ring systems were developed by Campiani and coworkers and also shown to act as EAAT inhibitors [78, 79]. While also nitrogen-containing heterocyclics, these compounds, like the PDC's, possess carboxylate groups directly attached to the five-membered ring. Interestingly, when tested against EAAT 3 (EAAC 1) expressed in HEK-293 cells, (4S,2'S)-2-[2'-aminopropionate]-4,5-dihydro-oxazole-4-carboxylic acid ((4S,2'S)-APOC) (Fig. 7c) was found not only to bind, but to act as a substrate of the transporter, exhibiting an I_{max} comparable to L-Glu and a K_m value of about 30 µM. Like AHTP and TDPA, the presumed α-amino and α-carboxylate mimics of L-Glu are incorporated

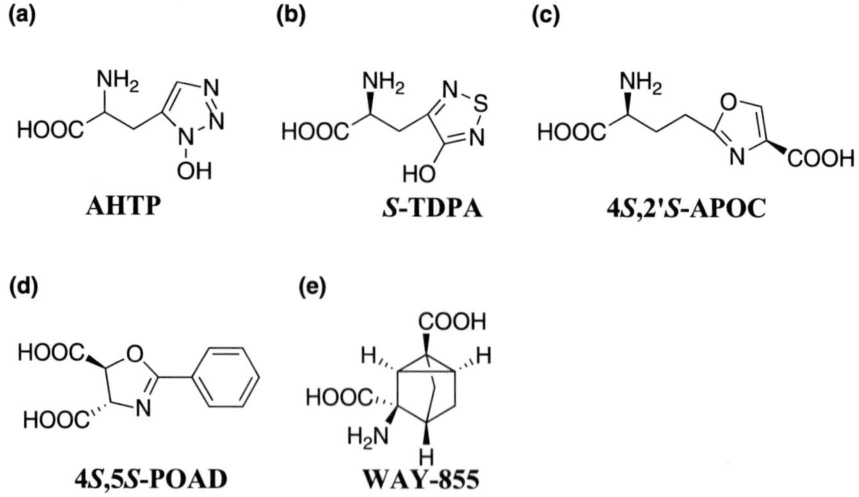

(a)   **(b)**   **(c)**

AHTP   **S-TDPA**   **4S,2'S-APOC**

(d)   **(e)**

4S,5S-POAD   **WAY-855**

Fig. 7 Heterocyclic and carbocyclic EAAT inhibitors

into APOC via a 2-amino-3-propionate group. If the linkage to the ring is increased by another methylene group, it results in a marked loss of activity. Interestingly, it appears as if the nitrogen of the oxazole ring can also function as an amino group mimic, similar to the PDC's, as (4S,5S)-2-phenyl-4,5-dihydro-oxazole-4,5-dicarboxylate (4S,5S-POAD) (Fig. 7d) inhibited EAAT 3 with a K_i value of 14 μM [78]. In contrast to 4S,2'S-APOC, however, 4S,5S-POAD acts as a non-substrate inhibitor.

Rather than a heterocyclic template, Dunlop and coworkers have also developed a tricycloheptane-based analogue, 3-amino-tricyclo[2.2.1.0[2.6]]heptane-1,3-dicarboxylate (WAY-855) (Fig. 7e), that is somewhat akin to the PDC analogues in its use of conformational restriction, with the exception that the α-amino acid group mimic is not embedded within the ring [80]. WAY-855 has attracted considerable attention because it displays a very strong preference for EAAT 2, compared to either EAAT 1 or 3. When tested in oocytes expressing EAATs 1–3, the analogue was reported to be a non-substrate inhibitor of all three subtypes that was 50-fold more potent at EAAT 2, where it exhibited an IC_{50} of about 1 μM.

2.3
Analogues as Probes of Substrate and Subtype Specificity

The combination of a growing library of analogues that potently bind to the EAATs, the ease with which the subtypes can be individually studied, and a crystallographic-based structural model of a homologous transporter have all had a dramatic impact on advancing our understanding of EAAT pharmacology. Two issues that have now progressed to the forefront of the field concern delineating the structure–activity relations that govern subtype selectivity and substrate activity. The goal of developing subtype selective inhibitors is especially applicable to functional studies aimed at determining if the various EAATs, particularly EAATs 1–3 differentially contribute to the physiological roles generally ascribed to uptake, including signal termination, transmitter recycling, and excitotoxic protection. To date the greatest progress has been made in developing inhibitors that exhibit a marked selectivity for EAAT 2. These compounds include DHK and L-*anti-endo*-3,4-PDC among the pyrrolidine dicarboxylates, WAY-213394 and WAY-213613 among the aspartamides, PivA-TBOA and CNB-TBOA among benzoylaspartates, and the tricyclohexane analogue WAY-855 [32, 48, 56, 80]. While not as much progress has been made with respect to the other subtypes, new lead compounds are beginning to emerge. In the instance of the neuronal transporter EAAT 3, L-β-*threo*-benzyl-aspartate and NBI-59159 have been shown to preferentially block EAAT 3 with IC_{50}/K_i values (0.8 and 0.1 μM, respectively) that are about 10 to 20 times lower than found with EAATs 1 or 2 [54, 81]. Both of these compounds function as non-substrate inhibitors. The triazole-based analogue AHTP is one of the first conformationally constrained analogues to exhibit a modest degree of selectivity for EAAT 1 ($\approx$ 3-fold), compared to EAATs 2 or 3, although it is relatively weak when compared to the potency of the more recently developed blockers ($IC_{50} \approx 100$ μM). Interestingly, substrate selectivity can also be used to distinguish EAAT 1 from EAAT 2. For example, L-serine-O-sulfate and 4-methylglutamate exhibit I_{max} values that are much more comparable to L-Glu at EAAT 1 ($\approx$ 100 and 80%, respectively), than at EAAT 1, where L-serine-O-sulfate is only a moderate substrate ($I_{max} \approx 50\%$) and 4-methylglutamate is a non-substrate inhibitor. As a majority of the analogues discussed above have only been assessed for activity at EAATs 1–3, considerably less is known about the specificity of EAATs 4 and 5. Initial studies on these transporters have, however, identified a few distinguishing properties also linked to substrate specificity. For example the longer glutamate homologues L-α-aminoadipate and L-homocysteic acid that exhibit little or no interactions with EAATs 1–3, serve as substrates of EAAT 4 with I_{max} greater than L-Glu itself. In contrast, the substrate specificity of EAAT 5, is readily differentiated from the other subtypes by β-THA and L-*trans*-2,4-PDC, which are partial or full substrates of EAATs 1–4, but non-substrate inhibitors of EAAT 5.

It is quite intriguing that subtle variations among the "R" groups within families of compounds such as the benzoylaspartates, benzylaspartates, and aspartamides may be central to developing subtype selective inhibitors. SAR-based pharmacophore modeling with inhibitors of EAAT 1-3 and, more recently, crystallographic results, are consistent with the conclusion that the primary functional groups of the endogenous substrates (i.e., α-amino, α-carboxyl and distal carboxyl groups), are likely occupying very similar positions within the transporter binding sites. In turn, this would suggest that the selectivity observed with the compounds described above may be due to the presence of lipophilic groups and the likelihood that the protein domains with which these groups interact vary among the subtypes. In particular, the structural data generated when the transporter (GLt$_{Ph}$) was crystallized with TBOA bound, suggest that the HP 2 loop thought to be involved in substrate gating may represent one of these lipophilic domains. However, the fact that a number of inhibitors, such as TBOA itself, act at all three EAATs and exhibit little or no selectivity, while the closely related PMB-TBOA and L-β-threo-benzyl-aspartate exhibit preferences for EAAT 2 and EAAT 3, respectively, indicates that it is not merely the presence or absence of a lipophilic R group that dictates specificity but probably subtle differences in its linkage, orientation and/or chemical composition.

The presence of the lipophilic R groups on these inhibitors also appears to be an important determinant in whether or not compounds act as alternative substrates for the transporters. Computational-based pharmacophore modeling with identified substrates demonstrated that these compounds not only share a very similar placement of amino and carboxylate groups, but also the space occupied by their respective carbon backbones [54]. Comparisons with other competitive inhibitors that lacked substrate activity allowed regions to be identified where steric bulk is tolerated within the binding site, but falls outside of the volume constrained by the substrates. In particular, there was considerable overlap in the regions occupied by the lipophilic R groups linked to the distal part of aspartate, such as present in the benzoylaspartates, benzy-laspartates, and aspartamides. The increased potency of these inhibitors led to the hypothesis that the interaction of these lipophilic R groups with specific protein domains enhanced binding, but also led to a loss of substrate activity. Significantly, crystallographic studies with TBOA added a structural explanation to this concept, as the R-group of TBOA was shown to interact with the HP 2 loop of GLt$_{Ph}$ and produce a configuration that was different from that observed in the "closed" orientation normally seen with the binding of aspartate [39]. Thus, the binding of TBOA (and its closely related analogues) may be enhanced because of its ability to interact with the HP 2 loop, but because this prevents the HP 2 loop from closing, the gating mechanism fails to operate and the inhibitor cannot be transported. It will be interesting to see if other non-substrate inhibitors, with smaller R-groups, (e.g., methyl-PDCs), can be utilized to define the limits of HP 2 loop movement during binding and translocation.

2.4
Analogues as Probes of Physiological and Pathological Roles

The EAATs are juxtaposed within the pathways of glutamate-mediated neurotransmission in such a manner as to potentially influence a number of variables, including: the concentration and time course with which vesicularly released glutamate interacts with synaptic EAA receptors, spillover of transmitter out of the synaptic cleft and access to extrasynaptic receptors, and the general maintenance of extracellular levels of glutamate below those which could trigger excitotoxic pathology. Considerable effort has been focused on trying to determine how these various roles are distributed and/or differ among transporter subtypes, cell populations, and specific circuits. Toward this goal the inhibitors and substrates discussed above have often been used in attempts to assess whether or not the EAATs significantly contribute to a particular activity. For example, the questions as to the respective contributions of uptake and diffusion to the termination of the excitatory signal has been investigated by examining if excitatory postsynaptic current (EPSC) properties are effected by the presence of EAAT inhibitors. Such studies have revealed a heterogeneity among synaptic connections, as transport blockers produced a slowing of the EPSC in some circuits (e.g., inputs to cerebellar Purkinje neurons), but not in others (between CA 3 and CA 1 hippocampal neurons) [82–85]. Comparisons among a number of preparations suggest that transport will have more of an influence on excitatory signaling where there is an increased likelihood that L-Glu can accumulate in the synaptic cleft. These relatively higher levels of L-Glu could be a consequence of increased release due to the presence of multivesicular release, increased density of release sites, or increased incidence of high frequency stimulation. Alternatively, any spatial characteristic that would make the cleft less accessible to the bulk CSF, such as glial ensheathment, would tend to lessen diffusion and favor accumulation. Examples of connections in which EAAT-mediated uptake has been reported to contribute to signal termination include synapses of the cerebellar climbing fibers, mossy fibers, retinal bipolar ganglion, and auditory brain stem [86–91].

The localization of EAATs around synapses also places these transporters in a position to regulate the amount of L-Glu that can either escape from or enter a particular synaptic cleft. In this manner, uptake can potentially serve as a control point in determining the extent to which L-Glu released from one synapse can activate EAA receptors present perisynaptically (i.e., spillover) or even in adjacent synapses (i.e., cross-talk). Once again, EAAT inhibitors have featured prominently in these studies and have been used to demonstrate that attenuating transporter activity can enhance the ability of L-Glu to reach and activate mGluRs at a distance. Examples include postsynaptic group I mGluR in the hippocampus and cerebellum, the group II mGluRs that mediate presynaptic inhibition on hippocampal neurons, and group III mGluRs on GABA terminals [92–94]. While it is more difficult to establish whether synaptic ionotropic receptors are being specifically activated by extrasynaptic L-Glu, studies suggest that this indeed is occurring and that the EAATs are influencing this signaling process in hippocampal CA1 cells, cerebellar stellate cells, and olfactory bulb mitral cells [95–98].

As will be discussed later in this review, the efflux of L-Glu through the System x_c^- (Sx_c^-) transporter rather than synaptic release may, in some brain regions, represent another physiological relevant source of extracellular L-Glu. The normal operation of Sx_c^- stoichiometrically couples the import of L-cystine with the export of L-Glu [99]. In addition to providing a needed amino acid precursor, evidence suggests that under some circumstances the resulting extracellular L-Glu provides for the tonic activation of presynaptic mGluRs that, in turn, regulate L-Glu release [100–102]. Even though the exchange of L-cystine and L-Glu are mediated by Sx_c^-, the EAATs play an integral role in both of these pathways. In the first, the EAATs maintain the L-Glu gradient necessary to drive the import of L-cystine, and in the second, it is the balance between the densities, location, and activities of both Sx_c^- and the EAATs that should ultimately determine if L-Glu accumulates to a sufficient level to activate the mGluRs.

As extracellular levels of L-Glu increase, the functional significance of the EAATs shifts from signal termination to excitotoxic protection. Both pharmacological and molecular manipulations of the EAATs indicate that there is an inverse relationship between transport activity and excitotoxic vulnerability. For example, co-administration of glutamate with transport inhibitors that exhibit little or no toxicity by themselves, such as L-2,4-PDC or β-THA, have been found to exacerbate the extent of the glutamate-mediated neuronal damage both in vitro and in vivo [103–106]. Subsequent studies also demonstrated that the pathological consequences of compromised EAAT activity was not necessarily dependent upon the inhibitors being co-administered with exogenous L-Glu. Thus, neuronal loss was also observed in long-term organotypic spinal cord slice cultures that were chronically exposed to EAAT inhibitors [107]. This study was of particular interest because of the reported loss of EAAT 2 in amyotrophic lateral sclerosis (ALS) (for review see [25]) and the observation that the motor neurons within these slices were found to be the most vulnerable to injury. Stereotaxic injections of TBOA directly into the hippocampus or striatum also produced increased extracellular levels of L-Glu, electroencephalographic activity indicative of seizures and neuronal damage [108]. In contrast, comparable administrations of L-2,4-PDC did not produce significant neuronal loss, suggesting that the substrate properties of an inhibitor (i.e., whether or not it can be cleared from the extracellular space, see above) can have a marked influence on its actions in vivo. The ability of an inhibitor to serve as a substrate also raises the possibility that the analogues may participate in the process of heteroexchange with intracellular L-Glu. This possibility was highlighted in experiments in which pathological conditions (e.g., typically energy depletion) lead to the disruption of the ion gradients used to maintain the L-Glu gradient and, consequently, promote its efflux through the reversed action of the transporter [48, 109–112]. Under such conditions, non-substrate inhibitors, such as TBOA and L-trans-2,3-PDC attenuate this reversed action, essentially trapping the transporter binding site in its external-facing configuration. Alternative substrates, on the other hand, can be translocated and allow a continued efflux of L-Glu as the transporter cycles between the external and internal compartments.

The pharmacological experiments described above are also consistent with studies in which the expression of the EAATs has been molecularly manipulated. Thus, neuronal pathology indicative of excitotoxicity was also observed following the in vitro or in vivo administration of antisense oligonucleotides for EAATs 1–3 [113, 114]. Mice that are homozygous deficient for Glt 1 (EAAT 2) show an increased vulnerability to cortical injury, hippocampal neurodegeneration, and die prematurely from severe spontaneous seizures [115]. GLAST (EAAT 1) knockout mice do not exhibit any overt neurodegeneration, but display motor problems and increased neuropathology following cerebellar injury [116]. Similarly, neurodegenerative pathology was not present in mice homozygous deficient for EAAC 1 (EAAT 3), although there is a significant decrease in locomotor activity [117]. More recent studies on these mice report an increased neurodegeneration with age, decreased neuronal levels of glutathione and an increased vulnerability to oxidative stress, consistent with a role of EAAT 3 in cysteine transport and glutathione production [118]. Although brain development appeared normal in each of these transgenic mice, a GLAST/GLT 1 (EAAT 1/EAAT 2) double knock out revealed multiple deficits in cortical and hippocampal organization, indicating that there may be compensation by other subtypes in the individual mouse models [119].

Taken together, these pharmacological and molecular studies highlight the neuroprotective role of the EAATs and, consequently, have prompted a therapeutic interest in agents that enhance glutamate uptake, rather than inhibit it. Unfortunately, traditional structure activity studies and rational analogue design strategies lend themselves more to the development of inhibitors and substrates, rather than activating agents acting at a distinct site. That being said, a few potential strategies have emerged to enhance transporter activity. With respect to potential allosteric modulators, arachidonic acid has been reported to inhibit the activity of EAAT 1, but increase EAAT 2 mediated uptake by increasing its affinity for L-Glu [120]. More recently, attention has focused on the identification of molecules capable of targeting and activating specific gene promoters regulating EAAT expression. In the case of GLT 1 (the rat homologue of EAAT 2) a blind screen of 1,040 FDA approved drugs and nutritionals conducted as part of the NINDS Drug Screening Consortium identified a series of 10 to 15 β-lactam and cephalosporin antibiotics as potent stimulators of GLT1/EAAT 2 protein expression and activity. In particular, the addition of ceftriaxone to organotypic rat spinal cord slice cultures for 5–7 days resulted in a 3-fold increase in both cell surface GLT1 expression, as determined by immunoblotting and transport activity quantified by uptake of radiolabeled L-Glu [121]. Additionally, ceftriaxone was able to prevent glutamate induced large ventral motor neuron degeneration resulting from the chronic blockade of glutamate transport by *threo-β*-hydroxyaspartate (THA) in organotypic spinal cord cultures. These findings point to possible therapeutic potential in neuropathologies where glutamate transport may be compromised elevating the risk of excitotoxicity.

3
System x_c^-: Cystine/Glutamate Exchanger

Much of the interest in the EAAT system has revolved around the process of uptake, that is the ability of these transporters to regulate extracellular L-Glu level by sequestering it intracellularly in glia or neurons. System x_c^- (Sx_c^-), on the other hand, has attracted attention because its accepted mode of operation provides a route for the export of glutamate from cells into the extracellular environment of the CNS and, consequently, access to EAA receptors. As Sx_c^- is an obligate exchanger, this efflux of L-Glu is linked with import of L-cystine (L-Cys$_2$), a sulfur-containing amino acid critical to a number of metabolic pathways, most notably the synthesis of glutathione (GSH) [122]. In this respect, both sides of this exchange reaction have significant implications within the CNS: the uptake of L-Cys$_2$ as a precursor in the maintenance of GSH levels for oxidative protection, and the efflux of L-Glu as a novel source of the neurotransmitter for excitatory signaling or excitotoxicity. The significance of these actions is reflected in the range of CNS processes to which Sx_c^- has recently been linked, including: oxidative protection [123], the operation of the blood–brain barrier [124], drug addiction [102], and synaptic organization [125].

3.1
Properties

Like most transporters, Sx_c^- was first identified (and named) on an activity dependent basis, where its specificity, ionic dependence, and exchange properties differentiated it from other transporters. Recent molecular studies reveal that Sx_c^- is part of the heteromeric amino acid transporter family (HATs; aka glycoprotein-associated amino acid exchangers) [126, 127]. These proteins are disulfide-linked heterodimers containing one of two type-II membrane *N*-glycoproteins termed "heavy chains" (e.g., 4F2hc or rBat) in combination with one of several different highly hydrophobic, non-glycosylated "light chains". In the instance of Sx_c^-, the heavy and light chains are 4F2hc ($\approx$ 80 kDa; aka CD98 or FRP1) and xCT ($\approx$ 40 kDa), respectively. The amino acid transport activity of the complex resides with the xCT subunit, which is thought to adopt a 12 transmembrane domain topology with the N and C termini found intracellularly. Thiol modification studies on xCT suggest that a re-entrant loop between TM 2 and 3, as well as regions of TM 8 may be involved in substrate binding and permeation [128, 129]. Interestingly, the xCT subunit has also been identified as the fusion-entry receptor for Kaposi's sarcoma-associated virus [130]. The 4F2hc subunit has a single transmembrane domain and is required for the trafficking of xCT to the plasma membrane [128, 131, 132].

Sx_c^- has been identified in a wide variety of CNS cells (e.g., astrocytes [99, 133, 134], microglia [135], retinal Muller cells [136], immature cortical neurons [137], and glioma cell lines [99, 138]), as well as fibroblasts, macrophages, hepatocytes and endothelial cells [124, 127, 139, 140]. The transporter is an obligate exchanger that mediates a 1 : 1 exchange of extracellular and intracellular amino acid substrates. In contrast to the EAATs, its activity is Na-independent and Cl-dependent, although it is still unclear if Cl^- ions are co-transported. As mentioned above, its primary mode of operation couples the import of L-Cys2 with the export of L-Glu, thereby allowing the L-Glu gradient generated by (EAATs) to serve as the driving force for L-Cys2 accumulation. Activity can be assessed with either radiolabeled L-Cys2 or L-Glu, with each acting as a competitive inhibitor of the other [99, 138]. On the basis of the influence of pH, it was concluded that both L-Glu and L-Cys2 are transported in an anionic form which, in turn, is also consistent with uptake being electroneutral [141].

3.2
Pharmacology

Initial pharmacological characterizations of Sx_c^- defined the substrate activities of L-Cys2 and L-Glu ($K_m \approx 50\,\mu M$) (Fig. 8a,b), established it as an obligate exchanger and identified several defining features of its specificity, including that: (i) L-aspartate is neither a substrate nor inhibitor, (ii) homocysteate is an effective inhibitor, (i.e., the binding site will tolerate a SO_3^- moiety in place of the distal COO^- functional group) and (iii) α-aminoadipate and α-aminopimelate are inhibitors (i.e., the binding site will accept longer homologues of L-Glu) [142]. Ironically, some of the earliest insights into the presence and pharmacology of Sx_c^- specifically within the CNS, emerged not from uptake studies, but from autoradiographic and membrane binding experiments. Thus, at a time when considerable effort was being devoted to the characterization of the EAA receptors, a chloride-dependent glutamate binding site identified in CNS preparations was suspected of being a novel receptor [143]. Subsequent studies, however, provided evidence that the observed binding was: likely associated with a transporter, could be blocked by L-Cys2, exhibited a distinct distribution, and was enriched on glial membranes [144–148]. As the properties and roles of Sx_c^- in other cell systems became more widely recognized, it became clear that much of this CNS binding was undoubtedly attributable to the Sx_c^- transporter.

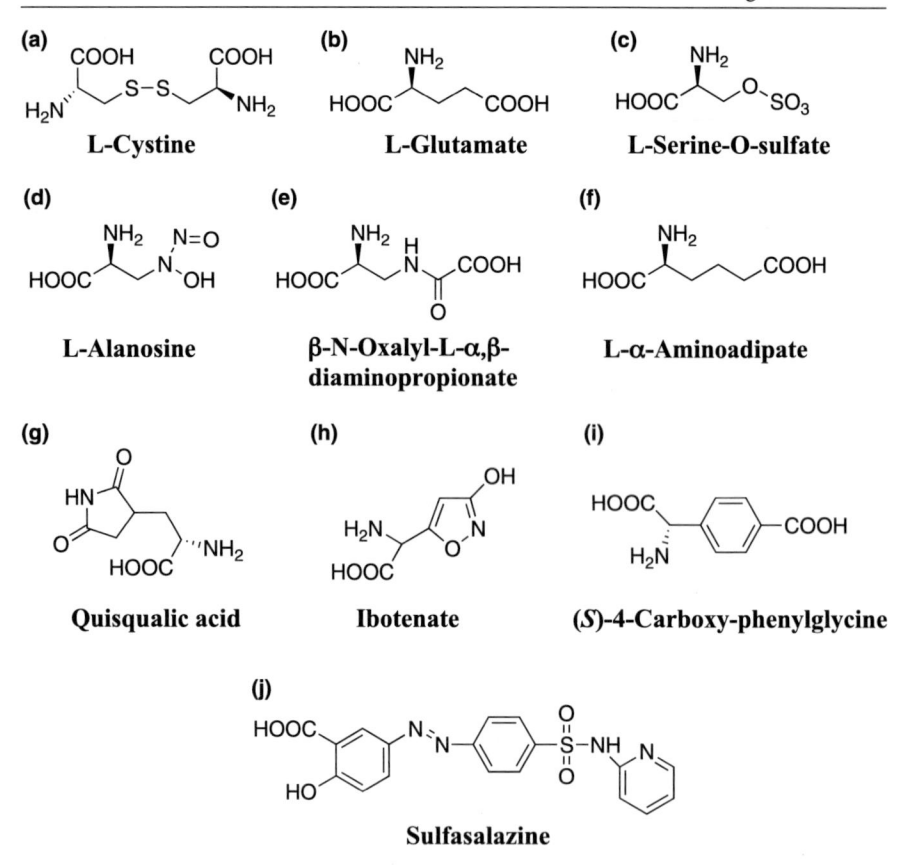

Fig. 8 System x_c^- inhibitors and substrates

Exploiting the finding that CNS-derived tumor cell lines exhibit high levels of Sx_c^- activity [99, 149–152] has allowed more thorough investigations of the structure–activity relationships (SARs) governing binding and translocation. Consistent with the inhibitory activity of L-homocysteate, the substrate-binding site accommodates the SO_3^- moieties of S-sulfo-L-cysteine and L-serine-O-sulfate ($K_i \approx 25\,\mu M$) (Fig. 8c), the SO_2^- group of L-homocysteine-sulfinate, but not the PO_3^{2-} group of L-serine-O-phosphate. Additionally, it appears as if an N-nitroso group can also serve as a γ-carboxylate mimic in the substrate binding site of Sx_c^-, as the antibiotic and antitumor drug L-alanosine (Fig. 8d) was identified as a competitive inhibitor in studies examining the transporter as a possible therapeutic target in the treatment of astrocytomas [153]. Assays demonstrating that the *lathyrus* toxin β-N-oxalyl-L-α,β-diaminopropionic acid (β-ODAP) (Fig. 8e) is also a competitive inhibitor about equipotent with L-Cys2, provides further evidence that the distal portion of L-Glu can be derivatized and still retain activity at the transporter [151]. Additionally, both L-alanosine and β-ODAP can serve as alternative substrates for the transporter (see below) and exchange with intracellular L-Glu when incubated with cells expressing Sx_c^-. The ability to bind longer homologues of L-Glu, such as α-aminoadipate (Fig. 8 f), is substantiated by the actions of S-carboxymethyl- and S-carboxyethyl-L-cysteine. However, there is a limit where increasing chain length to that of L-homocystine or L-djenkolate results in a marked loss of inhibitory activity [152].

The EAA ionotropic and metabotropic receptor agonists NMDA, KA, AMPA, L-AP4 and *trans*-ACPD were also all found to be essentially inactive as inhibitors, as were the Na^+-dependent EAAT inhibitors dihydrokainate and L-*trans*-2,4-PDC. While a lack of activity by these class-defining ligands clearly differentiate Sx_c^- from the other glutamate system proteins, a number of conformationally constrained EAA analogues well-recognized for their interactions with EAA receptors have been identified as Sx_c^- inhibitors. Probably the most prominent among these is quisqualate (QA) ($K_i \approx 5\,\mu M$) (Fig. 8g) [99, 149, 152], which also acts as an agonist at AMPA and mGluR receptors [1]. The action of QA at Sx_c^-, as well as the significance of this transporter in providing L-Cys$_2$ for glutathione synthesis, was illustrated by the work of Coyle, Murphy and colleagues who demonstrated that certain neuronal cell lines and immature cortical neurons were vulnerable to QA-mediated toxicity because of compromised glutathione synthesis and oxidative stress, rather than a typical EAA receptor-mediated excitotoxic mechanism [137, 149]. Other EAA analogues that are now realized to exhibit some cross-reactivity with Sx_c^-, include: (i) ibotenate ($K_i \approx 30\,\mu M$) (Fig. 8h) [152], an NMDA receptor [154] and non-selective metabotropic agonist [155], (ii) (RS)-4-Br-homoibotenate ($K_i \approx 20\,\mu M$) [152], an AMPA receptor agonist [156] (iii) (RS)-5-Br-willardiine [152], a kainate receptor agonist [157] and (iv) 4-S-carboxy-phenylglycine (4-S-CPG; $K_i \approx 5\,\mu M$) (Fig. 8i) [99, 149], a group I metabotropic receptor antagonist [158]. 4-S-CPG is of particular interest because it is one of the more potent inhibitors of Sx_c^- yet identified and because it has been shown to control the growth of glial tumors that express high levels of the transporter [159]. More detailed studies with other phenylglycines revealed that 4-R-CPG was 1,000-fold less potent than the S-enantiomer and that S-4-carboxy-3-hydroxyphenylglycine and S-3-carboxy-4-hydroxyphenylglycine were also effective inhibitors as judged by the ability to block Sx_c^--mediated release of L-Glu from tumor cells [160]. The lack of a carboxy group on the phenyl ring or the addition of a methyl group to the α-carbon also resulted in a marked loss of inhibitory activity.

Even though the cross-reactivities of these compounds with other EAA proteins render them somewhat less valuable as physiological probes (as well as raise questions as to agonist-induced responses attributed to specific receptors) the analogues listed above begin to shed light on the SAR's that govern the binding of L-Glu and L-Cys$_2$ to Sx_c^-. Thus, akin to what was observed with the EAATs, the requirement of an amino and carboxylate group in a classic L-α-amino acid configuration appears almost absolute. Greater flexibility, however, appears to be tolerated with respect to the position of distal carboxylate, which can be replaced with a variety of mimics. As many of the inhibitors entirely lack a distal carboxylate equivalent to the second amino group of L-Cys$_2$, it may be that this functional group plays a less important role. The distances separating the carboxylate groups of L-Glu and L-Cys$_2$, as well as in the configuration of charged groups in the conformationally restricted inhibitors (e.g., QA, 4-S-CPG, 3-S-4HPG) suggest that the binding site may tolerate some degree of variation with respect to placement of the distal acidic group. Indeed, given the strict requirement at the α-position, it is tempting to speculate that the same domains on the transporter are interacting with the proximal functional groups on all the inhibitors, while there may be more than one domain participating in the binding of the distal charges. Given the shorter length of L-Glu (and a number of the other inhibitors) relative to L-Cys$_2$, the binding domains for the two substrates may not be identical, but subsets of one another that exhibit the greatest overlap in the regions accommodating the α carboxylate and amino groups.

3.3
Substrate Specificity

As discussed with the EAATs, competition studies address the process of binding (evidenced by inhibition of uptake), but shed little insight as to whether an inhibitor is also a transportable substrate. Unlike the EAATs, however, uptake by Sx$_c^-$, is not electrogenic, which limits the utility of electrophysiological recordings to quantify substrate translocation. Advantageously, the fluorometric assay originally developed to examine the vesicular release of L-Glu from synaptosomes can be readily modified to follow Sx$_c^-$ activity [151, 152, 161]. Thus, the exchange of extracellular L-Cys$_2$ for intracellular L-Glu through Sx$_c^-$ can be quantified, in real time, by measuring the conversion of NADP$^+$ to NADPH as the extracellular L-Glu is rapidly metabolized by glutamate dehydrogenase (GDH) included in the assay mixture. Using this approach the competitive inhibitors discussed above can be further differentiated on the basis of substrate activity. Once again, it appears that subtle structural changes in ligands can substantially influence not only binding, but also substrate activity. For example, ibotenate and serine-O-sulfate exhibit comparable K_i values ($\approx 30\ \mu$M) for inhibiting the uptake of ^{3}H-L-Glu by Sx$_c^-$, yet ibotenate is transported as well as the endogenous substrate L-Cys$_2$, while the substrate activity of serine-O-sulfate is only half as much [152]. QA, which is one of the most potent inhibitors of Sx$_c^-$ ($K_i \approx 5\ \mu$M), exchanges with L-Glu at a rate of about 35% of that observed with L-Cys$_2$. The demonstration that QA can directly exchange for intracellular L-Glu through Sx$_c^-$ is also consistent with its proposed role in "QA-sensitization" observed in physiological brain slice preparation, where internalized QA can be subsequently released through Sx$_c^-$ and activate EAA receptors [162]. In contrast to these partial substrates, 4-S-CPG also blocks Sx_c^- with a K_i of $\approx 5\ \mu M$, but essentially fails to exchange with intracellular L-Glu. As appears to be the case with EAAT ligands, the most potent inhibitors identified tended to exhibit lower levels of substrate activity. This might reflect the fact that, as ligands are optimized for inhibition (and therefore binding), the compounds may be interacting strongly with domains that are distinct from those that participate in the translocation mechanism.

3.4
Analogues as Probes of Physiological and Pathological Roles

Sx_c^- is a particularly intriguing transporter from a functional perspective, because both the import of L-Cys$_2$ and the export of L-Glu are each associated with a distinct set of physiological roles within the CNS. Until only recently, most studies on Sx_c^- have primarily focused on its significance as a rate-limiting step in the provision of intracellular cysteine needed to maintain appropriate levels of the antioxidant glutathione (GSH). Given the sensitivity of the CNS to oxidative pathology it is not surprising that specific pathways have evolved to ensure neurons and glia have adequate capacities to produce GSH. Although both cell types require intracellular CysH to synthesize GSH, the extracellular precursor and route of entry necessary to provide this CysH appear to differ. Thus, a Cys$_2$/CysH shuttle has been proposed in which the CysH needed to maintain neuronal GSH levels is ultimately dependent upon the Sx_c^--mediated uptake of Cys$_2$ into astrocytes and its subsequent efflux from the cell as either CysH or GSH [122, 163–165]. The significance of Sx_c^- in GSH synthesis is underscored by the fact that QA-mediated toxicity in some cells is a consequence of blocking this transporter and the resultant oxidative stress, rather than the result of EAA receptor-mediated excitotoxicity [99, 149].

Within this protective context, Sx_c^- expression is a part of a number of different adaptive cellular responses (e.g., amino acid starvation, oxidative stress, toxic exposure) that are under transcriptional control regulated by genomic *cis*-elements. In the most thoroughly studied cases, exposure to electrophiles and/or increased oxidative stress activates transcription factors (e.g., Nrf2) that bind to electrophilic-responsive elements (EpRE)/antioxidant-responsive elements (ARE) and result in the upregulation of proteins presumed critical to detoxification and/or antioxidant defense mechanisms, including: GSH transferase, γ-glutamylcysteine synthetase, NAD(P)H: quinone reductase, heme-oxygenase 1 and Sx_c^- [166–169]. The transporter is also present (and inducible) at the blood–brain barrier, where it may serve as a point of entry not only for L-Cys$_2$, but also structurally related drugs and neurotoxins (e.g., β-L-ODAP, see above) [124, 151, 164, 170]. Surprisingly, the level of Sx_c^--mediated uptake typically reported to be present in primary cultures of astrocytes is remarkably low. This inconsistency was largely resolved when it was found that Sx_c^- activity is markedly up-regulated when the neonatal astrocyte cultures are differentiated with dibutyryl-cAMP and adopt a morphology more closely resembling that observed in vivo [134].

The Cys_2/CysH shuttle model, and its dependence upon Sx_c^- is, however, not without controversy. Indeed, it has been reported that the EAATs may provide another route of entry of Cys_2 into either neurons or glia [133, 171, 172]. Interestingly the EAATs, particularly neuronal EAAT 3, have also been postulated to transport CysH selectively into neurons [173, 174]. This role has received considerable support with the finding that EAAT 3/EAAC 1-null mice exhibit markedly lower levels of neuronal GSH, increased markers of oxidative damage, enhanced sensitivity to oxidative stress, and exhibit age-dependent neurodegeneration [118].

As an obligate exchanger, the import of L-Cys2 is coupled to the export of an equivalent amount of L-Glu. While this allows cells to utilize high intracellular concentrations of L-Glu as a driving force for the uptake of needed L-Cys2, it also provides a route for the accumulation of L-Glu in the extracellular space that, if not adequately regulated by the EAATs, could potentially trigger an excitotoxic response. Thus, in the two disorders where the evidence is strongest for Sx_c^- acting as a source for excitotoxic levels of L-Glu, CNS infection and glial tumors, it is not surprising that both of the cell types involved express enriched levels of the transporter. In CNS infection and inflammation, the cellular source of L-Glu are microglia, whose oxidative-based defense mechanisms necessitate high levels of Sx_c^- activity to provide the Cys_2/CysH necessary to maintain an adequate glutathione supply. Ironically, it thus appears that the excitotoxic pathology associated with a number of infections may actually be a secondary result of the migration of high numbers of microglia into an area and a subsequent release of L-Glu as these cells import the L-Cys2 integral to their protective roles [135, 175].

Similarly, astrocytoma cells express markedly higher levels of Sx_c^- [153, 160]. Numerous cellular and functional changes accompany the progression from low-grade astrocytomas to high-grade glioblastoma multiforme tumors. Of particular interest is the observation that the increase in Sx_c^- expression is also accompanied by a reduction in EAAT-mediated transport of glutamate [138]. Several studies utilizing human tumor cell lines and animal models now suggest these tumors employ the mechanism of excitotoxicity, in part by the release of excessive glutamate through Sx_c^-, to actively kill surrounding neurons in the peritumoral space to aid tumor expansion [138, 160, 176]. Epileptic seizures associated with brain tumors are also believed to result from this released glutamate. Further, the concurrent ability to acquire more L-Cys2 and maintain high levels of GSH allows the tumor to survive the necrotic biochemical environment engulfing the tumor. As a transporter, Sx_c^- may also impact the chemosensitivity and chemoresistance of a broad range of tumors through the uptake and efflux of anticancer agents [153, 159, 177]. The application of identified Sx_c^- inhibitors, (S)-4-CPG or sulfasalazine (Fig. 8j) to a series of glioma cell lines produced a marked reduction in L-Cys$_2$ uptake and intracellular GSH concentration whilst also inhibiting tumor cell growth and inducing caspase-dependent apoptotic cell death. Sulfasalazine was also demonstrated to suppress tumor growth in vivo. In addition to inhibiting the uptake of L-Cys$_2$, L-alanosine has been shown to induce cytotoxicity in lung and ovarian cancer cell lines as a consequence of its intracellular accumulation following uptake through Sx_c^-. Consistent with such a mechanism, inhibition of Sx_c^- with S-4-CPG decreased the cytotoxicity of L-alanosine. In contrast, the cytotoxic potential of geldanamycin was increased following inhibition of Sx_c^-, as the subsequent reduction in GSH level attenuated the detoxification of this drug. This suggests a multidrug strategy based on substrate and/or inhibitor activity may be a relevant therapeutic approach in treating tumors that needs to be further investigated.

Significantly, evidence is beginning to emerge that this export of L-Glu through Sx_c^- may be relevant to more than just pathological mechanisms and may actually represent a novel route of release through which L-Glu can activate extrasynaptic EAA receptors. Based primarily on microdialysis data, work by Kalivas, Baker and coworkers report that Sx_c^- appears to be a primary source of extracellular L-Glu in select brain regions, such as the striatum and nucleus accumbens [100–102]. Indeed in vivo microdialysis has revealed Sx_c^- is the primary source of non-vesicular extracellular glutamate outside the synaptic cleft and responsible for as much as 50–70% of the basal extracellular level in the nucleus accumbens (NAc). Consistent with this hypothesis, inhibition of Sx_c^- with 4-S-CPG reduced extracellular L-Glu level, as well as blocked the accumulation of L-Glu produced by inhibiting EAAT activity. This efflux of L-Glu is thought to regulate synaptic release (of both L-Glu and dopamine) through the tonic activation of extrasynaptic Group II mGluRs, which have been implicated in various forms of neuroplasticity and neurophyschiatric disorders. Of particular significance, these extracellular levels of L-Glu are reduced in the nucleus accumbens during cocaine addiction and withdrawal, an effect attributed to the decreased function of Sx_c^- [102]. Consistent with this model, the infusion of Cys$_2$ restored L-Glu levels and, importantly, treatment of rats with the CysH/Cys$_2$ prodrug N-acetyl-CysH prevented reinstatement in cocaine-addicted rats. This led to the conclusion that the increased susceptibility to relapse that accompanies withdrawal from cocaine addition is linked to the decreased activity of Sx_c^-, reductions in extracellular L-Glu levels, and consequent loss of mGluR-mediated regulation of excitatory transmission. Remarkably restoring Sx_c^- activity, presumably with increased substrate levels, prevents "cocaine-primed drug seeking" [102, 178]. This suggests that in addition to receptor-targeted approaches, agents directed at Sx_c^- may hold therapeutic value in the treatment of addiction.

This postulated role of Sx_c^- (and Cys$_2$) in regulating extracellular L-Glu level is, however, not without its controversies and complications. For example, very similar microdialysis studies in prefrontal cortex demonstrated that inhibition of Sx_c^- did not reduce extracellular L-Glu concentrations, yet did block the accumulation of L-Glu produced following the inhibition of EAAT activity [179]. All of these studies are also complicated by the fact that the Sx_c^- inhibitor most often employed was 4-S-CPG, which is also an mGluR1/5 antagonist. Ironically, during the course of the studies another mGluR1/5 antagonist (LY367385) was also found to be an Sx_c^- inhibitor [179]. In related studies by Atwell and coworkers, an L-Cys$_2$-mediated efflux of L-Glu through Sx_c^- was shown to be significant enough to activate non-NMDA in cerebellar slices and NMDA receptors in hippocampal slices [180, 181]. However, it was concluded that given both CSF levels of L-Cys$_2$ (typically in the 0.1–0.5 μM range) and its K_m values at Sx_c^- (typically reported in 100 μM range), it was unlikely that this mechanism contributed to tonic glutamate levels (and EAA receptor signaling) under normal physiological conditions. This conclusion, however, must be tempered somewhat by direct measurements of ^{35}S-L-Cys$_2$ uptake in slices of nucleus accumbens which yielded K_m values in the 2–4 μM range [102]. Further, an Sx_c^--mediated efflux of L-Glu was also shown capable of decreasing the synaptic release of L-Glu (decreased mEPSC and spontaneous EPSC frequency) as a consequence of presynaptic mGluR2/3 activation by levels of L-Cys$_2$ in the 0.1–0.3 μM range [178]. Lastly, recent studies in which homologues of Sx_c^- (e.g., *genderblind*) have been genetically eliminated in *Drosophila* support the participation of Sx_c^- in the regulation of extracellular L-Glu levels and further suggest a role of this extracellular L-Glu in iGluR desensitization and clustering [125].

Taken together, these findings suggest that Sx_c^- may indeed function to regulate extracellular L-Glu levels and set a tone at extrasynaptic mGluRs, although this contribution may be a function of the synaptic circuit being examined, the activity of Sx_c^- and EAATs, as well as extracellular Cys$_2$ levels. A particularly intriguing aspect of these variables concerns the balance set between these two transporters, with respect to kinetic properties, expression levels and localizations within the microenvironment of extrasynaptic receptors. Changes in any one of these properties, whether the result of development, plasticity, or pathology could impact excitatory signaling and/or excitotoxic vulnerability. Assessing the coordinate activity and functions of the EAATs and Sx_c^- will be dependent upon the continued development of potent, selective inhibitors and substrates in combination with more thorough understanding of the structure–activity relationships that govern binding and uptake.

Acknowledgements The authors are grateful to M.P. Kavanaugh, S.E. Esslinger, J.M. Gerdes, N. Natale, C.M. Thompson and P. Kalivas for their insightful discussions and comments. This work was supported in part by NINDS NS30570 and NCCR COBRE RR15583.

References

1. Balazs R, Bridges RJ, Cotman CW (2006) Excitatory amino acid transmission in health and disease. Oxford University Press, New York
2. Natale N, Magnusson K, Nelson J (2006) Curr Top Med Chem 6:823
3. Foster A, Kemp J (2006) Curr Opin Pharmacol 6:7
4. Waxman E, Lynch D (2005) Neuroscientist 11:37
5. Hynd MR, Scott HL, Dodd PR (2004) Neurochem Int 45:583
6. Olney JW (2003) Curr Opin Pharmacol 3:101
7. Rao SD, Weiss JH (2004) Trends Neurosci 27:17
8. Coyle J, Tsai G (2004) Psychopharmacol 174:32
9. Niswender CM, Jones CK, Conn PJ (2005) Curr Top Med Chem 5:847
10. Recasens M, Guiramand J, Aimar R, Abdulkarim A, Barbanel G (2007) Curr Drug Targets 8:651
11. Meldrum BS, Chapman AG (1999) Adv Neurol 79:965
12. Varney M, Gereau R (2002) Curr Drug Targets CNS Neurol Disord 1:283
13. Danbolt NC (2001) Prog Neurobiol 65:1
14. Borre L, Kavanaugh MP, Kanner BI (2002) J Biol Chem 277:13501
15. Gonzalez MI, Robinson MB (2004) Curr Opin Pharmacol 4:30
16. Diamond JS (2005) J Neurosci 25:2906
17. Kanner BI (2006) J Membr Biol 21:89
18. Dunlop J (2006) Curr Opin Pharmacol 6:103
19. Wadiche JI, Tzingounis AV, Jahr CE (2006) Proc Natl Acad Sci USA 103:1083
20. Farinelli SE, Nicklas WJ (1992) J Neurochem 58:1905
21. Choi DW (1994) Prog Brain Res 100:47
22. Rothman SM, Olney JW (1995) Trends Neurosci 18:57
23. Mattson MP (2003) Neuromolecular Med 3:65
24. Hara MR, Snyder SH (2007) Annu Rev Pharmacol Toxicol 47:117
25. Maragakis NJ, Rothstein JD (2004) Neurobiol Dis 15:461
26. Aarts MM, Tyminski M (2003) Biochem Pharmacol 66:877
27. Boillee S, Vande Velde C, Cleveland DW (2006) Neuron 52:39
28. Logan WJ, Snyder SH (1972) Brain Res 42:413
29. Storck T, Schulte S, Hofmann K, Stoffel W (1992) Proc Natl Acad Sci USA 89:10955
30. Pines G, Danbolt NC, Bjoras M, Bendahan A, Eide L, Koepsell H, Storm-Mathisen J, Kanner BI (1992) Nature 360:464
31. Kanai Y, Hediger MA (1992) Nature 360:467
32. Arriza JL, Fairman WA, Wadiche JI, Murdoch GH, Kavanaugh MP, Amara SG (1994) J Neurosci 14:5559
33. Arriza JL, Eliasof S, Kavanaugh MP, Amara SG (1997) Proc Natl Acad Sci USA 94:4155
34. Fairman W, Vandenberg RJ, Arriza JL, Kavanaugh MP, Amara SG (1995) Nature 375:599
35. Hediger MA, Romero MF, Peng JB, Rolfs A, Takanaga H, Bruford EA (2004) Pflugers Arch Eur J Physiol 447:465
36. Palacin M, Estevez R, Bertran J, Zorzano A (1998) Physiol Rev 78:969
37. Gegelashvili G, Schousboe A (1998) Brain Res Bull 45:233
38. Wadiche JI, Arriza JL, Amara SG, Kavanaugh MP (1995) Neuron 14:1019
39. Boudker O, Ryan RM, Yernool D, Shimamoto K, Gouaux E (2007) Nature 445:387
40. Zerangue N, Kavanaugh MP (1996) Nature 383:634
41. Wadiche JI, Amara SG, Kavanaugh MP (1995) Neuron 15:721

42. Otis TS, Kavanaugh M (2000) J Neurosci 20:2749
43. Yernool D, Boudker O, Jin Y, Gouaux E (2004) Nature 431:811
44. Haugeto O, Ullensvang K, Levy LM, Chaudhry FA, Honore T, Nielsen M, Lehre KP, Danbolt NC (1996) J Biol Chem 271:27715
45. Eskandari S, Kreman M, Kavanaugh MP, Wright EM, Zampighi GA (2000) Proc Natl Acad Sci USA 97:8641
46. Beliveau R, Demeule M, Jette M, Potier M (1990) Biochem J 268:195
47. Arriza JL, Kavanaugh MP, Fairman WA, Wu Y, Murdoch GH, North RA, Amara SG (1993) J Biol Chem 268:15329
48. Koch HP, Kavanaugh MP, Esslinger CS, Zerangue N, Humphrey JM, Amara SG, Chamberlin AR, Bridges RJ (1999) Mol Pharmacol 56:1095
49. Balcar VS, Johnston GAR (1972) J Neurobiol 3:295
50. Roberts PJ, Watkins JC (1975) Brain Res 85:120
51. Vandenberg RJ, Mitrovic AD, Chebib M, Balcar VJ, Johnston GAR (1997) Mol Pharmacol 51:809
52. Lebrun B, Sakaitani M, Shimamoto K, Yasuda-Kamatani Y, Nakajima T (1997) J Biol Chem 272:20336
53. Chamberlin AR, Koch HP, Bridges RJ (1998) In: Amara SG (ed) Neurotransmitter Transporters (Meth Enzymol), vol 296. Academic Press, San Diego, p 175
54. Esslinger CS, Agarwal S, Gerdes JM, Wilson PA, Davies ES, Awes AN, O'Brien E, Mavencamp T, Koch HP, Poulsen DJ, Chamberlin AR, Kavanaugh MP, Bridges RJ (2005) Neuropharmacology 49:850
55. Esslinger CS, Titus J, Koch HP, Bridges RJ, Chamberlin AR (2002) Bioorg Med Chem 10:3509
56. Dunlop J, McIlvain B, Carrick T, Jow B, Lu Q, Kowal DM, Lin S, Greenfield A, Grosanu C, Fan K, Petroski R, Williams J, Foster A, Butera J (2005) Mol Pharmacol 68:974
57. Shigeri Y, Shimamoto K, Yasuda-Kamatani Y, Seal RP, Yumoto N, Nakajima T, Amara SG (2001) J Neurochem 79:297
58. Shimamoto K, LeBrun B, Yasuda-Kamatani Y, Sakaitani M, Shigeri Y, Yumoto N, Nakajima T (1998) Mol Pharmacol 53:195
59. Shimamoto K, Sakai R, Takaoka K, Yumoto N, Nakajima T, Amara SG, Shigeri Y (2004) Mol Pharmacol 65:1008
60. Shimamoto K, Otsubo Y, Shigeri Y, Yasuda-Kamatani Y, Satoh M, Kaneko T, Nakagawa T (2007) Mol Pharmacol 71:294
61. Rudnick G (1997) In: Mea R (ed) Neurotransmitter transporters, structure, function and regulation. Humana Press, New Jersey, p 73
62. Tatsumi M, K. G, Blakely R, Richelson E (1997) Eur J Pharmacol 340:249
63. Alaux S, Kusk M, Sagot E, Bolte J, Jensen A, Brauner-Osborne H, Gefflaut T, Bunch L (2005) J Med Chem 48:7980
64. Greenfield A, Grosanu C, Dunlop J, McIlvain B, Carrick T, Jow B, Lu Q, Kowal DM, Williams J, Butera J (2005) Bioorg Med Chem Lett 15:4985
65. Dunlop J, McIlvain HB, Carrick T, Jow B, Lu Q, Kowal DM, Lin S, Greenfield A, Grosanu C, Fan K, Petroski R, Williams J, Foster A, Butera J (2005) Mol Pharmacol 68:974
66. Ishida M, Ohfune Y, Shimada Y, Shimamoto K, Shinozaki H (1991) Brain Res 550:152
67. Nakagawa Y, Saitoh K, Ishihara T, Ishida M, Shinozaki H (1990) Eur J Pharmacol 184:205
68. Nakamura Y, Kataoka K, Ishida M, Shinozaki H (1993) Neuropharmacology 32:833

69. Yamashita H, Kawakami H, Zhang Y, Hagiwara T, Tanaka K, Nakamura S (1995) Eur J Pharmacol 289:387
70. Dowd LA, Coyle AJ, Rothstein JD, Pritchett DB, Robinson MB (1996) Mol Pharmacol 49:465
71. Faure S, Jensen A, Maurat V, Gu X, Sagot E, Aitken D, Bolte J, Gefflaut T, Bunch L (2006) J Med Chem 49:6532
72. Jensen A, Brauner-Osborne H (2004) Biochem Pharmacol 67:2115
73. Johnston GAR, Kennedy SME, Twitchin B (1979) J Neurochem 32:121
74. Bridges RJ, Stanley MS, Anderson MW, Cotman CW, Chamberlin AR (1991) J Med Chem 34:717
75. Willis CL, Humphrey JM, Koch HP, Hart JA, Blakely T, Ralston L, Baker CA, Shim S, Kadri M, Chamberlin AR, Bridges RJ (1996) Neuropharmacology 35:531
76. Stensbol TB, Uhlmann P, Morel S, Eriksen BL, Felding J, Kromann H, Hermit MB, Greenwood JR, Brauner-Osborne H, Madsen U, Junager F, Krogsgaard-Larsen P, Begtrup M, Vedso P (2002) J Med Chem 45:19
77. Brauner-Osborne H, Hermit MB, Nielsen B, Krogsgaard-Larsen P, Johansen TN (2000) Eur J Pharmacol 406:41
78. Campiani G, De Angelis M, Armaroli S, Fattorusso C, Catalanotti B, Ramunno A, Nacci V, Novellino E, Grewer C, Ionescu D, Rauen T, Griffiths R, Sinclair C, Fumagalli E, Mennini T (2001) J Med Chem 44:2507
79. Campiani G, Fattorusso C, De Angelis M, Catalanotti B, Butini S, Fattorusso S, Fiorini I, Nacci V, Novellino E (2003) Curr Pharm Des 9:599
80. Dunlop J, Eliasof S, Stack G, McIlvain HB, Greenfield A, Kowal DM, Petroski R, Carrick T (2003) Br J Pharmacol 140:839–846
81. Dunlop J, Butera J (2006) Curr Top Med Chem 6:1897
82. Barbour B, Keller BU, Liana I, Marty A (1994) Neuron 12:1331
83. Isaacson JS, Nicoll RA (1993) J Neurophysiol 70:2187
84. Sarantis M, Ballerini L, Miller B, Silver RA, Edwards M, Attwell D (1993) Neuron 11:541
85. Wadiche JI, Jahr CE (2001) Neuron 32:301
86. Matsui K, Hosoi N, Tachibana M (1999) J Neurosci 19:6755
87. Kinney GA, Overstreet LS, Slater NT (1997) J Neurophysiol 78:1320
88. Overstreet LS, Kinney GA, Liu YB, Billups D, Slater NT (1999) J Neurosci 19:9663
89. Turecek R, Trussell LO (2000) J Neurosci 20:2054
90. Dzubay JA, Otis TS (2002) Neuron 36:1159
91. Higgs MH, Lukasiewicz PD (1999) J Neurosci 19:3691
92. Brasnjo G, Otis TS (2001) Neuron 31:607
93. Scanziani M, Salin PA, Vogt KE, Malenka RC, Nicoll R (1997) Nature 385:630
94. Semyanov A, Kullmann DM (2000) Neuron 25:663
95. Asztely F, Erdemli G, Kullmann DM (1997) Neuron 18:281
96. Carter AG, Regehr WG (2000) J Neurosci 20:4423
97. Isaacson JS (1999) Neuron 23:377
98. Diamond JS, Jahr CE (2000) J Neurophysiol 83:2835
99. Cho Y, Bannai S (1990) J Neurochem 55:2091
100. Baker DA, Shen H, Kalivas PW (2002) Amino Acids 23:161
101. Baker DA, Xi ZX, Hui S, Swanson CJ, Kalivas PW (2002) J Neurosci 22:9134
102. Baker DA, McFarland K, Lake RW, Shen H, Tang XC, Toda S, Kalivas PW (2003) Nat Neurosci 6:743
103. McBean GJ, Roberts PJ (1985) J Neurochem 44:247
104. Robinson MB, Djali S, Buchhalter JR (1993) J Neurochem 61:586

105. Amin N, Pearce B (1997) Neurochem Int 30:271
106. Velasco I, Tapia R, Massieu L (1996) J Neurosci Res 44:551
107. Rothstein JD, Jin L, Dykes-Hoberg M, Kuncl RW (1993) Proc Natl Acad Sci USA 90:6591
108. Montiel T, Camacho A, Estrada-Sanchez A, Massieu L (2005) Neuroscience 133:667
109. Anderson CM, Bridges RJ, Chamberlin AR, Shimamoto K, Yasuda-Kamatani Y, Swanson RA (2001) J Neurochem 79:1207
110. Koch HP, Chamberlin AR, Bridges RJ (1999) Mol Pharmacol 55:1044
111. Bonde C, Sarup A, Schousboe A, Gegelashvilli G, Zimmer J, Noraberg J (2003) Neurochem Int 43:371
112. Marcaggi P, Hirji N, Attwell D (2005) Neuropharmacology 49:843
113. Rothstein JD, Dykes-Hoberg M, Pardo CA, Bristol LA, Jin L, Kuncl RW, Kanai Y, Hediger MA, Wang Y, Schielke JP, Welty DF (1996) Neuron 16:675
114. Rao V, Dogan A, K. B, Todd K, Dempsey RJ (2001) Eur J Neurosci 13:119
115. Tanaka K, Watase K, Manabe T, Yamada K, Watanabe M, Takahashi K, Iwama H, Nishikawa T, Ichihara N, Kikuchi T, Okuyama S, Kawashima N, Hori S, Takimoto M, Wada K (1997) Science 276:1699
116. Watase K, Hashimoto K, Kano M, Yamada K, Watanabe M, Inoue Y, Okuyama S, Sakagawa T, Ogawa S, Kawashima N, Hori S, Takimoto M, Wada K, Tanaka K (1998) Eur J Neurosci 10:976
117. Peghini P, Janzen J, Stoffel W (1997) EMBO J 16:3822
118. Aoyama K, Suh SW, Hamby AM, Liu J, Chan WY, Y. C, Swanson RA (2006) Nat Neurosci 9:119
119. Matsugami T, Tanemura K, Mieda M, Nakatomi R, Yamada K, Kono T, Ogawa M, Obata K, Watanabe M, Hashikawa T, Tanaka K (2006) Proc Natl Acad Sci USA 103:12161
120. Zerangue N, Arriza JL, Amara SG, Kavanaugh MP (1995) J Biol Chem 270:6433
121. Rothstein J, Patel S, Regan M, Haenggeli C, Huang Y, Bergles DE, Jin L, Dykes Hoberg M, Vidensky S, Chung D, Toan S, Bruijn L, Su Z, Gupta P, Fisher P (2005) Nature 433:73
122. Kranich O, Dringen R, Sandberg M, Hamprecht B (1998) Glia 22:11
123. Shih A, Erb H, Sun X, Toda S, Kalivas P, Murphy T (2006) J Neurosci 41:10514
124. Hosoya K, Tomi M, Ohtsuki S, Takanaga H, Saeki S, Kanai Y, Endou H, Naito M, Tsuruo T, Terasaki T (2002) J Pharmacol Exp Ther 302:225
125. Augustin H, grosjean Y, Chen K, Sheng Q, Featherstone D (2007) J Neurosci 27:111
126. Palacin M, Nunes V, Jimenez-Vidal M, Font-Llitjos M, Gasol E, Pineda M, Feliubadalo L, Chillaron J, Zorzano A (2005) Physiology 20:112
127. Sato H, Tamba M, Ishii T, Bannai S (1999) J Biol Chem 274:11455
128. Gasol E, Jimenez-Vidal M, Chillaron J, Zorzano A, Palacin M (2004) J Biol Chem 279:31228
129. Jimenez-Vidal M, Gasol E, Zorzano A, Nunes V, Palacin M, Chillaron J (2004) J Biol Chem 279:11214
130. Kaleeba JA, Berger EA (2006) Science 311:1921
131. Nakamura E, Sato M, Yang H, Miyagawa F, Harasaki M, Tomita K, Matsuoka S, Noma A, Iwai K, Minato N (1999) J Biol Chem 274:3009
132. Palacin M, Kanai Y (2004) Eur J Physiol 447:490
133. Allen JW, Shanker G, Aschner M (2001) Brain Res 894:131
134. Gochenauer GE, Robinson MB (2001) J Neurochem 78:276
135. Piani D, Fontana A (1994) J Immunol 3578
136. Kato S, Ishita S, Sugawara K, Mawatari K (1993) Neuroscience 57:473

137. Murphy TH, Schnaar RL, Coyle JT (1990) FASEB J 4:1624
138. Ye Z, Rothstein JD, Sontheimer H (1999) J Neurosci 19:10767
139. Sasaki H, Sato H, Kuriyama M, K., Sato K, Maebara K, Wang H, Tamba M, Itoh K, Yamamoto A, Bannai S (2002) J Biol Chem 274:44765
140. Bannai S, Takada A, Kasuga H, Tateishi N (1986) Hepatology 6:1361
141. Bannai S, Kitamura E (1981) J Biol Chem 256:5770
142. Bannai S (1986) J Biol Chem 261:2256
143. Fagg GE, Foster AC, Mena EE, Cotman CW (1982) J Neurosci 2:958
144. Pin JP, Bockaert J, Recasen M (1984) FEBS Lett 175:31
145. Bridges RJ, Hearn TJ, Monaghan DT, Cotman CW (1986) Brain Res 375:204
146. Bridges RJ, Nieto SM, Kadri M, Cotman CW (1987) J Neurochem 48:001
147. Kessler M, Baudry M, Lynch G (1987) Neurosci Lett 81:221
148. Anderson KJ, Monaghan DT, Bridges RJ, Tavoularis AL, Cotman CW (1990) Neuroscience 38:311
149. Murphy TH, Miyamoto M, Sastre A, Schnaar RL, Coyle JT (1989) Neuron 2:1547
150. Waniewski RA, Martin DL (1984) J Neurosci 4:2237
151. Warren BA, Patel SA, Nunn PB, Bridges RJ (2004) Toxicol Appl Pharmacol 200:83
152. Patel SA, Warren BA, Rhoderick JF, Bridges RJ (2004) Neuropharmacology 46:273
153. Huang Y, Barbacioru C, Sadee W (2005) Cancer Res 65:7446
154. Ebert B, Madsen U, Johansen TN, Krogsgaard-Larsen P (1991) Adv Exp Med Biol 287:483
155. Brauner-Osborne H, Krogsgaard-Larsen P (1998) Eur J Pharmacol 5:311
156. Coquelle T, Christensen JK, Banke TG, Madsen U, Schousboe A, Pickering DS (2000) Neuroreport 21:2643
157. Wong LA, Mayer ML, Jane DE, Watkins JC (1994) J Neurosci 14:3881
158. Bedingfield JS, Kemp MC, Jane DE, Tse HW, Roberts PJ, Watkins JC (1995) Br J Pharmacol 116:3323
159. Chung WJ, Lyons SA, Nelson GM, Hamza H, Gladson CL, Gillespie GY, Sontheimer H (2005) J Neurosci 25:7101
160. Ye ZC, Sontheimer H (1999) Cancer Res 59:4383
161. Nicholls DG, Sihra TS, Sanches-Prieto J (1987) J Neurochem 1987:50
162. Chase LA, Roon RJ, Wellman L, Beitz AJ, Koerner JF (2001) Neuroscience 106:287
163. Wang XF, Cynader MS (2000) J Neurochem 74:1434
164. Guebel DV, Torres NV (2004) Biochim Biophys Acta 1674:12
165. Dringen R, Gutterer JM, Gros C, Hirrlinger J (2001) J Neurosci Res 66:1003
166. Sun X, Erb H, Murphy TH (2005) Biochem Biophys Res Commun 326:371
167. Sato H, Nomura S, Maebara K, Sato K, Tamba M, Bannai S (2004) Biochem Biophys Res Commun 325:109
168. Kim JY, Kanai Y, Chairoungdua A, Cha SH, Matsuo H, Kim DK, Inatomim J, Sawa H, Ida Y, Endou H (2001) Biochim Biophys Acta 1512:335
169. Ishii T, Itoh K, Takahashi S, Sato H, Yanagawa T, Katoh Y, Bannai S, Yamamoto M (2000) J Biol Chem 275:16023
170. Nagasawa M, Ito S, Kakuda T, Nagai K, Tamai I, Tsuji A, Fujimoto S (2005) Toxicol Lett 155:289
171. Bender AS, Reichelt W, Norenberg MD (2000) Neurochem Int 37:269
172. McBean GJ, Flynn J (2001) Biochem Soc Trans 29:717
173. Zerangue N, Kavanaugh MP (1996) J Physiol 493:419
174. Chen Y, Swanson RA (2003) J Neurochem 84:1332
175. Barger SW, Basile AS (2001) J Neurochem 76:846
176. Takano T, Lin J, G. A, Gao Q, Yang J, Nedergaard M (2001) Nat Med 7:1010

177. Gout PW, Buckley AR, Simms CR, Bruchovsky N (2001) Leukemia 15:1633
178. Moran M, McFarland K, Melendez RI, Seamans JK (2005) J Neurosci 25:6389
179. Melendez RI, Vuthiganon J, Kalivas PW (2005) J Pharmacol Exp Ther 314:139
180. Warr O, Takahashi M, Attwell D (1999) J Physiol 514.3:783
181. Cavelier P, Attwell D (2005) J Physiol 564:397

Top Med Chem (2009) 4: 223–247
DOI 10.1007/7355_2009_030
© Springer-Verlag Berlin Heidelberg
Published online: 25 March 2009

Glycine Transporters and Their Inhibitors

Robert Gilfillan (✉) · Jennifer Kerr · Glenn Walker · Grant Wishart

Schering Plough Corporation, Newhouse, Motherwell ML1 5SH, UK
Robert.gilfillan@spcorp.com

Abstract Glycine plays a ubiquitous role in many biological processes. In the central nervous system it serves as an important neurotransmitter acting as an agonist at strychnine-sensitive glycine receptors and as an essential co-agonist with glutamate at the NMDA receptor complex. Control of glycine concentrations in the vicinity of these receptors is mediated by the specific glycine transporters, GlyT1 and GlyT2. Inhibition of these transporters has been postulated to be of potential benefit in several therapeutic indications including schizophrenia and pain. In this review we discuss our current knowledge of glycine transporters and focus on recent advances in the medicinal chemistry of GlyT1 and GlyT2 inhibitors.

Keywords Glycine · Glycine transporter inhibitor · GlyT1 · GlyT2

Abbreviations

AMPA	α-Amino-3-hydroxy-5-methyl-4-isoxazole-propionic acid
DAT	Dopamine transporter
GlyR	Glycine receptor
GlyT1	Glycine transporter type-1
GlyT2	Glycine transporter type-2
hERG	Human ether-a-go-go related gene

LTP   Long term potentiation
NET   Norepinephrine transporter
NFPS N-[3-(4′-Fluorophenyl)-3-(4′-phenylphenoxy) propyl]sarcosine
NMDA N-Methyl-D-aspartate
PCP   Phencyclidine
SSRI   Selective serotonin reuptake inhibitor
VIAAT   Vesicular inhibitory amino acid transporter

1
Introduction

As the simplest amino acid, glycine plays a ubiquitous role in many biochemical and physiological processes including important neurotransmitter functions at both inhibitory and excitatory synapses within the mammalian central nervous system. The inhibitory glycinergic system located principally in the spinal cord and brain stem of the central nervous system exerts control on many fundamental physiological processes including motor coordination, respiratory control and perception of pain [1]. Glycine released from the terminals of interneurons activates post-synaptic glycine receptors (GlyR) resulting in Cl⁻ influx, hyperpolarisation of the neuronal membrane, and inhibition of neuronal firing. The excitatory role of glycine is focussed on its action as an obligatory co-agonist with glutamate at the ionotropic N-methyl-D-aspartate (NMDA) receptor [2]. Defining the origin of glycine in the region of the NMDA receptor complex is less certain than that of GlyRs but may stem from overflow at glycinergic synapses and release from neighbouring glial cells.

In common with many other neurotransmitters including neuro-active amino acids, flux and control of glycine levels in the vicinity of both these receptor types is likely to be mediated by specific high affinity glycine transporters designated GlyT1 and GlyT2. Advances in understanding the role of these transporters in normal physiology and disease has been made through the use of selective mutations and design of small molecule inhibitors targeting these membrane proteins.

2
Glycine Transporter Isoforms

GlyT1 and GlyT2 were identified following homology screening with complementary (c)DNAs for previously cloned Na⁺/Cl⁻ dependent neurotransmitter transporters [3–5] and both transporters exist as different isoforms resulting from alternative splicing or usage of different promoters. The human GlyT1 gene (SLC6A9), located on chromosome 1, can be expressed as three variants

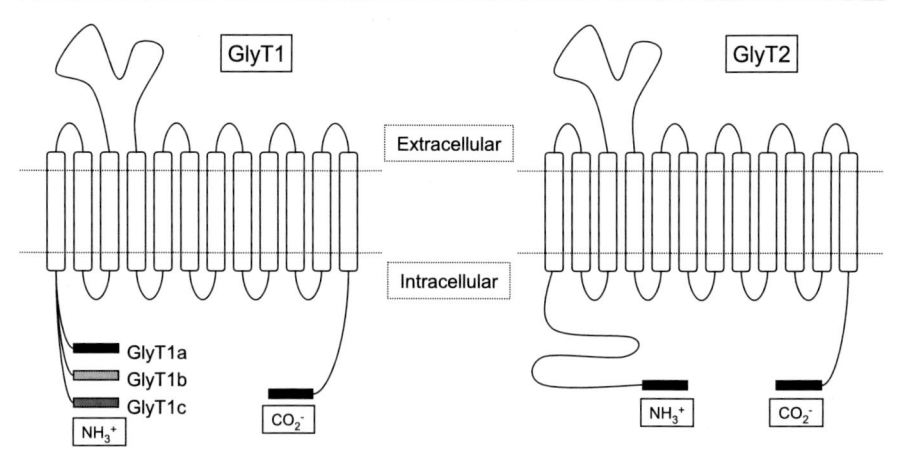

Fig. 1 Membrane topology of glycine transporters

(GlyT1a, b and c) which differ only in a restricted portion of the N-terminal region [6–8]. In addition, two variants of the C-terminal domain designated GlyT1d and e have been described for bovine tissue [9]. The human GlyT2 gene (SLC6A5) is found on chromosome 11 [10] and appears to exist as a single human isoform although at least three variants have been described for mouse [11] and two in rat [12]. In common with other members of this superfamily, glycine transporters are assumed to adopt a transmembrane topology with 12 transmembrane domains, six extracellular and five intracellular loops (Fig. 1). Both the amino- (N) and carboxy- (C) terminals are intracellular and play roles in directing appropriate trafficking of the protein to the cell membrane [13]. Glycosylation of asparagine residues on the large second extracellular loop also plays a role in membrane insertion [14], while di-leucine residues in the C-terminal tail of GlyT1 and a PDZ binding domain in GlyT2 are also implicated in correct targeting. The GlyT2 protein is unique in its class in that it possesses a long N-terminal domain comprising approximately 200 amino acids.

3
Distribution

Immunohistochemical and in situ hybridisation studies have been used to characterise the regional distribution and cellular localisation of glycine transporters [7, 15, 16]. GlyT1 is expressed primarily in glial astrocytes with significant presence in caudal areas including brain stem and spinal cord where it probably influences the functioning of the inhibitory glycinergic system. GlyT1 is also detected in most forebrain structures and evidence of close apposition to NMDA receptors [5] has been a significant driver supporting the quest for potent and selective GlyT1 inhibitors. Formerly, detection of neuronal expression of GlyT1 has been rare, with amacrine and bipolar cells in the retina featuring as the prime examples [17]. However, more recently, use of novel antibodies has also demonstrated the presence of GlyT1 protein in a subpopulation of glutamatergic terminals in rat forebrain [16]. GlyT2 is located on inhibitory glycinergic neurons distributed throughout the spinal cord, cerebellum and brainstem, and shows a high degree of overlap with the expression pattern of GlyR implying a major role for this transporter in the control of glycinergic neuronal activity [15].

4
Function

Both isoforms of the glycine transporter act as electrogenic carriers co-transporting Na^+ and Cl^- ions in concert with passage of their natural substrate across the plasma membrane. The stoichiometry associated with Na^+ transport has been shown to be 2 Na^+ ions per glycine molecule for GlyT1 but 3 Na^+ ions for GlyT2 [18]. This difference in coupling combined with the cellular localisation of the carriers helps explain their respective roles in controlling glycine concentrations and flux. The higher driving force afforded by the extra Na^+ allows GlyT2 to achieve millimolar intracellular glycine concentrations enabling filling of neuronal synaptic vesicles by low affinity vesicular transporters (VIAAT). The lower coupling ratio for GlyT1 suggests that this transporter will operate close to equilibrium allowing both uptake and release of glycine.

Considerable knowledge relating to function was gained through the production of gene knock-out mice [19–21]. Both GlyT1$^{-/-}$ and GlyT2$^{-/-}$ animals appear overtly normal at birth but die within one day and two weeks, respectively. GlyT1$^{-/-}$ mice fail to suckle, display an abnormal body posture and show poor motor responses including marked respiratory deficiency. GlyT2$^{-/-}$ progeny display different motor deficiencies including tremor, muscle spasticity and poor motor co-ordination. Similar phenotypes are observed in mice with mutations affecting GlyR function or expression (spastic, spasmodic and oscillator) and mutations within the human GlyT2 gene and the GlyR α1 subunit have been associated with startle disease [22] and hyperekplexia [23, 24]. Deletion of one gene does not affect expression of the other indicating a lack of compensation or adaptation, and also helping to delineate specific roles for either transporter. Deletion of GlyT1 appears to result in excessive stimulation of inhibitory GlyRs as respiratory activity in brain stem slices from GlyT1$^{-/-}$ neonates can be normalised by application of strychnine [20]. These observations indicate a major role for GlyT1 in regulating glycine concentrations at inhibitory GlyRs and also provide an animal model which mimics the symptoms observed in human glycine encephalopa-

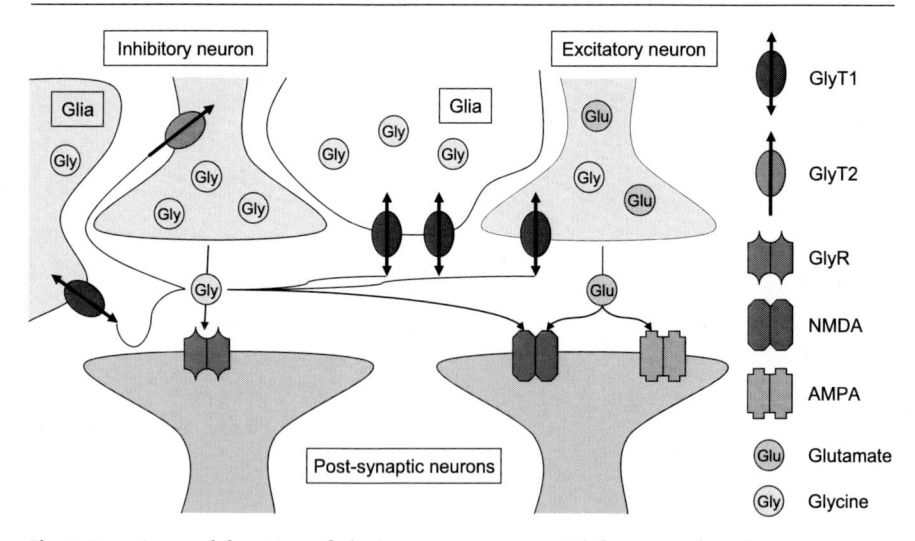

Fig. 2 Location and function of glycine transporters at inhibitory and excitatory neurons

thy. Evidence of altered NMDA receptor function was provided by Tsai et al. in 2004 [19], who observed an increased ratio of NMDA/AMPA receptor activity in hippocampal slices from $GlyT^{+/-}$ mice and also noted improved spatial retention in a water maze task.

The information yielded from knockout studies suggests that GlyT1 and GlyT2 have distinct roles related to glycinergic synapses. GlyT1 appears to function in the classical mode of removing glycine from the synaptic environment and ensuring high temporal fidelity of GlyR action. Neuronally located, GlyT2 may be outside the active zone of the synapse but plays a crucial role in driving refilling of synaptic vesicles for subsequent release upon neuronal stimulation. Figure 2 provides a depiction of the possible location and roles of GlyT1 and GlyT2.

5
Therapeutic Potential

Several examples of selective GlyT1 inhibitors including ALX 5407 (**4**, Fig. 5), Org 24 598 (**5**, Fig. 5) and SSR504734 (**26**, Fig. 12) have been shown, using microdialysis, to elevate extracellular glycine levels within brain structures and the spinal cord of rats [25–27]. On the basis of this mechanistic proof of principle, the pharmacology of these compounds has been explored using classical and novel paradigms to build the case for their use in the treatment of diverse disease states. However, the nature of binding of these novel ligands to the transporter may influence the observed efficacy and tolerability profiles. The prototypic GlyT1 inhibitor, sarcosine (**2**, Fig. 4), is a competitive substrate with low affinity for the transporter, whereas the first generation of high affinity amino acid-derived inhibitors (ALX 5407 and Org 24 598) are not transported and generally do not display competitive inhibitory kinetics [28]. There has been speculation that sarcosine-based inhibitors may act at GlyT1 in an irreversible manner leading to marked and prolonged adverse effects such as respiratory depression and akathisia-type behaviour [29]. ALX 5407 and its racemic equivalent NFPS have been shown to act in an essentially irreversible manner at GlyT1c expressed in a fibroblast cell line [30] and in *Xenopus laevis oocytes* expressing GlyT1b, respectively [31]. However, studies using rat forebrain membranes show rapid displacement of $[^3H]$-NFPS by the unlabelled ligand ($t_{1/2}$ of 28 min) indicating lack of covalent binding [28].

Recent non-sarcosine-based inhibitors synthesised by Sanofi are claimed to be reversible as defined by measurement of ex vivo glycine uptake [27, 32]. The kinetics of binding and related functional measures have recently been explored by Mezler et al. and illustrate an apparently clear separation between irreversible, non-competitive examples (NFPS, Org 24 598) and reversible, competitive compounds (SSR504734) [33]. Clinical experience with both types of compounds may resolve the question of preferred mechanism with respect to side effects and efficacy. In addition, the use of recently described novel radioligands $[^{35}S]$-ACPPB and $[^{11}C]$-GSK931145 will aid in the translation of in vivo occupancy and efficacy measures in pre-clinical and clinical settings [34, 35].

5.1
Schizophrenia

Despite the successful application of typical and atypical antipsychotic agents in the treatment of schizophrenia, in common with other psychiatric conditions, treatment regimes and outcomes retain many facets of unmet need. Current therapies are principally based on antagonism of dopamine receptors and exhibit efficacy in treating positive symptoms but fail to address negative symptoms and associated cognitive deficits. Observations that various glutamatergic antagonists including PCP and ketamine can reproduce the spectrum of schizophrenia-like symptoms in normal subjects and exacerbate psychotic symptoms in schizophrenic patients, led to the development of the hypoglutamatergic hypothesis. Pre-clinical investigations utilising glutamate antagonists such as PCP and MK801 to model aspects of the disease have been used to explore mechanisms aimed at addressing a glutamate-based deficit [36, 37]. Facilitation of NMDA receptor function by raising the local concentration of the natural co-agonist, glycine, forms the cornerstone of arguments promoting the development and use of GlyT1 inhibitors in the treatment of schizophrenia. Development of high affinity GlyT1 inhibitors followed the initial encouraging clinical observations that high doses of oral glycine demonstrate moderate improvements in negative symptoms when added to most conventional antipsychotic agents [38, 39]. The alternate natural co-agonist D-serine [40] and the low affinity GlyT1 inhibitor and substrate, sarcosine [41, 42], also show promise in improving negative symptomatology. However, a larger study examining efficacy in schizophrenic individuals of adjunctive glycine or D-cycloserine for negative symptoms and cognition failed to demonstrate significant improvement by either agent [43].

Cognition: In addition to addressing the negative symptoms in schizophrenia, increasing NMDA receptor activity and potentially, neuronal plasticity, is expected to improve learning and address attentional deficits observed in this condition. In vitro and limited in vivo studies confirm the enhancing effects of uptake inhibitors NFPS and CP-802,079 on NMDA function and long term potentiation (LTP) [44–46]. Selective deletion of GlyT1 within mouse forebrain resulted in improved mnemonic functions versus control animals [47, 48]. Procognitive effects of SSR504734 and SSR103800 have been proposed following positive effects observed in reversal of acute and neurodevelopmental latent inhibition deficits in rats [49]. Translation of these phenomena into improvements in clinically assessed cognitive domains is an eagerly sought objective in this area of drug development.

5.2
Pain

Activity of selective GlyT1 inhibitors NFPS and sarcosine has been demonstrated in mouse models of neuropathic pain [50]. Sites of action are model dependent but both spinal and supra-spinal sites are implicated and ablation of anti-allodynic activity by strychnine and siRNA versus GlyRα3 confirms the involvement of strychnine-sensitive glycine receptors. The role of GlyT1 in the control of spinal glycine levels and hypothesised effects on spinal NMDA receptors has also been demonstrated through the use of selective tools [25]. Profound anti-allodynic effects of selective GlyT2 inhibitors, Org 25 543 (**44**, Fig. 15) and ALX 1393 were also observed in these studies [51].

6
Medicinal Chemistry of Glycine Transporter Inhibitors

In comparison to the medicinal chemistry of monoamine transporter inhibitors, that of glycine transporter inhibitors is a relatively immature field. Nonetheless, it has now been over a decade since the first publications on novel, selective inhibitors appeared in the scientific literature and during the intervening years the number of patents and publications has continued to grow apace (Fig. 3). It is not surprising, given their significant therapeutic potential, that almost all of the glycine transporter inhibitors described in

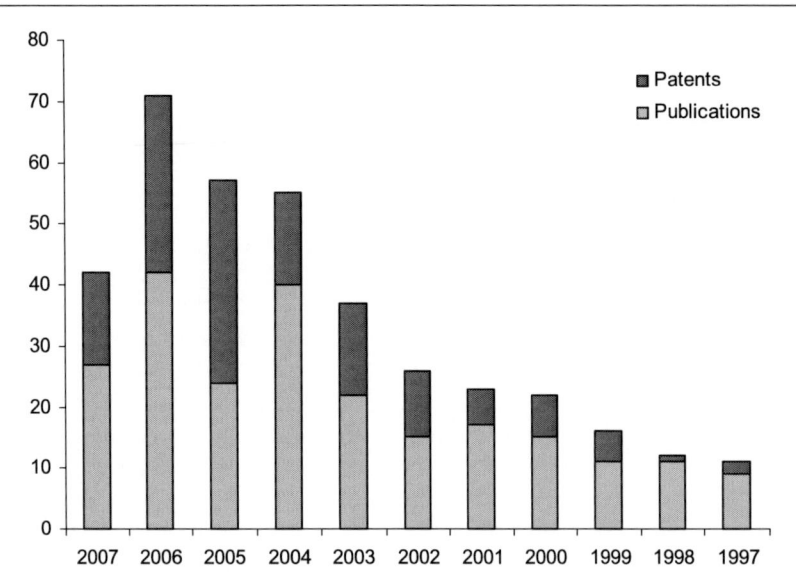

Fig. 3 Publications and patents by year. Data extracted from SciFinder, searching for the phrase "glycine transporter inhibitor"

patents and publications emanate from pharmaceutical company research. The inevitable downside of this is that, to date, only a limited amount of published data is available. Despite this, some clear trends have become apparent and these will be discussed in the next sections where we will look in detail at the medicinal chemistry of GlyT1 and GlyT2 inhibitors.

6.1
Amino Acid-Derived GlyT1 Inhibitors

The majority of work on glycine transporter inhibitors has focused on the discovery of selective GlyT1 inhibitors, and the first such reported selective inhibitor was sarcosine (**2**, Fig. 4) [52].

Sarcosine is an endogenous amino acid and is a competitive inhibitor of GlyT1 with negligible activity against GlyT2 [4, 6]. It is produced naturally via the metabolism of choline, primarily thought to occur in the liver and kidney, and in turn is further metabolised by glycine-N-methyltransferase to produce

1 Glycine

2 Sarcosine
GlyT1 IC$_{50}$ 37 μM

Fig. 4 The structures of glycine (**1**) and sarcosine (**2**)

glycine. In healthy subjects it is present in plasma at very low concentrations ($\sim$ 5 ng/ml) and has a relatively short half life (1.6 h) [53, 54]. Although sarcosine has been used in a number of clinical studies [41, 55, 56], it is a less than optimal drug molecule and the high doses required to elicit a clinical response (2 g per day) probably reflect its poor absorption/CNS penetration and low potency. It is therefore not surprising that there is currently a great deal of interest in the discovery of more potent and drug-like GlyT1 inhibitors.

The tactics employed in the discovery of the first novel GlyT1 inhibitors could be described as rational inasmuch as the compounds prepared were based on the appendage of lipophilic diaryl moieties to sarcosine or a related amino acid group. Alternatively, these compounds may be viewed as diaryl "drug-like" amines with an *N*-acetic acid appendage. Viewed either way, this simple tactic proved very effective and has resulted in the discovery of a wide range of amino acid-containing GlyT1 inhibitors. The first such compounds reported were ALX 5407 ((*R*)-*N*-[3-(4'-fluorophenyl)-3-(4'-phenylphenoxy) propyl]sarcosine, *R*-NFPS) [26] and Org 24 598 [57] (4 and 5, Fig. 5). It is interesting to note that both these compounds are derived from the SSRI fluoxetine (3), and that the glycine transporter belongs to the same Na^+/Cl^- family as the serotonin transporter. However, it would appear that this is purely coincidental as neither compound displays any inhibitory activity against serotonin or other related transporters. ALX 5407 is a potent GlyT1 inhibitor (IC_{50} 3–36 nM) with no appreciable activity at GlyT2 and, in contrast to sarcosine, is not a transporter substrate [58–60].

Org 24 598 is structurally closely related to ALX 5407 and it too is a potent (IC_{50} 7–126 nM) and selective inhibitor of GlyT1 [57, 60]. SAR around the Org 24 598 series showed that the homologous β- or γ-amino acid derivatives demonstrated significantly reduced GlyT1 inhibitory activity. However, substitution in the aromatic rings was better tolerated and, in particular, electron-withdrawing lipophilic substituents appeared to be preferred. The absolute stereochemistry was also shown to have an important bearing on GlyT1 inhibitory activity with the (*R*)-(–)-enantiomer, as is also the case with NFPS, proving to be the eutomer [57].

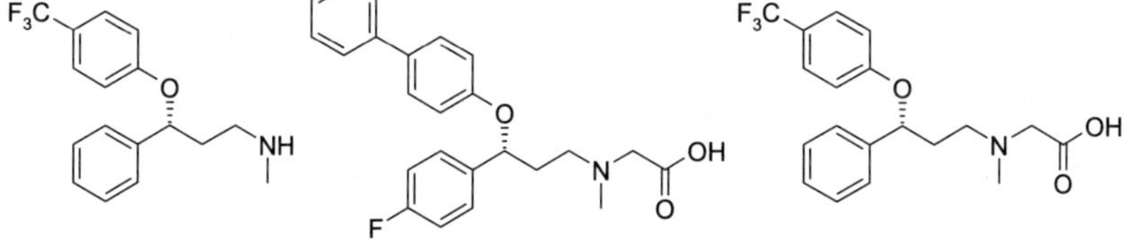

3 Fluoxetine

4 (*R*)-NFPS/ALX 5407
GlyT1 IC_{50} 3-36 nM

5 Org 24598
GlyT1 IC_{50} 7-126 nM

Fig. 5 Fluoxetine and the first reported amino acid derived GlyT1 inhibitors

In addition to identifying ALX 5407 as a potent and selective GlyT1 inhibitor, Trophix/NPS/Allelix also describe a number of other amino acid-containing analogues in the patent literature; these include diphenylmethylpiperidines such as **6** and a number of substituted α-amino acid derivatives such as **7, 8** and **9** (a mixed GlyT1/2 inhibitor) (Fig. 6) [61–64]. The fluoxetine scaffold was further exploited by Lowe and co-workers at Pfizer who describe a series of benzophenone ether derivatives [65], including compound **10** (Fig. 6) [59]. This was followed by a further two filings where the linking methylene ketone was replaced by an ether or a difluoromethylene group as in compounds **11** and **12**, respectively.

In a subsequent patent from Organon, a series of 1-phenyl-2-methylaminomethyltetralins are claimed [66]. GlyT1 activity is demonstrated for a number of examples with potencies in the range 50 nM–1 μM. In a similar fashion to within the Org 24 598 series, the GlyT1 inhibitory activity appears to be sensitive to the absolute stereochemistry and substitution pattern around the aromatic rings, with (–)-(**13**) given as the most potent example (GlyT1

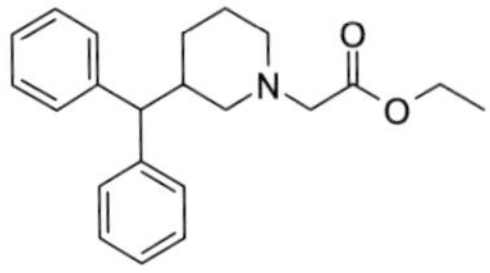

6 Trophix; WO1997045423

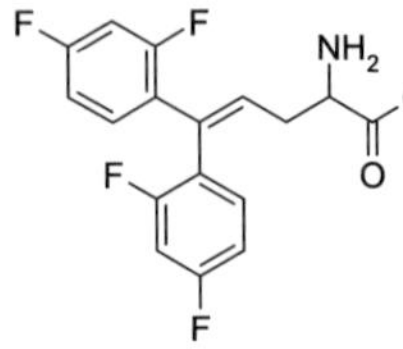

7 Allelix; US006103743

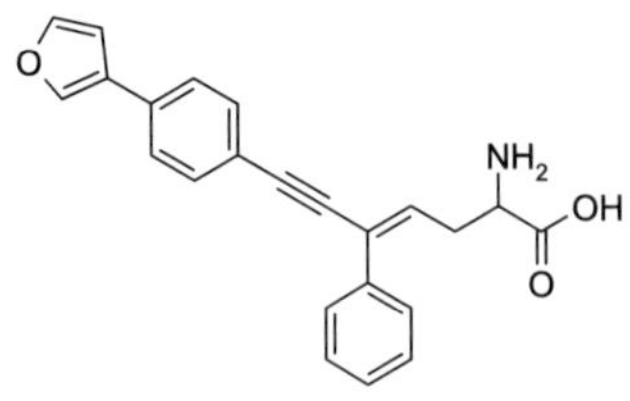

8 Allelix; US020169197

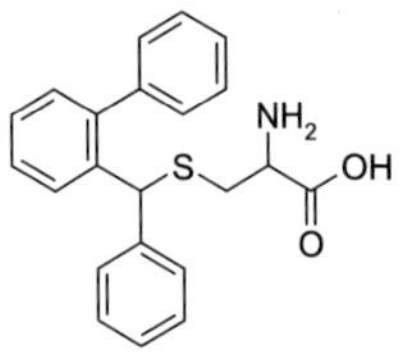

9 Allelix; US006166072

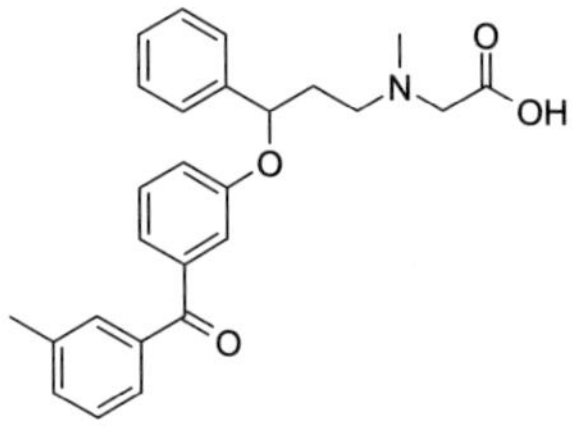

10 NPTS Pfizer; WO2002000602
GlyT1 IC_{50} 153 nM

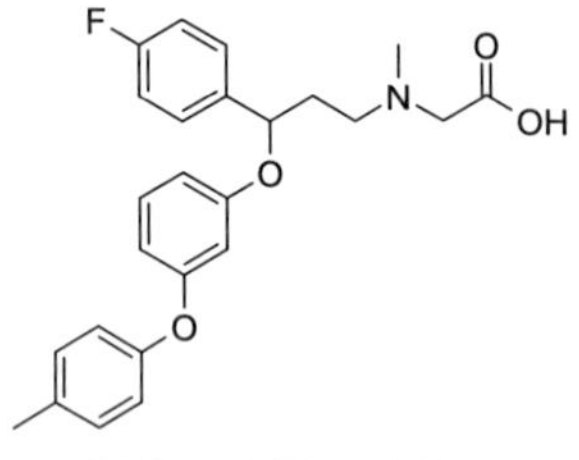

11 Pfizer; WO2003000646

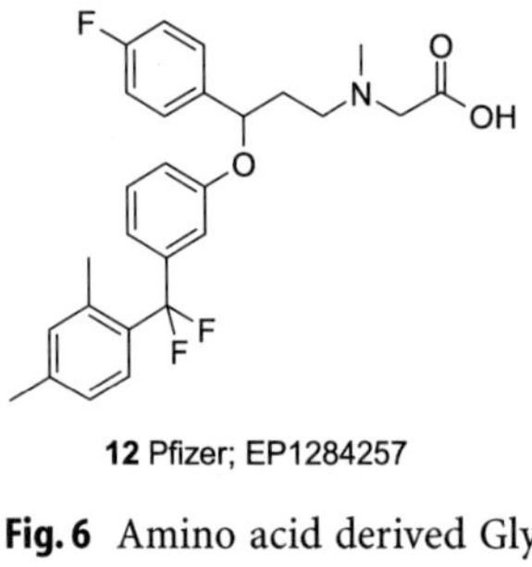

12 Pfizer; EP1284257

Fig. 6 Amino acid derived GlyT1 inhibitors

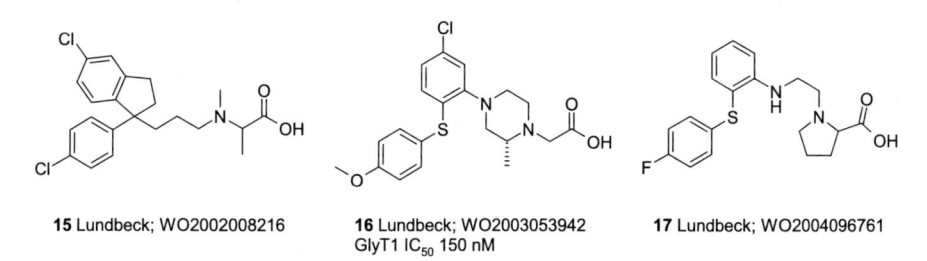

Fig. 7 Amino acid derived GlyT1 inhibitors from Organon

IC_{50} 50 nM). In another invention from Organon, a series of spiro ($2H$-1-benzopyran-2,4′-piperidine) derivatives are described [67]. Again, GlyT1 inhibitory activity appears to be sensitive to substitution around the aromatic rings and compound **14** is given as one of the more potent examples with a GlyT1 IC_{50} = 250 nM (Fig. 7).

The GlyT1 inhibitors from Lundbeck (Fig. 8) are based on indane derivatives such as compound **15** or diphenyl thioethers exemplified by compounds **16** and **17**. The SAR around compound **16** demonstrated that methyl substitution on the piperazine ring was essential for good inhibitory activity, with 2-(R)-methyl or 2-(R)-5-(S)-dimethyl substitution proving optimal [68]. Furthermore, it was shown that activity could be significantly modulated by varying the aromatic ring substituents. For example, replacement of the chloro-substituent in compound **16** by a methoxy group results in a > 10-fold reduction in activity. Compound **16** was shown to have excellent bioavailability (F = 100%) and was able to significantly elevate glycine concentrations by up to 140% over basal levels following microdialysis studies in rats.

Thomson et al. have described the SAR around a selection of Merck indandione derivatives including compound **18** (Fig. 9) [69]. They demonstrate that GlyT1 inhibitory activity is sensitive to aromatic ring substitution and, interestingly, they show that small substituents alpha to the acid (e.g. Me or Et) are tolerated, and that the distance between the first aromatic ring and basic nitrogen is important (optimal distance is four bond lengths).

Fig. 8 Amino acid derived GlyT1 inhibitors from Lundbeck

18 Merck
GlyT1 IC_{50} 0.95 nM

19 Lilly; WO2005100301
GlyT1 Ki 70 nM

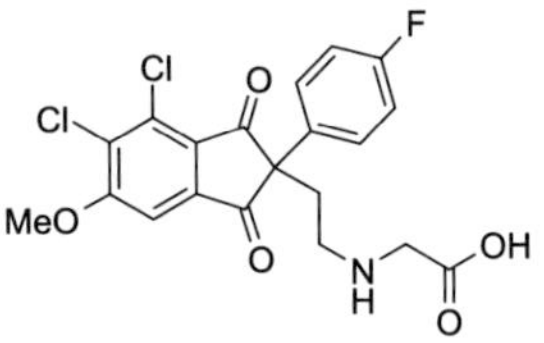

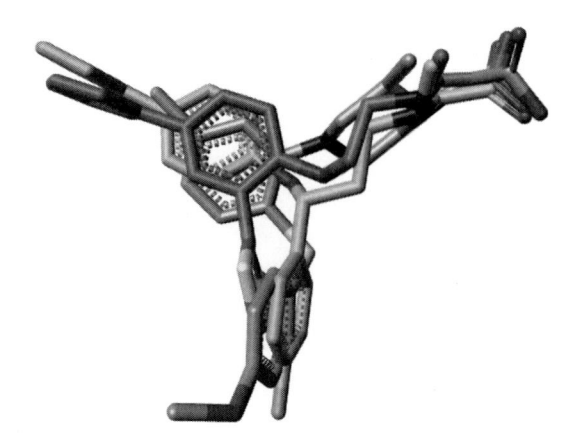

Fig. 9 Amino acid derived GlyT1 inhibitors from Merck and Lilly

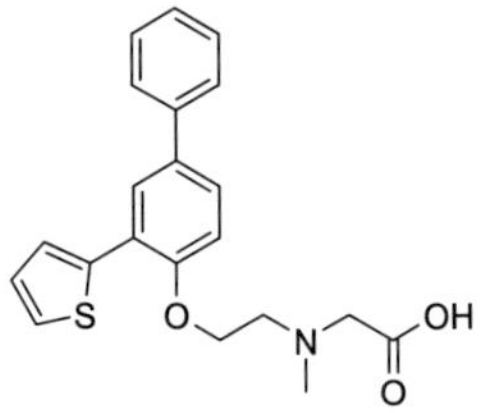

Compound **18** was also shown to be $> 1,000$-fold selective over a number of other targets including GlyT2 and hERG. Most recently, Walter et al. (Lilly) have demonstrated that there is still scope for further discoveries and innovation around these amino acid derivatives. They describe a series of 2-acyl- and 2-aryl biphenyl ethanolamine amino acid GlyT1 inhibitors [70]. The thiophene derivative compound **19** (Fig. 9) was one of several potent inhibitors described and, following subcutaneous dosing at 30 mg/kg, it produced a 196% increase in glycine CSF levels in rat, an effect similar to that observed with ALX 5407 [71].

What is apparent from all of the amino acid-containing GlyT1 inhibitors discussed here is that they all seem to fit a common pharmacophore consisting of the glycine amino acid motif and at least two aromatic hydrophobes. This is exemplified by the superposition of ALX 5407, **16** and **19** (Fig. 10), which clearly shows that the amino acid moieties can be aligned such that the two aromatic hydrophobes common to the majority of amino acid-derived GlyT1 inhibitors can occupy similar regions. A further hydrophobic pocket exists which is exploited by an additional aromatic hydrophobe for com-

Fig. 10 Superposition of ALX 5407 **4** (*cyan*), **16** (*orange*) and **19** (*magenta*)

pounds such as ALX 5407, **19** and many other examples. However, the additional aromatic moiety in this region does not appear to be an essential pharmacophoric feature since it can be replaced by a halogen in the case of the Lundbeck compound **16**.

6.2
Non-Amino Acid-Derived GlyT1 Inhibitors

As previously discussed, the optimisation of amino acid-based structural motifs has allowed for the identification and development of a series of potent and selective GlyT1 inhibitors. However, recent literature is suggesting a general shift away from the use of amino-acid derivatives as glycine uptake inhibitors, with over 40 patents having been published on non-sarcosine-based compounds since 2006. In addition, several candidates have entered clinical trials during this period and, whilst the structures of R-1678 (Roche, entered Ph II April 2008), SSR103800 (Sanofi, entered Ph I February 2008), and GSK1018921 (GlaxoSmithKline, entered Ph I August 2007) have not been disclosed, each of the parent companies has only patented structures which lack the sarcosine or a similar amino-acid motif. Furthermore, recent communications from Merck have revealed that a compound which is currently undergoing Ph I clinical trials for the treatment of schizophrenia (DCCyB) is indeed a non-sarcosine-derived GlyT1 inhibitor (**20**; GlyT1 IC_{50} 29 nM) (Fig. 11) [72].

The identification of non-sarcosine-containing molecules as glycine uptake inhibitors is not a particularly recent phenomenon, with Janssen Pharmaceuticals having patented such compounds as early as 1999 [73, 74]. The following examples **21–25** are a structurally diverse selection of only some of the earlier non-sarcosine-based GlyT1 inhibitors, all of which have been extensively reviewed previously and so will not be further discussed in this article (Fig. 11) [75–81]. Instead, this particular section of the review on glycine transporter inhibitors will focus on non-sarcosine-derived molecules and associated data published in the literature from 2006 onwards.

One of the first non-sarcosine-based glycine uptake inhibitors to be identified and advanced into development by Sanofi was SSR504734 **26** [82]. This potent GlyT1 inhibitor (GlyT1 IC_{50} 18 nM) has been the subject of a number of review articles which go on to discuss its extensive in vivo profile [76, 78–81]. Subsequent to the disclosure of this series in the patent literature between 2003–2005, Sanofi have since published a more recent patent application [83] extending the piperidine ring of **26** to cover a range of bicyclic amines as typified by compound **27**. In total, 31 examples are given with in vitro data being associated to only three compounds, e.g. **27** (GlyT1 IC_{50} 3 nM) (Fig. 12).

Since the initial disclosure of this series of aryl amides as potent and selective GlyT1 inhibitors, the interest surrounding this motif has mushroomed

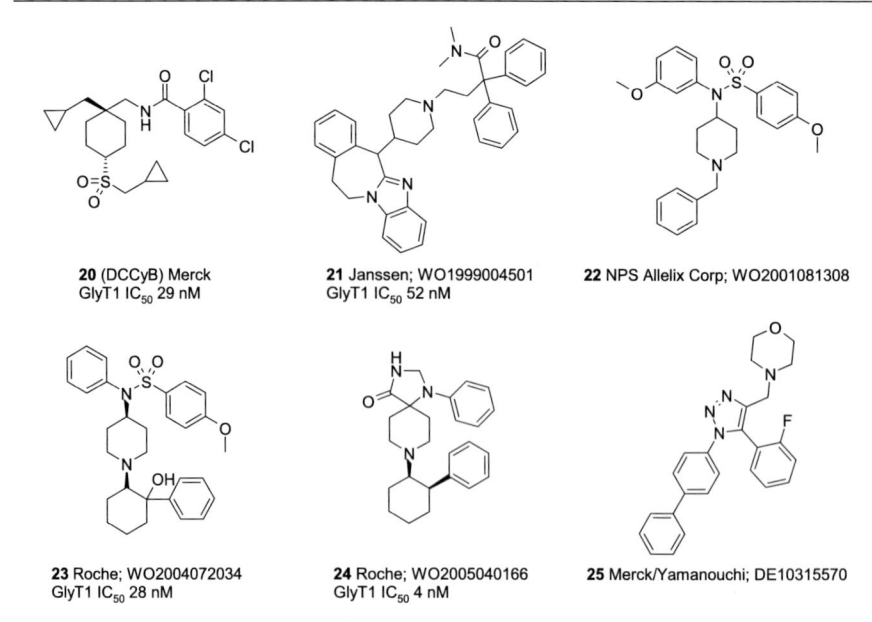

Fig. 11 Non amino acid derived GlyT1 inhibitors

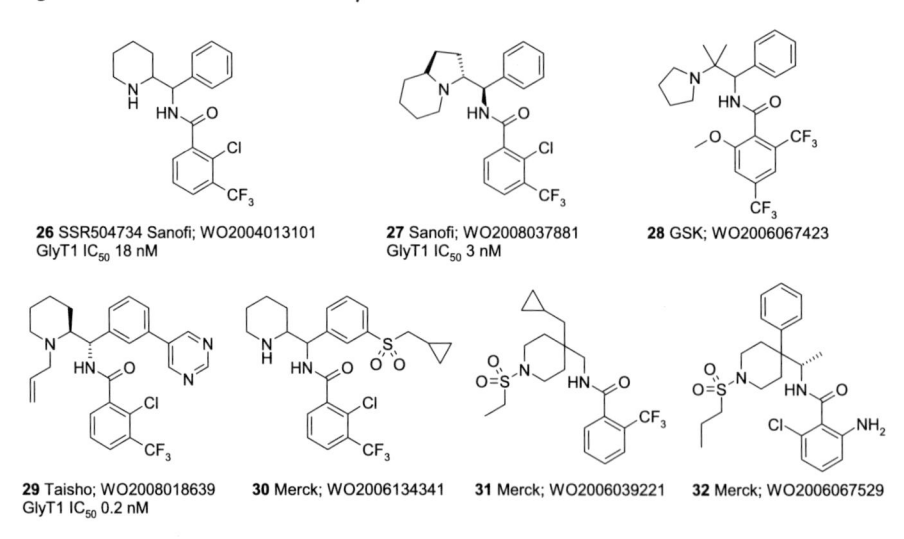

Fig. 12 Aryl amide derived GlyT1 inhibitors

such that numerous other competitors have also attempted to identify novel compounds within a structurally similar chemical series.

In particular, GSK have published a total of 12 patents (2006–2007) covering a set of aryl amides which are typified by compound **28** (Fig. 12) [84–95]. The examples given vary in substitution pattern around the aromatic ring, with heteroaryl replacements for the phenyl ring also being claimed. In order to differentiate their compounds from those of Sanofi, GSK have replaced the conformationally restricted piperidine unit associated with **26** and **27** with an open chain amine. Unfortunately, no biological data has been released for this series of compounds as yet, although GSK claim that the most active examples all have a GlyT1 $IC_{50} < 100$ nM.

Further structural modifications to the Sanofi core have also been made by Taisho, who claimed in a recent patent [96] to have further substituted the phenyl ring of **26** with a range of heteroaryl groups. There are 24 exemplified compounds within the patent which have associated in vitro GlyT1 activities, with potency ranging from 0.2 nM for compound **29** to 12.4 nM (Fig. 12).

Similarly, Merck have also developed a series of aryl amides which bear resemblance to the Sanofi compound **26**. Compound **30** is representative of a series of aryl sulfone-containing molecules which mainly vary in substitution around each of the phenyl rings (heteroaryl rings also being covered) and replacement of the piperidine moiety with alternative amines, aryl and alkyl groups (Fig. 12) [97]. No biological data is given for any of the 77 exemplified compounds. In addition, a second series of sulfone- and sulfonamide-containing molecules has been claimed which affords cover to the clinical candidate DCCyB (**20**) and typical examples **31** and **32** (Fig. 12) [98–105]. Although no biological data is contained within any of these six patents, a publication has revealed that the aryl-containing series was identified from an HTS campaign, the original hit being optimised to afford compound **32** which is a potent inhibitor of human (IC_{50} 2.6 nM), mouse (IC_{50} 5.1 nM) and rat GlyT1 (IC_{50} 2.1 nM) [106]. As such, this compound was further progressed to demonstrate both an increase in extracellular levels of glycine (340% of basal control levels at a dose of 10 mg/kg) and significantly enhanced prepulse inhibition (PPI) of the rodent acoustic startle response, a rodent behavioural model sensitive to antipsychotic treatment.

In what might be perceived to be an extension of the aforementioned aryl amide series, where the secondary nature of the amide group appears essential, both GSK and Roche have also developed a similar range of tertiary amido-aryl compounds as potent and selective GlyT1 inhibitors. A total of 196 exemplified compounds, as typified by **33** and which maintain the aryl sulfone whilst substituting the additional aryl groups for alkyl, thioalkyl and piperidine moieties as an example, are covered by three GSK patents (Fig. 13) [107–109]. No biological data is available although, once again, preferred compounds are claimed to have a GlyT1 $IC_{50} < 100$ nM.

Roche have also utilised a very similar scaffold to develop a range of GlyT1 inhibitors which are exemplified by **34** (Fig. 13) [110–112]. The biological activity of such compounds ranges from 6–500 nM, with substituted 5-membered heterocycles tending to exert higher activity than either 6-membered or fused heterocycles. Interestingly, however, a recent patent has emerged which discusses the crystalline forms of a single advanced stage compound possessing a pendant pyridine moiety [113]. A second

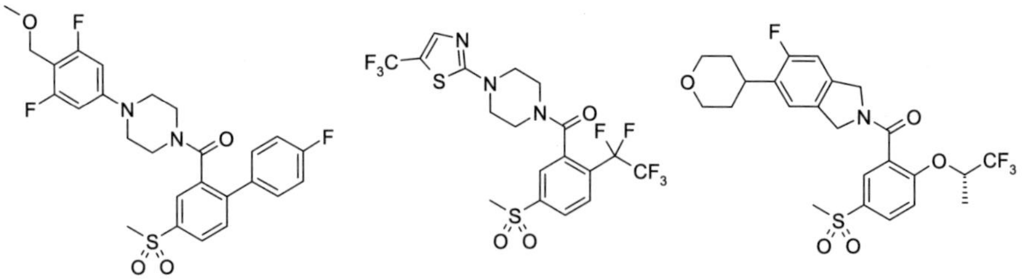

33 GSK; WO2006094842

34 Roche; WO2006072436
GlyT1 IC_{50} 6 nM

35 Roche; WO2006082001
GlyT1 IC_{50} 2 nM

Fig. 13 Aryl amide derived GlyT1 inhibitors

series of conformationally constrained amides also produced compounds with excellent GlyT1 potency as exemplified by **35** (Fig. 13) (GlyT1 IC_{50} 2 nM) [114–118].

In addition to developing potent GlyT1 inhibitors within the aryl amide series already mentioned, both GSK and Roche have also patented relatively similar cyclic amide motifs (Fig. 14). Although no biological data is available for the GSK series, a number of publications claim preferred compounds to have an activity of $IC_{50} < 100$ nM at GlyT1, with **36** being representative of the structures exemplified within these patents [119–124]. In turn, Roche have patented a series of compounds which are typified by **37** [125]. Biological activities range from $IC_{50} = 7$ nM for **37** to $IC_{50} = 82$ nM for a compound which has replaced the benzylamine motif with an alkyl amine.

Further examples of the diversity in chemical structure which appear to exert GlyT1 inhibitory activity are illustrated in the following additional examples by GSK, Roche and Pfizer (Fig. 14). Compound **38** is an example from the last in a series of patents published by GSK on this aryl morpholine unit [126]. Again, no biological activity has been quoted for any of the compounds exemplified within this series. However, GSK have recently published data describing the optimisation of one of their first non-sarcosine-derived chemical series [127]. High-throughput screening identified the aryl sulfonamide core which was further derivatised to afford compound **39**, a potent (GlyT1 IC_{50} 100 nM) and selective (GlyT2 $IC_{50} > 10$ μM) glycine uptake inhibitor. Although demonstrating poor in vitro microsomal stability, significant CYP2D6 inhibition and poor oral bioavailability in rats, compound **39** showed an acceptable PK profile following s.c. administration and, as such, may act as a valuable tool in further defining the role of non-sarcosine-based GlyT1 inhibitors in the treatment of schizophrenia.

More recently, Roche have published two patents on a series of aryl glycine amide derivatives as typified by **40** and **41**, with biological activities ranging from 15–369 nM for those compounds which are exemplified [128, 129]. Finally, Pfizer have published two recent patents, the second of which details the binding activities of no less than 3,340 compounds to the glycine trans-

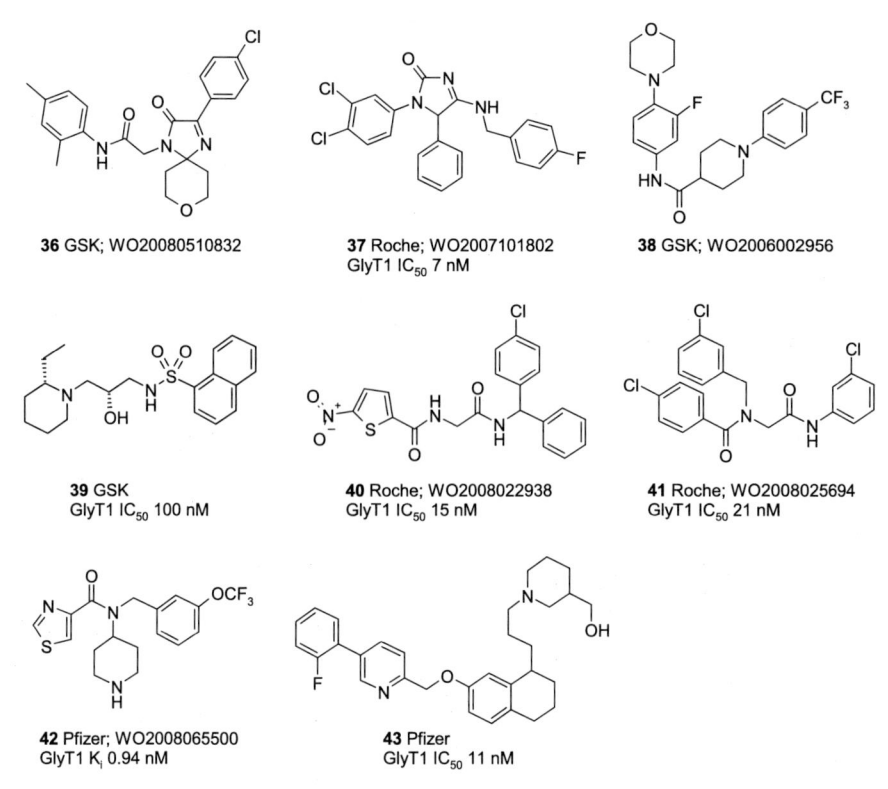

Fig. 14 Non amino acid GlyT1 inhibitors

porter, a typical example of which is compound **42** which displays a binding affinity of 0.94 nM [130, 131]. In addition, Pfizer have also published details of an HTS campaign and the subsequent optimisation of the original hit to afford compound **43** (GlyT1 IC$_{50}$ 11 nM) [132]. Unfortunately, the compounds in this series all suffered from issues relating to CYP2D6 inhibition, activity at the hERG channel and poor microsomal stability. As such, the series was abandoned.

In conclusion, a wide structural variety of both basic and neutral non-sarcosine-based ligands have now been patented as glycine uptake inhibitors. As such, and with the further addition of the sarcosine-based compounds, it is difficult to identify a general pharmacophore for GlyT1 inhibitory activity which may imply the presence of multiple binding sites on the glycine transporters. Regardless of their binding mode, with the advancement of several of these non-sarcosine-derived ligands into the clinic, and especially with the publication of structural data relating to the Merck Ph I candidate DCCyB, it is evident that such compounds are proving to be effective as potent and selective glycine uptake inhibitors which warrant further investigation.

6.3
GlyT2 Inhibitors

Compounds acting at this molecular target were initially investigated as potential treatments for pain and muscle spasticity but elucidation of the role of GlyT2 in vesicle refilling and the association of GlyT2 mutations with hyperkplexia cast doubt on their application and possible side effects [22]. Pre-clinical evidence of marked activity in several mouse models of neuropathic pain has recently been published for Org 25543 (44, Fig. 15), a selective GlyT2 inhibitor [133]. These data may revive interest in this mechanism as a route towards novel pain control therapy.

Consequently, inhibitors of GlyT2 are less well represented in the medicinal chemistry literature. However, as previously mentioned, potent and selective GlyT2 inhibitors have been identified. More specifically, researchers at Organon were the first to publish the structure of a potent and selective GlyT2 inhibitor, Org 25543 (44, Fig. 15; GlyT2 IC_{50} 16 nM; GlyT1 IC_{50} > 100 μM), which was identified following optimisation of an HTS hit [134, 135]. Around the same time, NPS also published information on a series of amino acid derivatives e.g. 45 (Fig. 15) which also showed a significant inhibition of GlyT2 [136]. These series of compounds are compared in a subsequent review by NPS [137]. In addition, Organon have also pub-

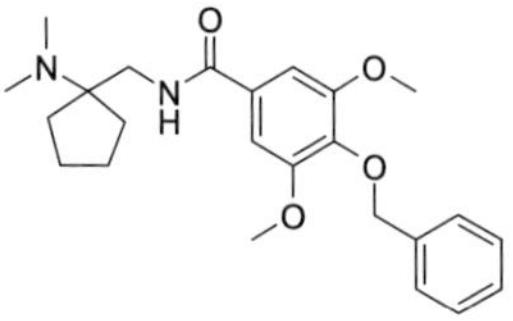

44 Org 22543; WO2003010132
GlyT2 IC_{50} 16 nM

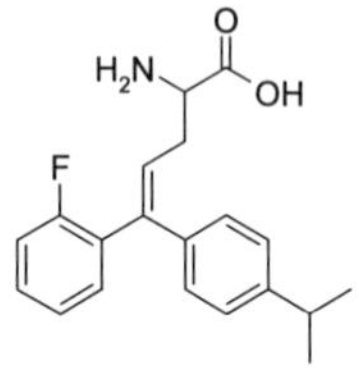

45 NPS
GlyT2 IC_{50} 290 nM

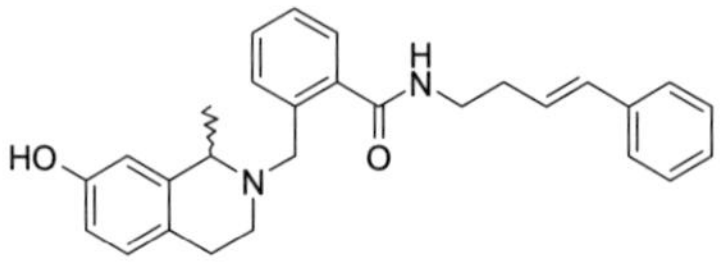

46 Organon; US0167119
GlyT2 IC_{50} 30 nM

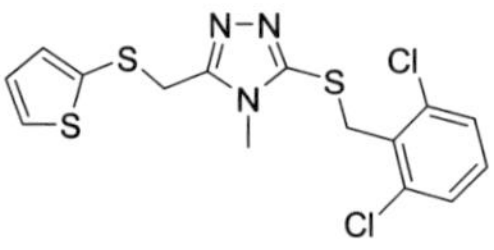

47 Telik; WO2002064135

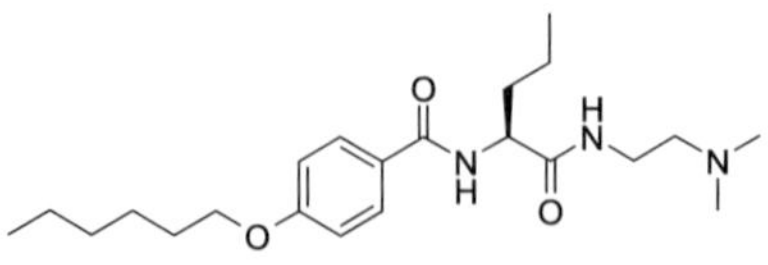

48 Tanabe; WO2006080477

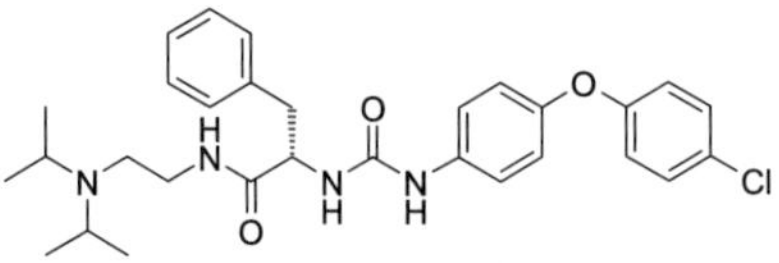

49 J&J; WO2005044810
GlyT2 IC_{50} 18 nM

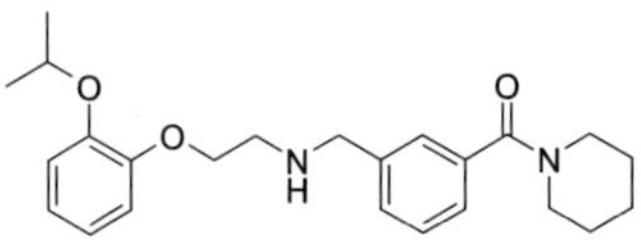

50 J&J; WO2005021525
GlyT2 IC_{50} 400 nM

Fig. 15 GlyT2 inhibitors

lished information on a series of 2-(aminomethyl)-benzamide-based GlyT2 inhibitors, with the most potent compound being **46** (Fig. 15) (GlyT2 IC_{50} 30 nM) [138, 139]. The geometrical arrangement of the hydroxyl and the amino functionalities in this chemotype was shown to be crucial for both activity and selectivity, in particular over other transporters such as NET and DAT.

Other companies who have shown an interest in this field of research through the publication of patent applications (without associated biological data) include Telik, who covered a range of heterocyclic inhibitors of GlyT2 such as **47** [140], and Tanabe, who patented a series of norvaline derivatives as typified by **48** (Fig. 15) [141].

Finally, Johnson & Johnson have published a series of papers and associated patents covering a range of α-, β-, and γ-amino acid derivatives [142–144], in addition to a set of benzoyl piperidine-based GlyT2 inhibitors [145–147]. In particular, compound **49** appears as the most potent amino acid derivative (GlyT2 IC_{50} 18 nM), with **50** affording significantly less activity from the aryl amide series (GlyT2 IC_{50} 400 nM) (Fig. 15).

7
Conclusion

Our knowledge and understanding of the role and function of glycine and specific glycine transporters has grown considerably over the last decade. This has been helped by the discovery of selective small molecule inhibitors of GlyT1 and GlyT2. At present, the potential development and clinical application of GlyT2 inhibitors seems uncertain. On the other hand, GlyT1 inhibition has received considerable attention owing to the attractiveness of GlyT1 as a drug target for indications such as schizophrenia and cognition. Two distinct structural classes of GlyT1 inhibitor have emerged: the amino acid-derived inhibitors such as ALX 5407 and Org 24598 and, latterly, the non-amino acid-derived inhibitors such as R-1678 and DCCyB. It remains to be seen whether or not the encouraging preclinical observations translate into clinical efficacy and what differences, if any, emerge between the two structural classes.

References

1. Kirsch J (2006) Cell Tissue Res 326:535
2. Johnson JW, Ascher P (1987) Nature 325:529
3. Guastella J, Brecha N, Weigmann C, Lester HA, Davidson N (1992) Proc Natl Acad Sci USA 89:7189
4. Liu QR, López-Corcuera B, Mandiyan S, Nelson H, Nelson N (1993) J Biol Chem 268:22802

5. Smith KE, Borden LA, Hartig PR, Branchek T, Weinshank RL (1992) Neuron 8:927
6. Kim KM, Kingsmore SF, Han H, Yang-Feng TL, Godinot N, Seldin MF, Caron MG, Giros B (1994) Mol Pharmacol 45:608
7. Adams RH, Sato K, Shimada S, Tohyama M, Püschel AW, Betz H (1995) J Neurosci 3:2524
8. Borowsky B, Mezey E, Hoffman BJ (1993) Neuron 10:851
9. Hanley J, Jones E, Moss S (2000) J Biol Chem 275:840
10. Morrow JA, Collie IT, Dunbar DR, Walker GB, Shahid M, Hill DR (1998) FEBS Lett 439:334
11. Ebihara S, Yamamoto T, Obata K, Yanagawa Y (2004) Biochem Biophys Res Commun 317:857
12. Ponce J, Poyatos I, Aragón C, Giménez C, Zafra F (1998) Neurosci Lett 242:25
13. Olivares L, Aragón C, Giménez C, Zafra F (1994) J Biol Chem 269:28400
14. Olivares L, Aragón C, Giménez C, Zafra F (1995) J Biol Chem 270:9437
15. Zafra F, Aragón C, Olivares L, Danbolt NC, Giménez C, Storm-Mathisen J (1995) J Neurosci 15:3952
16. Cubelos B, Giménez C, Zafra F (2005) Cereb Cortex 15:448
17. Menger N, Pow DV, Wässle H (1998) J Comp Neurol 401:34
18. Roux MJ, Supplisson S (2000) Neuron 25:373
19. Tsai G, Ralph-Williams RJ, Martina M, Bergeron R, Berger-Sweeney J, Dunham KS, Jiang Z, Caine SB, Coyle JT (2004) Proc Natl Acad Sci USA 101:8485
20. Gomeza J, Hülsmann S, Ohno K, Eulenburg V, Szöke K, Richter D, Betz H (2003) Neuron 40:785
21. Gomeza J, Ohno K, Hülsmann S, Armsen W, Eulenburg V, Richter DW, Laube B, Betz H (2003) Neuron 40:797
22. Rees MI, Harvey K, Pearce BR, Chung SK, Duguid IC, Thomas P, Beatty S, Graham GE, Armstrong L, Shiang R, Abbott KJ, Zuberi SM, Stephenson JB, Owen MJ, Tijssen MA, van den Maagdenberg AM, Smart TG, Supplisson S, Harvey RJ (2006) Nat Genet 38:801
23. Eulenburg V, Becker K, Gomeza J, Schmitt B, Becker CM, Betz H (2006) Biochem Biophys Res Commun 348:400
24. Shiang R, Ryan S, Zhu Y, Hahn A, O'Connell P, Wasmuth J (1993) Nat Genet 5:351
25. Whitehead KJ, Pearce SM, Walker G, Sundaram H, Hill D, Bowery NG (2004) Neuroscience 126:381
26. Atkinson BN, Bell SC, De Vivo M, Kowalski LR, Lechner SM, Ognyanov VI, Tham C-S, Tsai C, Jia J, Ashton D, Klitenick MA (2001) Mol Pharmacol 60:1414
27. Boulay D, Pichat P, Dargazanli G, Estenne-Bouhtou G, Terranova JP, Rogacki N, Stemmelin J, Coste A, Lanneau C, Desvignes C, Cohen C, Alonso R, Vigé X, Biton B, Steinberg R, Sevrin M, Oury-Donat F, George P, Bergis O, Griebel G, Avenet P, Scatton B (2008) Pharmacol Biochem Behav 91:47
28. Mallorga PJ, Williams JB, Jacobson M, Marques R, Chaudhary A, Conn PJ, Pettibone DJ, Sur C (2003) Neuropharmacology 45:585
29. Harsing LG Jr, Juranyi Z, Gacsalyi I, Tapolcsanyi P, Czompa A, Matyus P (2006) Curr Med Chem 13:1017
30. Atkinson BN, Bell SC, De Vivo M, Kowalski LR, Lechner SM, Ognyanov VI, Tham CS, Tsai C, Jia J, Ashton D, Klitenick MA (2001) Mol Pharmacol 60:1414
31. Aubrey K, Vandenberg R (2001) Br J Pharmacol 134:1429
32. Depoortère R, Dargazanli G, Estenne-Bouhtou G, Coste A, Lanneau C, Desvignes C, Poncelet M, Heaulme M, Santucci V, Decobert M, Cudennec A, Voltz C, Boulay D, Terranova JP, Stemmelin J, Roger P, Marabout B, Sevrin M, Vigé X, Biton B, Stein-

berg R, Françon D, Alonso R, Avenet P, Oury-Donat F, Perrault G, Griebel G, George P, Soubrié P, Scatton B (2005) Neuropsychopharmacology 30:1963

33. Mezler M, Hornberger W, Mueller R, Schmidt M, Amberg W, Braje W, Ochse M, Schoemaker H, Behl B (2008) Mol Pharmacol 74:1705–1715
34. Zeng Z, O'Brien JA, Lemaire W, O'Malley SS, Miller PJ, Zhao Z, Wallace MA, Raab C, Lindsley CW, Sur C, Williams DL Jr (2008) Nucl Med Biol 35:315
35. Passchier J, Murthy V, Catafau A, Gunn R, Gentile G, Porter R, Herdon H, Slifstein M, Rabiner E, Laruelle M (2008) J Nucl Med 49:129P
36. Javitt DC, Frusciante M (1997) Psychopharmacology 129:96
37. Karasawa J, Hashimoto K, Chaki S (2008) Behav Brain Res 186:78
38. Javitt DC, Zylberman I, Zukin SR, Heresco-Levy U, Lindenmayer JP (1994) Am J Psychiatry 151:1234
39. Heresco-Levy U, Javitt DC, Ermilov M, Mordel C, Horowitz A, Kelly D (1996) Br J Psychiatry 169:610
40. Heresco-Levy U, Javitt DC, Ebstein R, Vass A, Lichtenberg P, Bar G, Catinari S, Ermilov M (2005) Biol Psychiatry 57:577
41. Lane HY, Chang YC, Liu YC, Chiu CC, Tsai GE (2005) Arch Gen Psychiatry 62:1196
42. Lane HY, Liu YC, Huang CL, Chang YC, Liau CH, Perng CH, Tsai GE (2008) Biol Psychiatry 63:9
43. Buchanan RW, Javitt DC, Marder SR, Schooler NR, Gold JM, McMahon RP, Heresco-Levy U, Carpenter WT (2007) Am J Psychiatry 164:1593
44. Chen L, Muhlhauser M, Yang CR (2003) J Neurophysiol 89:691
45. Martina M, Gorfinkel Y, Halman S, Lowe JA, Periyalwar P, Schmidt CJ, Bergeron R (2004) J Physiol 557:489
46. Kinney GG, Sur C, Burno M, Mallorga PJ, Williams JB, Figueroa DJ, Wittmann M, Lemaire W, Conn PJ (2003) J Neurosci 23:7586
47. Singer P, Boison D, Möhler H, Feldon J, Yee BK (2007) Behav Neurosci 121:815
48. Yee BK, Balic E, Singer P, Schwerdel C, Grampp T, Gabernet L, Knuesel I, Benke D, Feldon J, Mohler H, Boison D (2006) J Neurosci 26:3169
49. Black MD, Varty GB, Arad M, Barak S, De Levie A, Boulay D, Pichat P, Griebel G, Weiner I (2008) Psychopharmacology (e-pub ahead of print)
50. Tanabe M, Takasu K, Yamaguchi S, Kodama D, Ono H (2008) Anaesthesiology 108:929
51. Morita K, Motoyama N, Kitayama T, Morioka N, Kifune K, Dohi T (2008) J Pharmacol Exp Ther (e-pub ahead of print)
52. McBain CJ, Kleckner NW, Wyrick S, Dingledine R (1989) Mol Pharmacol 36:556
53. Molina JA, Jimenez-Jimenez FJ, Gomez P, Vargas C, Navarro JA, Orti-Pareja M, Gasalla T, Benito-Leon J, Bermejo F, Arenas J (1997) J Neurol Sci 150:123
54. Glorieux FH, Scriver CR, Delvin E, Mohyuddin F (1971) J Clin Invest 50:2313
55. Lane H-Y, Huang C-L, Wu P-L, Liu Y-C, Chang Y-C, Lin P-Y, Chen P-W, Tsai G (2006) Biol Psychiatry 60:645
56. Tsai G, Lane H-Y, Yang P, Chong M-Y, Lange N (2004) Biol Psychiatry 55:452
57. Brown A, Carlyle I, Clark J, Hamilton W, Gibson S, McGarry G, McEachen S, Rae D, Thorn S, Walker G (2001) Bioorg Med Chem Lett 11:2007
58. Herdon HJ, Godgery FM, Brown AM, Coulton S, Evans JR, Cairns WJ (2001) Neuropharmacology 41:88
59. Lowe JA, Drozda SE, Fisher K, Strick C, Lebel L, Schmidt C, Hiller D, Zandi KS (2003) Bioorg Med Chem Lett 13:1291
60. Mallorga PJ, Williams JB, Jacobson M, Marques R, Chaudhary A, Conn PJ, Pettibone DJ, Sur C (2003) Neuropharmacology 45:585

61. Ognayanov VI, Borden L, Bell SC, Zhang J (1997) International Patent Application WO 1997 045 423
62. Bell SC, Da Silva K, Hopper A, Methvin I, Meade EA, Ognyanov VI, Slassi A (2000) US Patent Application US 6 103 743
63. Bell SC, De Vivo M, Hopper A, Methvin I, O'Brien A, Ognyanov VI, Schumacher R (2000) US Patent Application US 006 166 072
64. Egle I, Frey J, Isaac M (2002) US Patent Application US 0 169 197
65. Lowe JA (2002) International Patent Application WO 2002 000 602
66. Gibson SG, Jaap DR, Thorn SN, Gilfillan R (2000) International Patent Application WO 2000 007 978
67. Gibson SG, Miller DJ (2001) International Patent Application WO 2001 036 423
68. Smith G, Ruhland T, Mikkelsen G, Andersen K, Christoffersen CT, Alifrangis LH, Mørk A, Wren SP, Harris N, Wyman BM, Brandt G (2004) Bioorg Med Chem Lett 14:4027
69. Thomson CG, Duncan K, Fletcher SR, Huscroft IT, Pillai G, Raubo P, Smith AJ, Stead D (2006) Bioorg Med Chem Lett 16:1388
70. Man T, Milot G, Porter WJ, Reel JK, Rudyk HCE, Valli MJ, Walter MW (2005) International Patent Application WO 2005 100 301
71. Walter MW, Hoffman BJ, Gordon K, Johnson K, Love P, Jones M, Man T, Phebus L, Reel JK, Rudyk HC, Shannon H, Svensson K, Yu H, Valli MJ, Porter WJ (2007) Bioorg Med Chem Lett 17:5233
72. Wolkenberg SE (2008) MEDI-220, American Chemical Society National Meeting and Exposition. Philadelphia, PA. American Chemical Society, Washington, DC
73. Luyten WHML, Janssens FE, Kennis LEJ (1999) International Patent Application WO 1999 045 011
74. Luyten WHML, Janssens FE, Kennis LEJ (1999) International Patent Application WO 1999 044 596
75. Slassi A, Egle I (2004) Expert Opin Ther Pat 14:201
76. Lowe JA (2005) Expert Opin Ther Pat 15:1657
77. Harsing LG, Juranyi Z, Gacaslyi I, Tapolcsanyi P, Czompa A, Matyus P (2006) Curr Med Chem 13:1017
78. Lindsley CW, Shipe WD, Wolkenberg SE, Theberge CR, Williams DL, Sur C, Kinney GG (2006) Curr Top Med Chem 6:771
79. Lindsley CW, Wolkenberg SE, Kinney GG (2006) Curr Top Med Chem 6:1883
80. Hashimoto K (2006) Recent Pat CNS Drug Discov 1:43
81. Hashimoto K (2007) CNS Agents Med Chem 7:177
82. Dargazanli G, Estenne-Bouhtou G, Marabout B, Roger P, Sevrin M (2004) International Patent Application WO 2004 013 101
83. Dargazanli G, Estenne-Bouhtou G, Medaisko F, Renones MC (2008) International Patent Application WO 2008 037 881
84. Bozzoli A, Bradley DM, Coulton S, Gilpin ML, MacRitchie JA, Porter RA, Thewlis KM (2006) International Patent Application WO 2006 067 414
85. Bradley DM, Branch CL, Chan WN, Coulton S, Gilpin ML, Harris AJ, Lai JYQ, Marshall HR, MacRitchie JA, Nash DJ, Porter RA, Spada S, Thewlis KM, Ward SE (2006) International Patent Application WO 2006 067 417
86. Bradley DM, Branch CL, Chan WN, Coulton S, Gilpin ML, Harris AJ, Jaxa-Chamiec AA, Lai JYQ, Marshall HR, MacRitchie JA, Nash DJ, Porter RA, Spada S, Thewlis KM, Ward SE (2006) International Patent Application WO 2006 067 423
87. Bozzoli A, Branch CL, MacRitchie JA, Marshall HR, Porter RA, Spada S (2006) International Patent Application WO 2006 067 430

88. Bozzoli A, Branch CL, MacRitchie JA, Marshall HR, Porter RA, Spada S (2006) International Patent Application WO 2006 067 437
89. Branch CL, Marshall H, MccRitchie J, Porter RA, Spada S (2006) International Patent Application WO 2007 080 159
90. Coulton S, Gilpin ML, Porter RA (2006) International Patent Application WO 2007 113 309
91. Anderton CL, Clapham D, Keel TR, Kindon LJ (2007) International Patent Application WO 2007 147 831
92. Coulton S, Gilpin ML, Porter RA (2007) International Patent Application WO 2007 147 834
93. Coulton S, Gilpin ML, Porter RA (2007) International Patent Application WO 2007 147 836
94. Gentile G, Herdon HJ, Passchier J, Porter RA (2007) International Patent Application WO 2007 147 838
95. Coulton S, Gilpin ML, Porter RA (2007) International Patent Application WO 2007 147 839
96. Sekiguchi Y, Okubo T, Shibata T, Abe K, Yamamoto S, Kashiwa S (2008) International Patent Application WO 2008 018 639
97. Blackaby WP, Huscroft IT, Keown LE, Lewis RT, Raubo PA, Street LJ, Thomson CG, Thomson J (2006) International Patent Application WO 2006 134 341
98. Blackaby WP, Castro Pineiro JL, Lewis RT, Naylor EM, Street LJ (2006) US Patent Application US 0 276 655
99. Blackaby WP, Castro Pineiro JL, Lewis RT, Naylor EM, Street LJ (2006) International Patent Application WO 2006 131 711
100. Blackaby WP, Castro Pineiro JL, Lewis RT, Naylor EM, Street LJ (2006) International Patent Application WO 2006 131 713
101. Blackaby WP, Fletcher SR, Jennings A, Lewis RT, Naylor EM, Street LJ, Thomson J (2006) International Patent Application WO 2006 067 529
102. Lindsley CW, Wisnoski DD, Wolkenberg SE (2006) International Patent Application WO 2006 039 221
103. Burnds HD, Hamill, TG, Lindsley CW (2007) International Patent Application WO 2007 041 025
104. Hallett D, Lindsley CW, Naylor EM, Zhao Z, Theberge CR, Wolkenberg SE, Nolt BM (2007) International Patent Application WO 2007 053 400
105. Blackaby WP, Lewis RT, Naylor EM (2007) International Patent Application WO 2007 060 484
106. Lindsley CW, Zhao Z, Leister WH, O'Brien J, Lemaire W, Williams DL, Chen T-B, Chang RSL, Burno M, Jacobson MA, Sur C, Kinney GG, Pettibone DJ, Tiller PR, Smith S, Tsou NN, Duggan ME, Conn PJ, Hartman GD (2006) ChemMedChem 1:807
107. Bradley DM, Branch CL, Brown BJ, Chan WN, Coulton S, Dean AW, Doyle PM, Evans B, Gilpin ML, Gough SL, MacRitchie JA, Marshall HR, Nash DJ, Porter RA, Stasi LP (2006) International Patent Application WO 2006 094 840
108. Bradley DM, Branch CL, Chan WN, Coulton S, Dean AW, Doyle PM, Evans B, Gilpin ML, Gough SL, MacRitchie JA, Marshall HR, Nash DJ, Porter RA, Ward SE (2006) International Patent Application WO 2006 094 842
109. Bradley DM, Branch CL, Brown BJ, Chan WN, Coulton S, Dean AW, Doyle PM, Evans B, Gilpin ML, Gough SL, MacRitchie JA, Marshall HR, Nash DJ, Porter RA, Stasi LP (2006) International Patent Application WO 2006 094 843
110. Jolidon S, Narquizian R, Norcross RD, Pinard E (2006) US Patent Application US 0 149 062

111. Jolidon S, Narquizian R, Norcross RD, Pinard E (2006) International Patent Application WO 2006 061 135
112. Jolidon S, Narquizian R, Norcross RD, Pinard E (2006) International Patent Application WO 2006 072 436
113. Bubendorf A, Deynet-Vucenovic A, Diodone R, Grassmann O, Lindenstruth K, Pinard E, Rohrer FE, Schwitter U (2008) International Patent Application WO 2008 080 821
114. Jolidon S, Narquizian R, Norcross RD, Pinard E (2006) US Patent Application US 0 128 713
115. Jolidon S, Narquizian R, Norcross RD, Pinard E (2006) US Patent Application US 0 160 788
116. Jolidon S, Narquizian R, Norcross RD, Pinard E (2006) US Patent Application US 0 167 023
117. Jolidon S, Narquizian R, Norcross RD, Pinard E (2006) International Patent Application WO 2006 082 001
118. Jolidon S, Narquizian R, Norcross RD, Pinard E (2007) International Patent Application WO 2007 147 770
119. Dean AW, Porter RA (2007) International Patent Application WO 2007 014 762
120. Coulton S, Marshall H, Nash DJ, Porter RA (2007) International Patent Application WO 2007 104 775
121. Coulton S, Marshall H, Nash DJ, Porter RA (2007) International Patent Application WO 2007 104 776
122. Branch CL, Marshall H, Nash DJ, Porter RA (2007) International Patent Application WO 2007 116 061
123. Ahmad NM, Lai JYQ, Andreotti D, Marshall HR, Nash DJ, Porter RA (2008) International Patent Application WO 2008 092 876
124. Ahmad NM, Lai JYQ, Marshall HR, Nash DJ, Porter RA (2008) International Patent Application WO 2008 092 877
125. Jolidon S, Narquizian R, Norcross RD, Pinard E (2007) International Patent Application WO 2007 101 802
126. Bozzoli A, Branch CL, Marshall H, Nash DJ (2006) International Patent Application WO 2006 002 956
127. Rahman SS, Coulton S, Herdon HJ, Joiner GF, Jin J, Porter RA (2007) Bioorg Med Chem Lett 17:1741
128. Jolidon S, Narquizian R, Pinard E (2008) International Patent Application WO 2008 022 938
129. Jolidon S, Narquizian R, Pinard E (2008) International Patent Application WO 2008 025 694
130. Michardy SF, Lowe JA (2006) International Patent Application WO 2006 106 425
131. Lowe JA, Sakya SM, Sanner MA, Coe JW, McHardy SF (2008) International Patent Application WO 2008 065 500
132. Lowe J, Drozda S, Qian W, Peakman M-C, Liu J, Gibbs J, Harms J, Schmidt C, Fisher K, Strick C, Schmidt A, Vanase M, Lebel L (2007) Bioorg Med Chem Lett 17:1675
133. Morita K, Motoyama N, Kitayama T, Morioka N, Kifune K, Dohi T (2008) J Pharmacol Exp Ther 326:633
134. Caulfield WL, Collie IT, Dickins RS, Epemolu O, McGuire R, Hill DR, McVey G, Morphy JR, Rankovic Z, Sundaram H (2001) J Med Chem 44:2659
135. Morphy JR, Rankovic Z (2003) International Patent Application WO 2003 010 132

136. Isaac M, Slassi A, Da Silva K, Arora J, MacLean N, Hung B, McCallum K (2001) Bioorg Med Chem Lett 11:1371
137. Isaac M, O'Brien A, Slassi A, Da Silva K, Arora J, MacLean N, Hong B, McCallum K, Bell S, Mead E, Devivo M, Schumacher R, Lechner S, Hopper A, Orgnyanov V, Tehim A (2003) PharmaChem 2:38
138. Ho K-K, Appell KC, Baldwin JJ, Bohnstedt AC, Dong G, Guo T, Horlick R, Islam KR, Kultgen SG, Masterson CM, McDonald E, McMillan K, Morphy JR, Rankovic Z, Sundaram H, Webb M (2004) Bioorg Med Chem Lett 14:545
139. Ho K-K, Baldwin JJ, Bohnstedt, Kultgen SG, McDonald E, Guo T, Morphy JR, Rankovic Z, Horlick R, Appell KC (2004) US Patent Application US 0 167 119
140. Laborde E, Villar HO (2002) International Patent Application WO 2002 064 135
141. Harada N, Hikota M (2006) International Patent Application WO 2006 080 477
142. Wolin RL, Venkatesan H, Tang L, Santillan A, Barclay T, Wilson S, Lee DH, Lovenberg TW (2004) Bioorg Med Chem 12:4477
143. Wolin RL, Santillan A, Barclay T, Tang L, Venkatesan H, Wilson S, Lee DH, Lovenberg TW (2004) Bioorg Med Chem 12:4493
144. Barclay TK, Santillan A, Tang LY, Venkatesan H, Wolin RL (2005) International Patent Application WO 2005 044 810
145. Wolin RL, Santillan A, Tang L, Huang C, Jiang X, Lovenberg TW (2004) Bioorg Med Chem 12:4511
146. Huang CQ, Lovenberg TW, Santillan A, Tang LY, Wolin RL (2005) US Patent Application US 0 049 239
147. Huang CQ, Lovenberg TW, Santillan A, Tang LY, Wolin RL (2005) International Patent Application WO 2005 021 525

Subject Index